S. HRG. 103–787

S. 1086, THE TELECOMMUNICATIONS INFRASTRUCTURE ACT OF 1993

AF817435

HEARINGS

BEFORE THE

SUBCOMMITTEE ON COMMUNICATIONS

OF THE

COMMITTEE ON COMMERCE, SCIENCE, AND TRANSPORTATION UNITED STATES SENATE

ONE HUNDRED THIRD CONGRESS

FIRST SESSION

JULY 14 AND SEPTEMBER 8, 1993

Printed for the use of the Committee on Commerce, Science, and Transportation

U.S. GOVERNMENT PRINTING OFFICE

70–334cc WASHINGTON : 1994

For sale by the U.S. Government Printing Office

Superintendent of Documents, Congressional Sales Office, Washington, DC 20402

ISBN 0-16-046068-9

COMMITTEE ON COMMERCE, SCIENCE, AND TRANSPORTATION

ERNEST F. HOLLINGS, South Carolina, *Chairman*

DANIEL K. INOUYE, Hawaii
WENDELL H. FORD, Kentucky
J. JAMES EXON, Nebraska
JOHN D. ROCKEFELLER IV, West Virginia
JOHN F. KERRY, Massachusetts
JOHN B. BREAUX, Louisiana
RICHARD H. BRYAN, Nevada
CHARLES S. ROBB, Virginia
BYRON L. DORGAN, North Dakota
BOB KRUEGER, Texas

JOHN C. DANFORTH, Missouri
BOB PACKWOOD, Oregon
LARRY PRESSLER, South Dakota
TED STEVENS, Alaska
JOHN McCAIN, Arizona
CONRAD BURNS, Montana
SLADE GORTON, Washington
TRENT LOTT, Mississippi
JUDD GREGG, New Hampshire

KEVIN G. CURTIN, *Chief Counsel and Staff Director*
JONATHAN CHAMBERS, *Republican Staff Director*

SUBCOMMITTEE ON COMMUNICATIONS

DANIEL K. INOUYE, Hawaii, *Chairman*

ERNEST F. HOLLINGS, South Carolina
WENDELL H. FORD, Kentucky
J. JAMES EXON, Nebraska
JOHN F. KERRY, Massachusetts
JOHN B. BREAUX, Louisiana
JOHN D. ROCKEFELLER IV, West Virginia
CHARLES S. ROBB, Virginia

BOB PACKWOOD, Oregon
LARRY PRESSLER, South Dakota
TED STEVENS, Alaska
JOHN McCAIN, Arizona
CONRAD BURNS, Montana
SLADE GORTON, Washington

CONTENTS

LIST OF WITNESSES

APPENDIX

S. 1086, THE TELECOMMUNICATIONS INFRASTRUCTURE ACT OF 1993

WEDNESDAY, JULY 14, 1993

U.S. SENATE,
SUBCOMMITTEE ON COMMUNICATIONS OF THE
COMMITTEE ON COMMERCE, SCIENCE, AND TRANSPORTATION,
Washington, DC.

The subcommittee met, pursuant to notice, at 9:30 a.m. in room SR–253, Russell Senate Office Building, Hon. Daniel K. Inouye (chairman of the subcommittee) presiding.

Staff members assigned to this hearing: John D. Windhausen, Jr., staff counsel, and Kevin M. Joseph, professional staff member; and Regina M. Keeney, minority senior staff counsel, and Mary P. McManus, minority staff counsel.

OPENING STATEMENT OF SENATOR INOUYE

Senator INOUYE. This morning's hearing will explore S. 1086, the Telecommunications Infrastructure Act of 1993. This bill was sponsored by Senator Danforth and myself. The purpose of this hearing is to aid the committee in understanding how this bill will affect the telecommunications market, and to discuss improvements to the legislation.

Although there are a number of disagreements about the issues involved in this bill, there is one issue on which there is no disagreement: Everyone agrees that this Nation needs to promote the growth and development of an enhanced high quality, universally available telecommunications network.

Rural residents believe that an enhanced network will give them access to more advanced health care, and provide greater opportunities for economic growth. Inner city residents want access to a diversity of information at low cost. Handicapped persons believe that telecommunications can compensate for their disabilities. Business users want more sophisticated telecommunication services that will enable them to communicate around the world, to support their global competitive efforts.

In short, virtually every consumer group, every participant in the telecommunications industry, and every State and local government is demanding that Congress pay greater attention to the potential economic and social benefits of advanced telecommunication services.

The administration has indicated an enhanced telecommunications infrastructure as one of its top priorities. And this bill, the Telecommunications Infrastructure Act, is our response to this national call for coordinated infrastructure policy.

Of course, disagreement abounds as to how to achieve this goal. There are those who would have the Federal Government pay for the installation of certain technologies. Under today's fiscal constraints, Government funding on this massive scale is unrealistic. Others would choose to deregulate the telecommunications industry. While deregulation can, in certain cases, promote substantial consumer benefits, most observers agree that blanket deregulation does not serve the interests of the consumers; or, in the long run, the industry.

This bill chooses a more moderate course that relies on the private sector to make investments in the telecommunications infrastructure, based upon consumer demand. The bill we are here to consider would recognize that Government can provide a guiding hand, by encouraging the private sector to invest in a way that ensures that all consumers may obtain the benefits of these new services.

The principal means chosen to promote infrastructure development under this bill is to promote competition to local telephone services. Clearly, competition has worked for telephone equipment, and competition is working for long distance services. Consumers now benefit from cheaper prices and more innovative products and services, because market forces operate in these two sectors. It is now time that consumers of local telephone services should obtain the same benefits of competition that are enjoyed by long distance customers and buyers of telephone equipment.

In return for opening the local telephone market to competition, this bill permits the telephone companies to enter the market for cable television service. While I have had my doubts about the wisdom of allowing the telephone companies to provide cable services in the past, there is no question that greater competition for cable service would promote consumer welfare.

There are a number of other provisions in this bill, that deserve extensive scrutiny. These issues are not easy to digest, and require a great deal of further analysis. Mr. Chairman, your comments, please.

OPENING STATEMENT OF SENATOR HOLLINGS

The CHAIRMAN. Today's hearing is on S. 1086, introduced by Senators Danforth and Inouye. The sponsors of this bill intend for it to promote competition among the local phone companies in order to spur greater investment in the Nation's telecommunications infrastructure. The bill also would allow the telephone companies to enter the cable television business. The bill does not let the Bell Companies into long distance or manufacturing. This legislation clearly raises several important issues about the future of our communications network.

I agree that competition is a worthy goal. Competition in a free market usually results in growth and development. However, I have seen the downside of just relying on competition as well—airline deregulation is one such example. Furthermore, I have seen the problems with deregulation for the sake of deregulation. For the past 12 years, we have had two different administrations that tried to deregulate everything, with no regard for the harm to the public interest from markets where effective competition does not or will not exist. We need to keep in mind the interests of consumers throughout the country in an efficient and affordable communications network.

Our telecommunications industry is one of the most successful industries in the world. For years, the telephone companies have provided Americans with quality, dependable telephone service at affordable rates. Telephone service always has been guaranteed on a universal basis no matter where an individual lives or what level of income that person earns. It is available on a ubiquitous, nondiscriminatory basis for one reason—because we have imposed this obligation on the local telephone company, and not because of the free market. The Government's agreement with the local phone companies is simple—if a company provides telephone service to everyone who wants it, the Government will grant it special status with a local franchise. The system has served all of our constituents well.

Nevertheless, the world is changing. Technology has provided the capability and economic incentives for companies to compete with the local telephone companies. The need and opportunities for increased communications services grow almost daily, it seems. These interests have led to the introduction of S. 1086 and this hearing today. I look forward to the testimony of the witnesses.

Thank you, Mr. Chairman.

Senator INOUYE. Thank you, Mr. Chairman. I am pleased that we have assembled this morning a diverse array of expert witnesses, so that we may receive their input on how this bill will affect the market for telecommunication services. And I appreciate the willingness of these witnesses to cooperate with the committee in exploring these issues, and I look forward to this testimony.

I am pleased now to call upon the vice chairman of this committee, Senator Packwood.

OPENING STATEMENT OF SENATOR PACKWOOD

Senator PACKWOOD. That is the first time I knew that, Mr. Chairman.

Senator INOUYE. On my committee you are vice chairman.

Senator PACKWOOD. You learn something new every time you go home. On the trip during the July recess, in talking with a physicist, he told me about an optical illusion. And that optical illusion is that on any kind of an incline, if you are standing at one end of it, it seems less steep looking down at it than it does standing at the other end looking up at it.

So that when anyone talks about a level playing field, it depends upon which end of the field they are standing on. And if they are standing at the bottom, it looks much more tilted against them than if you are standing at the top looking at it tilted for you. Which probably hinders us in attempting to define what a level playing field is. Because all of the parties and participants who are interested, are standing in different locations.

I think we know what we mean, in terms of local telephone service. We, in essence, would say, "Have at it. We will let you all in. We will let the cable companies in, we will let the MCI's and the AT&T's in. We will let the local Bells in. We will let everybody in, to compete on a level playing field."

And yet, some of the participants would say, "Well, it is not level, if some of us have to provide universal service and others do not. It is not level, if some of our service is regulated, and we provide a disproportionate amount of that service; and others who are cherrypicking," as the argument would be, "are not regulated, and do not have to provide service. Therefore, it is not a level playing field."

I think the bill that the chairman and Senator Danforth have introduced is a good start. My State of Oregon, I think in this session of the legislature, will enact a bill directing the public utility commission to open up local phone service to competition. The bill has passed both the house of representatives and the State senate and it should be signed by the Governor within the next 2 weeks.

So, we all start out with the same goal. Whether we set out with the same definition of a playing field is another matter. And I look forward, Mr. Chairman, to these hearings, to see if we can conceivably reach agreement on that definition.

Senator INOUYE. Thank you very much. Senator Stevens, do you have a statement?

Senator STEVENS. Well, I see the cosponsor is here. I would be happy to yield. I would yield to Senator Danforth.

Senator DANFORTH. I do not have a prepared statement, Mr. Chairman. I appreciate the holding of the hearing, and the expeditious way that you are handling this; your great leadership of what I think is one of the really important issues, the big issues that is before this Congress. So, I thank you very much.

OPENING STATEMENT OF SENATOR STEVENS

Senator STEVENS. Mr. Chairman, I agree with Senator Packwood. This is a good start. And I look forward to these hearings. As in the case of the bill that you had, that I raised the spectrum issue on, I intend to raise several issues on this one.

One of the major issues, in my opinion, is: How can the FCC do all these things we are telling them to do, in the timeframe that it must be done in order to keep up with this technological revolution that is going on, this tumbling technology that is taking over the telecommunications area?

I think we need some innovative funding concepts involved in this bill, that will enable the FCC to have the money, and not be subject to the appropriations or the tax policies that have to be passed periodically to keep them having the funding to do the job.

I also think that we have got to take a look at the, instead of leveling the playing field, determining the rate of access to the field by various portions of the industry. Because I do not think they can all get on the field at the same time, without tilting it one way or the other. And in my opinion, it is the definition of access that is going to really have to be worked out, to make this a fair bill.

But I look forward to working with you. I hope to be a cosponsor of it, by the time it goes to the floor. Thank you.

Senator INOUYE. Thank you. Senator Lott.

OPENING STATEMENT OF SENATOR LOTT

Senator LOTT. Thank you, Mr. Chairman. I was just looking over the witnesses and the panels that we are going to have today. This is certainly going to be most interesting; I am looking forward to it.

And I want to welcome these panelists in advance, and commend Senators Danforth and you, Mr. Chairman, for the work that you have already done in trying to develop a framework that will stimulate investment in these information highways of the future. I personally think this is the most dynamic part of the future and our economy, in the United States.

So, it is an exciting area that is going to provide just mind-boggling developments in long distance learning, health care, increased economic efficiency, all numbers of technology; and in particular, I think it is going to be helpful to a State like mine where, through the technology we are going to develop here, we can bring better teaching and better health care to our rural and poor areas.

So, it is an exciting field, and I am just delighted that you are having this hearing, and the work that has been done. I look forward to hearing the panel.

Senator INOUYE. Thank you. Senator Gorton.

OPENING STATEMENT OF SENATOR GORTON

Senator GORTON. Mr. Chairman, I, too, think that this is a vitally important hearing. The direction of creating more competition, in one of the most dynamic of all of our technological fields, is a worthy one. The complexities are intense, and we are at the beginning of that upward climb, as Senator Packwood put it, ourselves, of a learning experience to try to do this job right.

Senator INOUYE. Senator Burns.

OPENING STATEMENT OF SENATOR BURNS

Senator BURNS. Thank you, Mr. Chairman. I have a prepared statement I would like to enter in the record, if I might at this time.

Senator INOUYE. Without objection.

[The prepared statement of Senator Burns follows:]

PREPARED STATEMENT OF SENATOR BURNS

Mr. Chairman, let me commend you and Senator Danforth for your leadership in introducing this infrastructure bill and holding expeditious hearings. As you both well know, I was talking about telecommunications and information infrastructure long before it was popular to do so. As a result, it is with a great deal of pride and satisfaction for me to see the Senate Commerce Committee turn its attention to this important national policy issue.

Today in America, we are witnessing the dawn of a new era—a digital, multimedia era. As a result of breathtakingly rapid technological developments in the computer software and hardware, consumer electronics, and cable television and telecommunications industries, a true revolution in the delivery of entertainment, information, transactional and telecommunications services is at hand. Through a confluence of interests, this digital, multimedia technology revolution will bring together a broad cross-section of industries that have heretofore considered themselves unrelated.

Over the balance of this decade and into the 21st Century, this digitization phenonenon will revolutionize the communications industry, have profound implications for the consumer electronics, entertainment and computer industries, and change our way of life forever.

The U.S. Congress and new Clinton/Gore Administration has a golden opportunity to be America's new high-tech pioneers—an opportunity to explore the new American frontier of high-tech telecommunications and computers that will be unleashed through the deployment of hair-thin, glass strands of fiber optic cable and the crackling of radio spectrum frequencies.

By taking bold, forward looking actions to accelerate the deployment of advanced telecommunications and information networks—a National Information Infrastructure—we could markedly improve our international competitiveness posture and dramatically spur domestic economic growth, productivity and job creation. Furthermore, through advanced educational, health care and other social services made possible with advanced telecommunications and information technology, we can establish a quality of life for all Americans which is unparalleled in our nation's previous history. And through increased use of "telework" or "telecommuting" we will reduce our reliance on foreign sources of oil, have a positive impact on environmental concerns and help parents deal with child care and other family concerns.

Yet, as America stands at this critical crossroads—the dawn of a new, revolutionary era in high technology, entertainment, information and telecommunications—we continue to operate under an antiquated regulatory scheme built for a world in which order and stability were the watch words rather than this new digital, Information Age which should accommodate and accelerate innovation and change. Progress is being stymied and hobbled by a morass of legal and regulatory barriers which continue to segment and balkanize the telecommunications and information industries into protective enclaves.

Legislationn designed to aggressively move America forward in developing a feature-rich, state- of-the-art National Information Infrastructure would include, among other matters: (1) opening the local telephone exchange to greater competition in conjunction with greater regulatory freedom and flexibility for the local telephone companies to offer advanced services; (2) expanding and redefining universal service; (3) promoting regulatory symmetry so that entities providing similar services are regulated or de-regulated in a similar manner; (4) developing interconnection and interoperability standards and protocols so that different networks and systems interconnect and work as one; (5) protecting privacy and intellectual property rights; (6) crafting tough, enforceable safeguards to ensure that local information providers have access.to and reasonable rate for the information highways; (7) creating a competitive bidding system for radio spectrum assignments; (8) permitting greater flexible spectrum use so that spectrum is made available without imposing crimped, narrow limits on what it will be used for (by doing so, services will evolve far beyond original expectations as the cellular success story clearly demonstrates). In addition, we could propose specific legislation dealing directly with distance learning, telemedicine and telecommuting.

I look forward with great anticipation to today's testimony. Again, thank you, Mr. Chairman.

Senator BURNS. I want to congratulate you and Mr. Danforth for coming forward with this situation. I am certainly glad that, as Senator Gorton would say, that we are beginning. It seems like we began about 4 years ago, when we started talking about competition in the field, and offering ideas that would expand, or to lay the groundwork or the framework of a national infrastructure for telecommunications.

This is not a new subject for us, as you know that Senator Gorton and I had 1,200 last year that went along the same lines, but had some firewalls. And I want to congratulate Senator Danforth and Senator Inouye, how they have taken this issue and have studied it over time, and now have come to realize that, yes, this is something that has to be done for the good of this country.

I have long advocated this, because we have already started in my State of Montana, in the areas of distance learning, in telemedicine, in the areas that were never gone into before or used before; because of a vision that we had for our own State of Montana, that this industry covers great distances. And it is important to rule America in the areas of learning and education, and in rural health.

But it is also a necessity for our inner cities, where we have declining tax bases, where we have to share teaching resources and medical resources for the people, and we have to stimulate minds to break those bonds that hold us sometimes in areas where we just do not want to be held.

So, I congratulate you, and I hope to be a cosponsor of this legislation, too, before it is all over. There are some things that we can fix, and we are going to talk about. But I congratulate our chairman and our ranking member, for bringing this issue to the forefront. No other, I do not think there is any other issue before this Nation, has as much vision for it, for infrastructure, than this one does. So, I congratulate you, and thank you very much.

Senator INOUYE. Thank you very much. I think it would be appropriate to acknowledge at this point that Senator Burns has been the prod, that has brought us to this point here. And we appreciate that very much.

At this time, I would like to have a 6-minute video show, prepared by the Computer Systems Policy Project. It will give us, in a capsule form, what we are considering today.

[A video was shown.]

A VOICE FROM THE AUDIENCE. Hello, I am John Sculley, chairman of Apple Computer, and also chairman of the Computer Systems Policy Project. The video that you are about to see demonstrates how a national information infrastructure will benefit all Americans: Education, health care, scientific research, communications, and business and industry.

Mr. SCULLEY. During the growth of the Industrial Age, national links were built to move people, goods, and raw materials. Now, America is entering the Information Age, and a new infrastructure is needed. Just as interstate highways were crucial to our postwar development, national data links are necessary for our growth in the 21st century.

The answer is a new infrastructure connecting the U.S. information and computing resources, a national information infrastructure.

Just pull it out, and fast forward to those things.

I think we can handle that.

This is a computer visualization, showing the airflow structure of a thunderstorm, an example of the new visual forms of data handled by today's computers, an example of the power of today's supercomputers; and the supercomputer of today will be the personal computer of tomorrow.

In a recent demonstration, a high-speed data link connected an Illinois supercomputer to desktop computers on a stage in Boston. Two scientists, separated by one-half a continent, shared the supercomputer, to analyze the structure of a galaxy.

Computer experts have developed future scenarios, showing how the national information infrastructure could deliver high performance computing power to industry; for U.S. scientists and engineers; to students, no matter how far they might be from learning centers; to researchers, already finding problems growing too complex for their own local resources.

If I try to address of how much computer power, flexibility is required to study and understand the human body, it is very easy to say, there is no upper limit. And that has drawn our group and other groups into the use of supercomputers. We cannot afford to have them in our own backyard. We need the best.

The national information infrastructure could bring the advantage and power of high performance computing, now accessible to only a few, into any office or laboratory in America.

This is the medical library at the national Institutes of Health, one of our hundreds of extensive information resources. The information infrastructure would make libraries like this accessible, instantly, Nationwide.

Future scenarios show how data could be delivered in quickly understandable forms, like audio and video. Emergency medical personnel could have instant access to life-saving medical data. Doctors could collaborate across the networks from any location, urban or rural; access data on complex cases; share CAT scans and other diagnostic information.

Business also could find national trend data, or stay abreast of Government regulations. Picture files and video would be instantly available for communication and education. The information infrastructure would help us realize the full value of our stored information resources, and open them to use, nationwide.

Excuse me, Bill Gilbert is calling back.

Television and computer technologies are merging, to support collaboration across any distance. There is an enormous amount of information in screens like these, and it takes a high-speed network to transmit it. But once the infrastructure is in place, the Nation has a new, more effective way of working. Of course, it would have instant impact in the communication business; it would also greatly improve business communication, and academic research.

The information infrastructure would make the country information efficient; giving people in all walks of life access to the advice and data they need; in forms they could immediately understand and use. That is the power in the national information infrastructure. It delivers information in the many forms computers now use every day.

There is no place in the world that could have more than we could have; and have a little rural town called Hickman, with less than 2,000 people, with a high school of less than 260. And we can have the same things that anybody could have anywhere, plus get to live out in the country and breathe clean air.

Test projects with networks already show real promise in education. Linked computers bring rural students information resources they could not get any other way. Video links bring gifted teaching into several widespread classrooms simultaneously. National Geographic's Kids Network creates computer collaboration for elementary school science students. All of these programs are working to bring rich information resources closer to students.

We need to prepare teachers to use the many resources that the information infrastructure will bring to the classrooms of the near future. A student in Minneapolis could use the Shakespeare Archives of the Folger Library in Washington, DC. All the resources in the U.S. Geological Survey could bring 3-dimensional computerized maps and simulations into an open classroom.

With the new uses of wireless communication, students of any age or need would have access to new kinds of help. Elderly Americans, just starting to reach out today, would have access to a whole world of communication, all from their homes; better monitoring for health and security; easy, intimate communication with family and friends.

The national information infrastructure would provide America's students and elderly with information in forms they need today. This is a country very rich in information resources. The information infrastructure will make these resources more accessible; easier to use; and more productive. They will create an environment for nationwide collaboration.

Industry, academia, and Government are already cooperating. It is a beginning. Now, we need to build a real momentum to make this vision a reality. The national information infrastructure is critical for America's future.

For all Americans to realize these benefits, a national information infrastructure must be built. The computer assistance policy project calls upon the new administration and the new Congress to work with us and with other industries to make the information infrastructure a national priority and building it a national technology challenge, to dedicate the Government resources needed to coordinate this effort and to work with industry on key public issues, to pursue an ambitious research agenda including demonstration projects to provide the foundation for the information infrastructure.

By working in partnership, industry and Government can build this infrastructure, the key component toward dramatically improving how Americans live and work, how we educate ourselves and our children, and how we build industry and how we do business. We need this infrastructure to bring us into the 21st century. It is 1993, and we have to start building it today.

Thank you for watching, and I hope you join us in looking forward to the future.

[End of videotape.]

Senator INOUYE. Now, we will proceed with the live show. Our first witness is the president of the Council on Competitiveness, the project director of the 21st Century Information Infrastructure Project, Mr. Daniel Burton. Before we proceed, May I call on Senator Exon for a statement?

OPENING STATEMENT OF SENATOR EXON

Senator EXON. Mr. Chairman, thank you very much. I certainly want to congratulate you, and I am looking forward to the testimony by the witnesses today on this extremely important matter that we are wrestling with here. If it were not for your leadership, Mr. Chairman, we would not be at the point here of beginning some important discussions, I think, which are going to have a great deal to do with what happen in the future and the whole telecommunications industry.

Certainly, you and Senator Danforth have sparked debate, meaningful debate, much needed debate, on the future structure of America's telecommunications industry. No other sector of our economy holds more hope for jobs creation, productivity improvement, education enhancing the American way of life than what the subcommittee is tackling here today.

In the next century, quick, comprehensive, and affordable access to information and creation of infrastructure to facilitate that access will revolutionized American life much in the same way the assembly line revolutionized American life in the days of Henry Ford.

The challenge is to maximize the benefits and opportunity for the American consumer, the economy, and the worker with the minimum expense to the American taxpayer. Certainly, I congratulate the authors of S. 1086, and again, I emphasize that the debate that we are starting here today is extremely important. However, at this juncture I have not yet been in position to endorse their legislation. It seems to me that some changes are needed to be made. But once again, I emphasize the debate is extremely important in the view of this Senator.

It is interesting to me that all sides in the information debate say that they are the ones who represent fair competition. But none can agree on what fair competition really is. The bottom line, Mr. Chairman, it seems to me is what is best for the American consumer who is going to pay the bills, whatever type of an infrastructure we have in telecommunications in the future.

As I listen to today's testimony, I will continue to search for a formulation of policy which represents a grand compromise, if you will, that is obviously necessary in the new information age, one that will promote fair competition but assure that rural and small communities have access to advanced technology.

Senator Grassley and I have introduced legislation specifically aimed at helping rural and smaller phone systems share in technology advancements, and have appreciated the support of many members of this committee in that effort.

I look forward, Mr. Chairman, to working with you and my other colleagues to find that compromise and to move America ahead to a brighter telecommunications future.

Thank you, very much.

Senator INOUYE. Thank you very much, Senator Exon. Senator Rockefeller.

OPENING STATEMENT OF SENATOR ROCKEFELLER

Senator ROCKEFELLER. Mr. Chairman, I will put my statement in the record.

It is, as I gather previous ones have been, of a general nature, and so obviously I am interested in what happens in rural areas and I want to see us move forward strongly on information highways and everything else. And so it is a carefully balanced statement, one that would not shed much light, which is what I presume this hearing is all about.

Senator INOUYE. Your statement will be made a part of the record.

[The prepared statement of Senator Rockefeller follows:]

PREPARED STATEMENT OF SENATOR ROCKEFELLER

I wish to commend Chairman Inouye and Senator Danforth for authoring and introducing the Telecommunications Infrastructure Act.

This legislation has important goals. It is intended to promote competition in local telephone services and to increase investment. Its authors hope that the new investment will contribute toward development of the new telecommunications technologies associated in popular discussion with the idea of the "information superhighway."

If we can find a way to stimulate such development, it could be an important element in reviving the American economy. It could start a new wave of innovation and growth, in a manner comparable to what railroad building did for our economy in the 19th century. Of particular interest to me are such possibilities as more widespread access to health care and educational services and expanded business opportunities.

The legislation raises complex issues regarding competition and regulation, State and Federal oversight, consumer interests and business interests, the differences between urban and rural markets, and the options for ensuring that new technology is encouraged and made widely available.

Over the last decade we have grown more sophisticated about markets and deregulation. We know that where it is possible, competition remains key for economic growth and consumer welfare, but that deregulation must be undertaken carefully, especially in rural areas.

I look forward to the discussion of this bill. I look forward to working with all concerned to see if we can answer the questions and solve the problems raised by this legislation. Our goal is to open up for the American economy and the American people the benefits of a new age of innovation in telecommunications, while preserving the achievements already attained.

Senator INOUYE. Senator Breaux.

OPENING STATEMENT OF SENATOR BREAUX

Senator BREAUX. Well, thank you, Mr. Chairman. I will just be very brief. I see we saw a film on what technology can do. That is great and wonderful. The question is really who gets to do it. I mean, that is what we are really talking about. I mean, which segment of the industry gets to do what, and at what time and under what circumstances and what rules.

I am almost at the point of becoming convinced that we are never going to figure this thing out. I mean, we are tinkering around trying to provide a little competition here, but not too much. And a little competition over here but not too much. I mean, I am even to the point of becoming frustrated with it and saying let everybody compete. If you want to manufacture, go out and manufacture. If you can do it, do it.

If you can compete on the local telephone service, by golly, get in there and do it. If you think you can provide long distance service and have the equipment and the technology and the know-how, go to it and let us make it work. I think that we are all wrestling with a great deal of frustration on how to provide competition but restrict it, which is sort of contradictory in terms. And I think that is why we have such a difficult problem here, and I want to commend Senator Danforth and our chairman for really trying to make a real effort at this and seeing how frustrating it is.

But you all have certainly made a monumental effort of trying to provide the type of competition I have just suggested. Perhaps it is time to just say let everybody compete. I mean, that is what America is about.

Thank you.

Senator INOUYE. Thank you very much. Mr. Burton.

STATEMENT OF DANIEL BURTON, PRESIDENT, COUNCIL ON COMPETITIVENESS; AND PROJECT DIRECTOR, 21ST CENTURY INFORMATION INFRASTRUCTURE PROJECT

Mr. BURTON. Thank you, Mr. Chairman. I am very pleased to be with the subcommittee today and pleased to discuss the views of the Council on Competitiveness on America's information infrastructure. The council is a private, nonprofit, nonpartisan coalition of chief executives from U.S. business, U.S. labor, and U.S. academia. It has one overriding objective: to improve the ability of American companies and workers to succeed in world markets in such a way that we build rising standards of living here at home. We think that information infrastructure is critical to that task.

In January 1993, the council launched a project on 21st century information infrastructure that was unique, unique in that it brought together all of the major players who are building and actually using much of this infrastructure. Participating in this project are cable companies, regional Bell operating companies, long distance carriers, cellular phones, computer hardware, software, and service companies, broadcasters, publishers, labor unions, and universities.

These groups have very different and sometimes competing agendas, but I think it is very noteworthy that they are all willing and committed to come together to try and develop the private sector consensus that is so important to informed and motivated public policy. And I would posit that a year or even 6 months ago it was very difficult to get this kind of momentum and this kind of commitment in the private sector to coming up with a consensus position which will make your jobs here in Congress that much easier.

My comments this morning very much reflect the consensus views of the council's advisory committee, reflected in the report we released in May entitled "Vision for a 21st Century: Information Infrastructure." We do not have all the answers, and there are many aspects of your bill that I would like to reserve comments on until we have really had a chance to develop a deeper consensus. But I think we have developed a framework and a perspective at the Council on Competitiveness that represents an important industry consensus.

We had hoped to have either John Young, former CEO of Hewlett Packard or Charles Vest, the President of MIT, who are cochairing our effort with us here today. Time and distance made that impossible. We then investigated the possibility of using information infrastructure to link them up with this committee so that you could talk to them and interact with them, even over distance.

Unfortunately, while the technology exists to do this, the infrastructure is not readily available. We did not have enough leadtime to pull together the consulting services necessary, to do the necessary paperwork or to get the equipment and the video, to say nothing of the cost. And so instead of having my two chairmen here, I have a laptop computer and I have a television.

Now, had we had the infrastructure, one of these mechanisms, or perhaps another one that we have not yet manufactured, would be able to plug into a socket, readily link up to Mr. Young and Mr. Vest in California and Massachusetts, and provide an opportunity for this committee to interact very directly with them and answer some of the key problems that you are dealing with. I think this is an example of the fact that the technology exists, the demand is there. In fact, given enough time we can string it all together, but it does not exist in a timely, cost-effective flexible, accessible way that I think the American public and in fact this committee is looking for. Hopefully, in the future we will have a system so that in a hearing like this we can simply bring a device, plug it into the wall, and this committee will be able to easily interact with John Young or Charles Vest or whoever you think is appropriate.

I would like to focus on a couple of key questions that I think are behind the debate on information infrastructure today. First of all, where do we stand? The fusion of telecommunications and computing technology is setting the stage for an advanced information infrastructure that will literally transform the economy. Many of the expensive elements of this infrastructure are already in place, such as long distance fiber networks and computers. So are millions of potential users.

What is in short supply is the ability of different parties to work together to define and deploy key elements and interfaces. While many of our competitors have chosen to rely on centralized Government planning to accelerate the deployment of information infrastructure, the United States has relied on demand-driven competitive policies and market forces, tempered by some regulatory restraints. It is essential that we preserve this strength.

In doing so, however, and I think the committee recognizes this, we must not ignore the need for public policy reform. I would posit that the big question facing the American political system today, however, is not Government planning versus decentralized markets, as some would characterize it in an international context. The United States has reaped enormous benefits from its market-oriented competitive strategy, and will undoubtedly continue to do so. Instead, the policy challenge before us is to find the appropriate balance of regulation, competition, and cooperation.

The second question I would like to address is what is our vision? The information infrastructure of the 21st century, in fact, can be pretty easily defined. There is a consensus view that is widely shared: "The information infrastructure of the 21st century will enable all Americans to access information and communicate with each other easily, reliably, securely, and cost effectively, in any medium, voice, data, image, or video, anytime, anywhere. This capability will enhance the productivity of work and lead to dramatic improvements in social services, education, and entertainment."

The information infrastructure can be divided into four basic parts. First, a set of widely accessible and interoperable communications networks; second, digital libraries information data bases and services; third, information appliances and computing systems; and, fourth, trained people who can build, maintain, and operate these resources. Each of these components is dependent on the others, and it is important to think of the infrastructure in terms of a system.

The communications system is usually the piece of the infrastructure that receives the most attention. Interestingly enough, here, too, there is very broad consensus about what it is. The communications system of the future, like the network of today, will be made of wire, fiber, and radiowaves, or encompass all of these media. It will be a diverse collection of variable bandwidth digital interoperable interactive commercially provided organic networks that are easy to use and universally accessible.

The networks will be diverse because different networks have different needs and characteristics. They will be variable bandwidths because they must carry not only voice but also image, video, and high speed data. They will be digital because of the ease and accuracy with which digital information can be stored, transferred, moved, and managed. They will be interoperable because different networks need to communicate easily. They will be interactive because of the need for point-to-point communication. They will be commercially provided because a competitive marketplace ensures the widest range of services and products to the greatest number of users at affordable prices and because private capital is needed for the investment. They will be organic because they will develop at different times and places, depending upon market needs. They will be easy to use so that they can serve the needs of the broad public. And they will be universally accessible because every American should be able to utilize them.

The next question I would like to address is what is the role of Government? We believe that Government's task is to help articulate the national need for an advanced information infrastructure, to serve as a catalyst and coordinator, to bring U.S. telecommunications regulations in line with new requirements and market opportunities, and to make sure that its diverse R&D programs are well managed and designed to advance practical applications. The best way to enable the creation of an advanced information infrastructure is to let competitive forces work constructively in the marketplace.

Before I conclude, I would like to address one more question, which is where do we go from here? We believe that more than any technology program, the regulatory framework in the United States will determine the face of America's information infrastructure. In many respects, the U.S. regulatory framework, particularly in telecommunications, is a holdover from the 1930's. America's legal and regulatory bodies are fragmented and have not kept up with advances in technology.

The best role for Government is the one that is least intrusive. Rather than try to pick winners among technologies or choose one technology platform, the Government should let market forces determine winners and losers. Government regulations should aim to create a playing field that allows for all interested parties to exploit market opportunities as competitiveness issues warrant.

The policy context, therefore, should be industry-neutral and technology-neutral so that it promotes those applications that are efficient and affordable for individual consumers and companies of all sizes. Moreover, it should be structured so that it encourages systematic private sector investment. The immediate objective should be to make users comfortable with the technology and applications that the current network can readily deliver.

In conclusion, from a competitiveness perspective, I would like to emphasize that we believe the stakes are very high. Advances in technology and the rise of international competition have made knowledge the new currency of the global economy. Superior knowledge is what gives nations an edge in world markets and allows them to create jobs and maintain high standards of living for their citizens, issues related to economic growth. Jobs and competitiveness are therefore independently linked to questions about the information infrastructure that makes it possible to generate and share information.

To a very large extent, information infrastructure will determine the comparative advantage of nations in the Information Age. In order to capitalize on a growing reliance on information technology, we need to strive for a system that will make us more innovative and more productive than other nations. Ultimately, it is the ability to harness technology in the form of practical applications that will determine international leadership in this field.

Because it is impossible to predict the exact path that technology will follow or the products and services that consumers will demand, flexibility in market forces must be the hallmark of the U.S. approach.

That is the conclusion of my prepared statement. I would also like to ask that our vision statement which elaborates on some of these themes be included in the record.

[The prepared statement of Mr. Burton follows:]

PREPARED STATEMENT OF DANIEL F. BURTON, JR.

I am pleased to be able to discuss with you the views of the Council on Competitiveness on America's information infrastructure. The Council is a private, non-profit, non-partisan coalition of chief executives from U.S. business, labor and academia. It has one overriding objective—to improve the ability of American companies and workers to compete in world markets while building a rising standard of living at home.

In January 1993, the Council launched a project on 21st Century Information Infrastructure that brings together the diverse parties who are actually providing and using the infrastructure. Participating in the project are cable companies, the regional Bell operating companies,long-distance carriers, cellular firms, computer hardware, software and service companies, banks, broadcasters, publishers, labor unions and universities. These groups have different—often competing—interests, but they are committed to working with the Council to develop the broad-based private sector consensus that is so crucial to informed public policy.

My comments this morning will reflect the consensus views of the Council's advisory committee on information infrastructure. We do not have all the answers, and there are many aspects of your bill that I would like to reserve comments on until the Council has had time to develop a deeper consensus on some of the critical issues. But we have developed a perspective and a framework that represents an important industry consensus.

I had hoped that John Young or Charles Vest, Co-chairs of our Advisory Committee could also be with us here today. However, time and distance made that impossible. We investigated the possibilities of linking them to this hearing via video conferencing facilities. Unfortunately, while the technology exists to accomplish this, the infrastructure is not readily available. We did not have enough lead time to establish the appropriate connections and rent the equipment for a video conference—to say nothing of the cost. This is just one simple example of how a national information infrastructure could have allowed us to overcome the barriers of time and distance.

While many of the pieces of the infrastructure exist today, they are not interoperable or readily available, nor are they flexible and easy to use. In the future, the commonplace items here on the table with me—the TV and a laptop computer—or some new device yet to be manufactured, will provide multimedia capabilities so that I will be able to bring a device into this hearing room, plug it in, turn the switch, and you will be able to see and talk with John and Chuck.

1. Where Do We Stand?

The fusion of communications and computing is setting the stage for an advanced information infrastructure that will literally reshape the economy. Many of the expensive elements, such as long-distance fiber networks and computers, are already in place. So are millions of potential users. What is in short supply is the ability of different parties to work together to define and deploy key elements and interfaces.

While many of our competitors have chosen to rely on centralized government planning to accelerate the deployment of such an infrastructure, the United States has relied on demand-driven, competitive policies and market forces, tempered by some regulatory restraints. It is essential to preserve this strength.

In doing so, however, we must not ignore the need for public policy reform. The policy issue facing the United States is not centralized government planning versus decentralized markets. The nation has reaped enormous benefits from its market-oriented, competitive strategy and will undoubtedly continue to do so. Instead, the policy challenge is to find the appropriate balance of regulation, competition and cooperation.

2. What is Our Vision?

The vision for the future of America's information infrastructure can be easily described and is widely shared.

The information infrastructure of the 21st century will enable all Americans to access information and communicate with each other easily, reliably, securely and cost-effectively in any medium—voice, data, image or video—anytime, anywhere. This capability will enhance the productivity of work and lead to dramatic improvements in social services, education and entertainment.

Information infrastructure can be divided into four parts: 1) a set of widely accessible and interoperable communications networks, 2) digital libraries, information databases and services, 3) information appliances and computing systems that are easy to use, and 4) trained people who can build, maintain and operate these resources. Each of these is dependent on the others.

The communications system usually receives most of the attention. Here, too, there is broad consensus. The communications system of the future, like the network of today, will be made of wire, fiber or radio waves, or encompass all of these media. It will be a diverse collection of variable bandwidth, digital, interoperable, interactive, commercially provided, organic networks that are easy to use and universally accessible:

- Diverse because different networks serve different needs and have different characteristics;
- Variable bandwidth because they must carry not only voice, but also image, video and high-speed data;
- Digital because of the ease and accuracy with which digital information can be stored, transferred, moved and managed;
- Interoperable because different networks must be able to communicate easily;
- Interactive because of the need for point-to-point communication;
- Commercially provided because a competitive marketplace ensures the widest range of services and products to the greatest number of users at affordable prices and because private capital is needed for the investment;
- Organic because they will develop at different times and places depending on market needs;
- Easy to use so that they can serve the needs of the broad public; and
- Universally accessible because every American should be able to utilize them.

3. What Is the Role of Government?

Government's task is to help articulate the national need for an advanced information infrastructure, serve as a catalyst and coordinator, bring U.S. telecommunications regulations in line with new requirements and market opportunities, and make sure that its diverse R&D programs are well managed and designed to advance practical applications. The best way to enable the creation of an advanced information infrastructure is to let competitive forces work constructively in the marketplace.

4. Where Do We Go From Here?

More than any technology program, the regulatory framework will determine the face of America's information infrastructure. In many respects, the U.S. regulatory framework, particularly in telecommunications, is a hold-over from the 1930s. America's legal and regulatory bodies are fragmented and have not kept up with advances in technology. The best role for government is the one that is least intrusive. Rather than try to pick winners among technologies or choose one technology platform, the government should let market forces determine winners and losers. Government regulations should aim to create a playing field that allows for all interested parties to exploit market opportunities, as competitive issues warrant.

The policy context should be industry-neutral and technology-neutral so that it impartially promotes those applications that are efficient and affordable for individual consumers and companies of all sizes. Moreover, it should also be structured so that it encourages systematic private sector investment. The immediate objective should be to make users comfortable with the technology and applications that the current network can readily deliver.

CONCLUSION

The stakes are high. Advances in technology and the rise of international competition have made knowledge the new currency of the global economy. Superior knowledge gives nations an edge in world markets and allows them to create jobs and maintain high standards of living for their citizens. Issues related to economic growth, jobs and competitiveness are therefore intimately linked to questions about the information infrastructure that makes it possible to generate and share information.

To a large extent, information infrastructure will determine the comparative advantage of nations in the information age. In order to capitalize on our growing reliance on information technology, we need to strive for a system that will make us more innovative and productive than other nations. Ultimately, it is the ability to harness technology in the form of practical applications that will determine international leadership in this field. because it is impossible to predict the exact path that technology will follow, or the products and services that consumers will demand, flexibility and market forces must be the hallmark of the U.S. approach.

["Vision for a 21st Century Information Infrastructure," by the Council on Competitiveness may be found in the committee's files.]

Senator INOUYE. Thank you very much, Mr. Burton. Obviously, as Senator Breaux and Senator Stevens have noted, there is an air of frustration here because, first, we are not experts, none of us are. We do not know anything about the high technology involved. We are users and consumers for the most part, and we are concerned about cost.

Throughout your statement you used the phrase "cost-effective." Who do you suggest should bear the cost?

Mr. BURTON. Well, I think throughout my statement I also used the words "market forces" and "competition," and it is our view that market forces and competition will provide the lowest cost structure and the widest level of applicability. And, therefore, that is the principle that we endorse.

Senator INOUYE. Would that mean that those with the most readily available resources would be the dominant forces?

Mr. BURTON. We have not said that specifically. I think in addition to focusing on a market solution, we have also recognized that the user must come first. And that is not simply large business users, that is the broad American public. And questions of ease of access and affordability are absolutely essential and have to be addressed as part of the broader package.

Senator ROCKEFELLER. But you have not answered the chairman's question in any way, shape, or form. I think he wants an answer from you.

Mr. BURTON. Can you repeat the question, then, Mr. Chairman? I think I should also state that what I am representing here and really the extent that I feel that I can comment——

Senator ROCKEFELLER. The question that he asked was, those that have the resources, will they then be able to dominate? That was his question. You did not answer it.

Mr. BURTON. Right. I do not know that I can answer that. I think the views that I am representing are the consensus views. We have had this project underway for 6 months. I think we will be able to address issues like that more specifically if you give us 6 more months.

There are large users today who are already establishing their own infrastructures, their own network, their own systems. That is going forward if you look at a lot of large businesses, for example, and private concerns.

The issue is, How do you continue that momentum and broaden it so that the large American public also has ready access to the system?

Senator INOUYE. We have many other questions, but obviously as you can see this is the best turnout we have ever had on this subcommittee, and so time is of the essence.

I would like to call upon the author of the measure, Senator Danforth.

Senator DANFORTH. Thank you, Mr. Chairman. Mr. Burton, you described Government's role and as it should be. You also said that the current regulatory framework we are now existing under is a relic of the 1930's. I assume from that that what we are doing now is not right. I mean government and governmental policy is now serving as an impediment to the advancement of telecommunications. Is that not correct?

Mr. BURTON. Yes, we believe that some regulatory reforms are necessary.

Senator DANFORTH. And would they not have to be fairly dramatic regulatory reforms, in your opinion?

Mr. BURTON. There are a few fundamental issues which would have to be addressed. How do you shape the competitive environment and how do you guarantee universal access or broad access are two key questions.

The concern that we have heard expressed in the broad committee is that this infrastructure will evolve; it is evolving; we will have one. The key is how quickly it is evolving and how we can accelerate the deployment of an advanced information infrastructure.

Senator DANFORTH. Well, under the present situation Government is an impediment, is it not, in that it is basically saying to various industries you cannot compete in one area or another? I mean, obviously if we move to competition there are various transitional problems that have to be worked out. You have to be able to assure, for example, universal access.

But is it not fair to say that under the present system really we are not favoring competition, and therefore we are doing the opposite of what you testified we should be doing?

Mr. BURTON. Yes. We are not favoring competition as much as we need to.

Senator DANFORTH. You obviously have not testified pro or con with respect to the specific legislation that is before us. Are you familiar with the legislation?

Mr. BURTON. I am familiar with the broad contours of the legislation. As I mentioned in my opening comments what we are trying to do is really something that has not been done heretofore, which is bring together all of the major players in the private sector and broker a consensus on some key issues.

What we have done to date is put together a vision statement. We are currently looking at some of the critical issues. You and the chairman should be congratulated for putting forth a piece of legislation which really tries to get at some of the tough issues. Those are the same issues the council will be addressing in the future.

Senator DANFORTH. Let me just ask you one other question. This is obviously a very big subject area that we are working with in this legislation. The decisions that will be made, if we make decisions, are going to be ones that will be very controversial.

One approach to controversy and difficulty is to say, "Well, let us let inertia govern and just really do nothing, just maintain the situation as it is now." My own view is that we should get on with it; that while this is a big subject the issues are fairly well known, this is not a new matter. This is an area that has been debated and dealt with and addressed by courts and so on for a very long period of time.

Will you counsel the kind of caution that would end up leading nowhere, or do you think that we are pretty well in a position now where we can get on with the business of addressing the policy questions? And if there are problems in what we are doing now, actually changing those?

Mr. BURTON. It is very important to address the policy questions, and if there are problems to change them. I would posit that what has changed much of the landscape is technology. Today suddenly there are possibilities with the infrastructure that were not possible and not economically feasible on a broad basis before.

What we have is a technology that is forcing industries and markets to gallop ahead, and it is very important that the regulatory environment try and frame a situation which will allow for a very productive and efficacious deployment of that technology.

Senator DANFORTH. Thank you, Mr. Chairman.

Senator INOUYE. Thank you. Senator Packwood.

Senator PACKWOOD. When the chairman asked about those industries with resources he may have been driving at the argument raised by some, including I think AT&T, that if you had a level playing field immediately you would tilt automatically in favor of the Bells because they have an existing infrastructure. Is that true or not?

Mr. BURTON. It depends on who you talk to. I am not the person to give you the definitive answer. If you talk to the Bells, I think they are looking from the top of that. If you talk to some other companies, they are looking at it from the bottom of the incline, and so you get very different perspectives.

Senator ROCKEFELLER. But I think he asked you for your view. I mean, we are well aware there are different views.

Mr. BURTON. Again, the council's view is a consensus view, and that is the view that I am representing. And I apologize if I am not able to give you the precise and specific questions that you are posing. But I think our comparative advantage is very much the private sector consensus.

The key question with opening markets and open competition is really a question of pace and who gets into the action when. It is an all or nothing. It is how you phase it and how you establish it so that all parties have a fair play.

Senator PACKWOOD. I take that by staging you literally mean that. If we do not level the playing field immediately we may have to do it over a period of years to allow everybody time to get ready for the access.

Mr. BURTON. Yes. As far as the council is concerned if we do this tomorrow—first of all, I am not sure that there is a consensus for a solution to do that. And there are several thorny questions, universal access being one of them, that are important to build into the calculation. Therefore it is going to be a process of evolution.

Senator PACKWOOD. Next question. Assuming you level the playing field, and I sense you are not very enthusiastic about Government regulation, do you think lack of regulation and a free marketplace, given a level playing field, will more likely serve more needs than government regulation? Do I correctly state roughly the council's position?

Mr. BURTON. Yes. The principle of the council is that open competition is going to accelerate this. I think that the regulatory issue, however—again, it is not black or white.

Senator PACKWOOD. I understand. Do you worry that given that lack of regulatory structure, that all of the contestants into the fray may first try to pick off the most valuable sectors of the market, and that universal service, especially if it is less remunerative, may suffer?

Mr. BURTON. That is a concern, and that is why this committee and others have focused on universal service and broad access. The private sector recognizes the importance of that issue as well.

Senator PACKWOOD. I do not know if the council has a position and maybe you do as you have a bias against regulation. Did we move in the wrong direction when we partially reregulated the cable industry?

Mr. BURTON. Neither the council nor I have a position on that. [Laughter.]

Senator PACKWOOD. I think we did. I have no other questions, Mr. Chairman. Thank you.

Senator INOUYE. Thank you. Senator Stevens. Senator Lott.

Senator LOTT. No questions, Mr. Chairman.

Senator INOUYE. Senator Gorton.

Senator GORTON. At this point, obviously, the council does not endorse S. 1086?

Mr. BURTON. We do not have a position on S. 1086.

Senator GORTON. Does it endorse any of the elements in S. 1086?

Mr. BURTON. What the council has endorsed are some principles which I would be happy to share with you. There are basically six of them, which are as close as we have gotten to resolving some of the issues around which legislation must be built.

The first principle is that the user comes first. The main goal of public policy, we think, should be to provide users with the maximum possible levels of choice, service, and mobility with due respect to privacy, confidentiality, and security as well as fair and open access.

The other principles—affordability and ease of use are essential. Flexibility and responsiveness to market demand are key to success. The regulatory process must be evenhanded in terms of being industry neutral and technology neutral. In fact, a coordinated systems approach is important. So, we are looking at R&D programs, our procurement policies, and legislation at the same time.

And finally we think that it is important to take into account international concerns about connectivity and competition.

Senator GORTON. Well, I suspect you might find a broad, perhaps unanimous agreement on that set of principles here in this committee. We do not write principles, we write statutes, and if you are going to be of any help to us you have got to move beyond those principles and deal with the particular statutory proposals.

Mr. BURTON. Yes.

Senator GORTON. If you have these additional 6 months, are you going to be in a position in which you will give us specific recommendations with respect to this bill, either in its present form, subtractions from it or additions to it?

Mr. BURTON. Yes. We hope very much to comment on specifics that will be helpful to this committee and others, and let me just tell you how our process is working.

We put together our group in January. We put out our vision statement in May. So, we spent 4 months really getting everyone together, agreeing on a common vision, a common perspective, a common set of key issues. We are now in the process of hitting the stone wall of trying to pull the consensus together on specific proposals, propositions, priorities.

I think we will continue that process really over the next year and release statements focusing on specific policy provisions. We are also looking at a series of demonstration projects because we think in addition to policy it is very important to demonstrate in real terms that this stuff actually works, it has real applications, people are going to understand it, and there is a clear demand for it. And we will be moving on those issues really for the year and a half, and so hope to have a whole series of very specific comments for you.

Senator GORTON. Thank you. Thank you, Mr. Chairman.

Senator INOUYE. Thank you. Senator Stevens.

Senator STEVENS. No questions.

Senator INOUYE. Senator Rockefeller.

Senator ROCKEFELLER. Mr. Chairman, I will ask a question but I am not sure the gentleman will be able to answer it because I am not sure he knows the bill.

You use words like, "America's legal and regulatory bodies are fragmented and have not kept up with advances," and "the best role for Government is the one that is least intrusive. Government regulations should aim to create a playing field that allows for all interested parties to exploit market opportunities," et cetera.

And then in response to a question from Senator Packwood which was on the mark, you said that is of concern. Sort of like being of counsel, I guess. And my question—you know, we did this in airline deregulation, and we sort of said, go ahead and have at it, let us let things work. And I watched in West Virginia as American Airlines, United Airlines, and Eastern Airlines all cleared out within 2 months. And I fly around commercially on propeller-driven things with rather low maintenance.

So, my question to you is from the sort of philosophical principle view that you appear to view this short another 6 months, what do you see is the difference? Why would not rural areas be picked off, as Senator Packwood said, or just left out because the city areas and more populous areas would be far more profitable? Why would this not be airline deregulation all over again?

Mr. BURTON. First of all, let me say that I understand and can relate to your frustration about wanting more specific answers.

I would posit that what we represent, which we hope will be a major contribution, is a consensus. That consensus is forming. We think it has come a long way even in the past 3 or 4 months, and we hope to give you more specific answers in the future.

So, I am not ducking your questions, I am really just trying to present where we think a broad cross section of the private sector stands at present.

Clearly, when you get into questions of wide scale deregulation, broad public access is a key concern. Clearly, issues like cherry-picking off the best and most lucrative markets are ones which many business concerns would go to immediately. That is where the profits are, that is what motivates business activity.

And so I think there is a role for the regulatory environment in making sure that questions of universal access are fairly considered and built into the whole process of deregulation that we have underway.

I cannot give you a more specific answer than that at this point, but I think we share your concerns. The private sector is on record as sharing those concerns, and that is an issue that we hope to address, that we are currently addressing, and we hope to give you more specific responses to in the future.

Senator ROCKEFELLER. Thank you, Mr. Chairman.

Senator INOUYE. Senator Exon.

Senator EXON. Mr. Chairman, I would just like to follow up on the general tone of Senator Rockefeller's questioning, and it was encompassed in my opening remarks. Can you be any more specific, Mr. Burton, as to what your solutions would be?

We have pretty well established, I think, that there is concern here that we could find ourselves in this rush to deregulation back into a situation that we found ourselves in with the airline deregulation. It is very good for the people that live in New York, and San Francisco, and Oakland, and Atlanta, and Dallas, and not so good for the small communities of the country. Do you have any suggested frameworks that we should be looking at in that area?

Go back a little bit further, way back before airline deregulation. I think it is generally considered that we would not have in the rural communities across this Nation, which are still very important in all States—would not have electricity had it not been for the Rural Electrification Association. Do you see something of that nature as a possible solution to the concern that some of us have in this particular area?

Mr. BURTON. I am loath to continue frustrating the committee, but we do not have a specific framework or a specific program by which we can deregulate and guarantee universal access. I will be happy to consider that and take into consideration any specific questions that this committee would like to pose to us.

Senator EXON. Thank you, Mr. Burton. Thank you, Mr. Chairman.

I guess the point is that those of us from States that have large rural areas are very much concerned about this, and I hope everyone fully understands that we do feel that should be addressed, and the fact that there is no framework at the present time expressed by those whom we are relying on for a lot of advice and counsel indicates that we are going to have to give that a little bit more attention.

Thank you, Mr. Chairman.

Senator INOUYE. Thank you, Senator.

Senator Breaux.

Senator BREAUX. Thank you, Mr. Chairman. Thank you, Mr. Burton.

You make the following statement in your council's paper that you have produced, and it says, "More than any technology program, the regulatory framework will determine the fate of America's information infrastructure. The best role for the Government is one that is least intrusive."

That is really disturbing. I think that if the future of America's infrastructure and communication is being determined by a regulatory program as opposed to the type of technology we have, I think that is a very important statement, and it is one that gives me a great deal of concern that we are going to let the regulatory framework determine how fast we advance, as opposed to letting technology determine how fast we are able to provide services.

With that as a statement from the council, would you recommend that we do away with the modified MFJ judgment? I mean, that would take away the regulatory framework and let technology determine how fast we can proceed.

Mr. BURTON. You are not going to be surprised by my answer. I cannot tell you to take away the MFJ or to leave it in place. We do, however, believe that in fact the regulatory environment has had an extraordinarily powerful impact on the way that the infrastructure is evolving.

Senator BREAUX. Is it a positive impact, or a negative impact?

Mr. BURTON. Well, I think at this point it is hindering the accelerated employment of the infrastructure, and that is what we are concerned about.

Senator BREAUX. Is it correct to say that what you are saying is that we have the technology but that is being hampered by the regulatory framework that we operate under?

Mr. BURTON. Yes. Much of the technology exists.

Senator BREAUX. We could do a lot more if we were not restricted by arbitrary rules and regulations.

Mr. BURTON. The regulatory environment could do a lot more to accelerate deployment of the infrastructure.

Senator BREAUX. Our challenge is to find out how to do that.

Mr. BURTON. Yes, sir. We hope to help you in that process, even though we still have a ways to go.

Senator BREAUX. You apparently have the same problem. I think the frustration, Senator Rockefeller, is pointing out apparently—I mean, do you not represent almost all the players in this business?

Mr. BURTON. The purpose was to represent the major players in this business, and we really have tried to get them together. We are meeting on a very regular basis with a full committee. We then have a subcommittee on competition policy, another one on public access issues, and another one on demonstration projects.

Senator BREAUX. But is that not as close to an impossible task as you could find?

Mr. BURTON. It is difficult, surely. I think, however, if you look at this industry, over the past year, given the great advances that are being made technologically and what some of our foreign competitors are doing, there is hopefully some new opportunity for some consensus. It is our intent to develop that consensus and provide you with it here on the Hill.

Senator INOUYE. Thank you. Before I call upon Senator Robb, I would like to note that there is a vote in process at this moment. After Senator Robb has completed, we will call a short recess.

Senator Robb.

Senator ROBB. Thank you, Mr. Chairman, and I will be very brief. I think I have some reasonable expectation of the responses to some of the questions to this point. Let me see if I can be even a bit more general than some of my colleagues.

A relatively well-known Virginian, Mr. Jefferson, is purported to have said at one point or another that that Government governs best that governs least, or something to that effect. Would that general philosophy guide your approach to this problem? Without any specific reference to even eliminating MFJ or taking the Government out of the regulatory environment altogether, but would ultimately a level playing field, in your judgment, be defined by a lack of Government participation?

Mr. BURTON. We believe that the policy which will accelerate the deployment of the infrastructure most rapidly is the one that will move toward freer market competition.

Now, that is not to say that we expect for all Government regulations to be completely eliminated. Given the long history, that is probably unrealistic, but we think a reliance on market forces and letting the technologies, the industries, and the applications be sorted out in the marketplace is the best path for the United States.

Senator ROBB. Mr. Chairman, I think we have probably exhausted this particular topic in terms of what we are going to get at least at this particular time, and with the vote already in progress I will yield back any time so that we may cast our votes and return.

Senator INOUYE. Thank you very much, Senator Robb.

Mr. Burton, thank you very much for your participation this morning. I realize the answers are not easily forthcoming, but I believe that is the nature of this measure before us. It is a very difficult one. We hope we will be able to resolve some of the problems we have to face. With that, once again, thank you very much.

Mr. BURTON. Thank you, Mr. Chairman.

Senator INOUYE. Before I call the recess, will Mr. Crowe, Mr. Roberts, Mr. Ware, Mr. Coe, Mr. Niggli, Mr. McBee, and Mr. Frischkorn take their places to be ready when we return, and with that the hearing will stand in recess.

[A brief recess was taken.]

Senator INOUYE. On our first panel, we are pleased to have with us the chief executive officer of the Metropolitan Fiber Systems of Omaha; Mr. James Q. Crowe, the president of Comcast Corp. of Philadelphia, PA; Mr. Brian L. Roberts, the manager of Garden

Valley Telephone Co. of Erskine, MN; Mr. Lawrence C. Ware, a director of Merchant Acceptance Strategy of Visa International; Mr. Kermit Coe, the senior vice president of Entenergy Corp. of Little Rock, AR; Mr. Michael Niggli, the vice chairman of United States Telephone Association of Washington; and Mr. Gary McBee, and a representative of the Telecommunications Industry Association, Mr. Allen "Mike" Frischkorn, Jr.

Gentlemen, I thank you all for your participation this morning. May I first call upon Mr. Crowe?

STATEMENT OF JAMES Q. CROWE, CHIEF EXECUTIVE OFFICER, METROPOLITAN FIBER SYSTEMS

Mr. CROWE. Chairman Inouye, Senators, I am James Crowe, chairman and chief executive officer of MFS Communications Co. We are a new company that the FCC lists as the largest competitive local provider of phone services. Over the last 3 or 4 years we have raised and committed nearly $750 million to install fiber optic facilities in local exchanges in 14 metropolitan areas.

While I freely admit that we have strong views, they are at least views founded on practical experience with many of the issues dealt with in this measure, and I am honored today to enthusiastically support the bill. I believe it can be a vital part of a national communications policy and has implications for American jobs, American competitiveness, and our standard of living for decades to come.

Today, our Nation leads the way in the race to realize the promise of the age of information. However, the race is far from over, and around the world nations are moving quickly to harness the power of the private sector to improve their communications systems and compete with us more effectively.

These nations have learned from America that when technology and markets change, innovation comes from competitive markets, not from regulated monopolies. I believe another clear lesson of American history is that when technology and markets change rapidly, important innovation often comes from the hothouses of entrepreneurial companies, not large, mature, slow-moving enterprises.

This has certainly been the case in our computer industry and our software industry, where icons of American business have fallen, to be replaced by vital new enterprises, and America is the winner in this process. And now, in the communications industry, the irresistible forces of changing technology and changing markets are creating opportunities for today's new entrants, who will certainly be tomorrow's success stories.

Even though companies like my own have found opportunities to compete in a narrow range of services, typically those which are regulated by the procompetitive FCC, many cities and States cling to a monopolistic vision of communications promoted for decades by the consolidated Bell System.

This misguided view allows exclusive preferential treatment of local monopolies, and as a result, progress in upgrading our communications infrastructure is held hostage to a patch quilt of municipal and State regulation. This unacceptable situation is obvious, I believe, to the drafters of the bill, and they seek to rectify this situation.

The benefits of more open competition can be clearly seen in the long-distance industry. Competition has spurred the almost continuous replacement of older technologies by advanced systems. Vast sums have been spent improving the quality of service, sums provided by the private sector to build the interstate highways, if you would, of our Nation's communications system. Contrast this with the local exchange, the on-and-off ramps of the network.

Until recently, the baby Bells have refused to modernize their antiquated copperplate without a virtually guaranteed return on that investment. They have demanded over and over that they be allowed to enter new markets such as manufacturing and long distance, even though their monopoly controlled bottleneck facilities, coupled with a clear history of predatory practices, makes this unwise public policy until customers have a choice of local providers.

What little progress has been made in upgrading their facilities has come only when forced by competition from companies like MFS. There is more than a touch of irony in the fact that the baby Bells, with enormous cashflows extracted from ratepayers by a legally sanctioned monopoly, have chosen to invest this cash in upgrading the telephone systems of foreign countries while private capital is eager and unable to invest here in America. My company and many others are ready to invest in this American opportunity if given a chance.

As drafted, the bill contains many of the measures necessary to ensure the benefits of competition are realized. However, I would respectfully submit that two changes, both, I believe, consistent with the bill's intent, would greatly improve the opportunities for competition to flourish.

First, the bill should be expanded to mandate access for all competitors to utility-owned facilities such as poles, conduits, and access to buildings. Today, monopoly phone companies, cable companies, and other utilities either refuse such access outright or impose outrageously high fees for the use of these bottleneck facilities installed by ratepayer dollars.

Second, this measure should end discriminatory assessments of franchise fees and rights-of-way charges by municipalities. Currently, companies like MFS pay fees as high as 14 percent of total revenues to municipalities while monopoly phone companies generally pay nothing. Both these modifications are more fully developed in my written testimony.

I greatly appreciate the opportunity to testify in strong support of the bill. The good news is that the Nation's goal of universal access to a vast array of information can be promptly accomplished by private capital. Companies such as MFS stand ready to continue to make the investment necessary to improve not only the information highways but the on-and-off ramps to the network.

This bill will spur investment by removing antiquated and monopolistic barriers to competition. When enacted, competition can be counted on to drive technological advancement and industrial creativity and to help ensure America leads the world into the Information Age.

Thank you.

[The prepared statement of Mr. Crowe follows:]

PREPARED STATEMENT OF JAMES Q. CROWE

Mr. Chairman and members of the Subcommittee, I am James Q. Crowe, Chairman of the Board of Directors and Chief Executive Officer of MFS Communications Company, Inc. ("MFS"). I am honored to testify today in support of Senate Bill 1086 ("S. 1086"). My company's own short history evidences the many benefits of competition and innovation in the local telecommunications market and the compelling need for a comprehensive national policy promoting market entry in all facets of telecommunications. My presence on this panel is itself confirmation that local telephone service is decidedly not a natural monopoly and demonstrates that when antiquated and obsolete legal barriers to local telecommunications competition are eradicated, the private sector will respond forcefully with the investment and innovative services now required by the American people.

MFS commends Senators Inouye and Danforth for recognizing in S. 1086 the fallacy of the natural monopoly myth for local telephone service and for taking the bold and decisive action necessary to encourage open entry for all telecommunications services. The bill would remove obstacles to competition by mandating meaningful interconnection and interoperability and by encouraging number portability, reciprocal compensation and other competitive safeguards, including, most importantly, the removal of outdated state and local legal barriers to competition. The bill would also achieve the delicate, and often difficult, balance between unleashing competition through legislative directives and imposing restrictive government standards, which could have the unintended effect of stymieing advanced technology and natural market segmentation. Along with our overall support for the pro-competitive measures in S. 1086, MFS proposes amendments to the bill that, consistent with the bill's overall thrust, would eliminate two prohibitive barriers to a truly competitive local telecommunications environment. As discussed below, these amendments would: (1) prohibit telephone companies and other utilities from charging discriminatory rates for the use of conduit and pole attachments among similarly situated telecommunications carriers; and (2) prohibit municipalities from discriminating among telecommunications carriers in levying franchise fees and right-of-way charges.

THE NFS COMPANIES

MFS is a leading provider of telecommunications services for the business community and the nation's largest competitive local telecommunications service provider, offering a wide variety of sophisticated voice and data services in competition with local telephone companies. In addition, MFS has forged the path for competition in the local telecommunications market. In four years, MFS' subsidiary, MFS Telecom, has grown from a company with a single fiber optic network in Baltimore to a multifaceted telecommunications service provider with almost 600 employees and fiber optic networks in fourteen major metropolitan areas. Over that same period, the companies' sales have, on average, doubled every six months.

Consistent with the underlying themes of S. 1086, the rapid development of MFS and other competitors demonstrates that technological innovation, enhanced network reliability, improved service quality and future job growth are most often generated by small and medium sized entrepreneurial companies—not by large telecommunications providers wedded to a blind adherence to regulated monopoly. However, the true potential of competition cannot be realized until all barriers to competition are removed.

Indicative of the responsiveness and market leadership of entrepreneurial telecommunications entrants, MFS recently introduced several new service offerings specially tailored to meet the unique telecommunications needs of small and medium sized businesses and the increasingly complex high speed computer networking needs of both large and small business users. Many of these services and customer applications were not offered by the local exchange carriers prior to their introduction by MFS. Specifically, MFS' subsidiary, MFS Intelenet, provides advanced telecommunications services to small and medium sized businesses as a single source for comprehensive telecommunications services, making available to smaller users quality and pricing levels that are comparable to those available to larger communications users. MFS Intelenet began providing this comprehensive telephone service in New York City earlier this month.

Through its subsidiary, MFS Datanet, MFS recently introduced an innovative service which permits businesses to transport data between buildings, across town or across the nation by connecting local area networks ("LANs") with telecommunications facilities that transmit data at "native LAN speeds." These services allow personal computers and workstations on one LAN to communicate with personal computers and workstations on another LAN at the full, uncompromised speed at which computers on the same bAN can communicate, allowing for the development of a wide array of new productivity-enhancing multimedia applications. MFS Datanet's services are offered at a variety.of data transmission speeds to allow customers to choose the level which best satisfies their needs. In addition to high-speed LAN interconnection, MFS Datanet offers data telecommunications for mainframe computer channel extensions, high-speed remote file backup, file transfer and a variety of other purposes. The favorable response these services have received from the business community shows that, given proper competitive signals, private investors will ensure the development of information highways in ways that most efficiently and effectively meet users' needs.

Further, MFS, through its Network Technologies subsidiary, provides network systems integration services. In addition to constructing MFS' fiber optic networks, Network Technologies offers network systems integration services and turn-key facilities management services. These network integration projects include remote interactive learning facilities ("distance learning"), combined cable television-telephone networks and intelligent vehicle highway systems.

COMPETITION HAS DEBUNKED THE MYTH OF A NATURAL MONOPOLY IN LOCAL TELECOMMUNICATIONS SERVICES

If S. 1086 accomplishes one thing, it conclusively shows there are no inherent or technical barriers to full-fledged local competition. Yet, historically, many legislators and regulators believed the reverse. Contrary to popular belief, the history of the local exchange telecommunications market is not one impressed with the concept of a natural monopoly. Rather, in many respects, it is the history of entrepreneurial companies and innovation. The industry began not with large established companies but with the then-embryonic AT&T and other entrepreneurial upstarts. As with computer technology and other growth industries, important commercial innovation often comes from the hothouse of entrepreneurial companies, not large mature enterprises. Government's greatest contribution to these companies often is eliminating unnecessary and outmoded market-constricting regulations and permitting these entities to reach their full market potential. It is a lesson that sometimes seems lost on the entrenched telephone monopolies and many state and local telecommunications policymakers, but one that fortunately is recognized by the drafters of S. 1086.

To understand the economic bankruptcy underlying much state and local telecommunications regulation, it is important to understand that the Bell System was not built on a legally sanctioned "natural monopoly." Rather, it was devised nearly one hundred years ago by the actions of two men: the President of AT&T, Theodore Vail, and J.P. Morgan, its financial controlling partner. Morgan used control of capital markets to assist Vail in persuading independent telephone companies to sell out to the Bell System. Refusal to interconnect independent telephone companies to the Bell-controlled long distance network was a favorite tactic. (Indeed, refusal to interconnect has remained a persistent theme in telecommunications history.) This conduct continued until 1913 when the Bell System agreed to cease these practices in return for de facto governmental ratification of the Bell System's monopoly in long distance and in its local service territories. The Bell System subsequently embraced profit regulation (rate-of-return regulation) by the government in return for a monopoly franchise. This local monopoly concept was quickly memorialized and replicated nationwide in state public utility laws and other statutes.

In the 1970s, the Bell System, now firmly entrenched, began to fail satisfying customer needs for two interrelated reasons: silicon and market segmentation. As technology exploded, in the form of silicon-based integrated circuits and optical fiber, as communications needs began to migrate from commodities like switched voice minutes and began to segment and statify, the Bell Systems' muscle-bound, bureaucratic nature became an impediment to progress. The inability of the Bell System (or, for that matter, any single entity) to respond rapidly to changes in technology and customer demand for telecommunications services provides indisputable evidence that neither the long distance market nor the local exchange market is a natural monopoly. Simply put, when customers' needs and technology change, society is far better served by a diversity of choices delivered by a competitive market. S. 1086 recognizes this economic reality and, if enacted, will make the promise of telecommunications plurality and diversity a reality.

Indeed, economic history clearly demonstrates that, when markets and technology change rapidly, customer demand will almost inevitably result in competition, even if opposed by the entrenched forces of the status quo. Thus, demand for more cost effective long distance service and the advent of microwave transmission technology led to competition in interexchange services even though at the time regulatory bodies provided scant encouragement to MCI and other "upstarts." Similarly, the introduction of optical technology and customer demand for secure, high bandwidth services led directly to the start of the limited competition now present in local special access and private line services.

These same forces—rapidly changing markets and technologies—are still at work and, in fact, the pace of change is accelerating. Telecommunications markets are stratifying. Residential users want mobility and more diverse and personally tailored communications and entertainment. In business and government markets, a sea change is occurring. Men and women, machine and information are forming an interconnected web across the entire nation. When the switched network outside buildings can support transmission speeds that so far have been limited to local area networks inside buildings, the nation will benefit from advances in learning and productivity that we are just beginning to explore. As S. 1086 correctly recognizes, these forces—stratifying customer needs, technological advancements and product innovation—over time certainly will overwhelm any attempts to sustain the regulatory sanctioned exclusive franchises held by inefficient, technically outmoded local exchange companies that have never been driven by the needs of the market.

As the drafters of S. 1086 clearly understand, competition in the local exchange market is both inevitable and desirable. As in the long distance market, I expect that we will soon see, following enactment of this bill, a considerable number of new entrants employing diverse technologies to serve specific customer needs. It will include many players, each organized more around markets rather than around regulatory fiat. This time, however, if legislation and regulation permits, future industry leaders will be positioned more to serve customers' needs than geographic, technical or legal precedent. More than any other measure in recent memory, S. 1086 is an important step towards this inevitable reshaping of the industry.

PRIVATE CAPITAL, NOT RATEPAYER OR TAXPAYER DOLLARS, WILL FUND THE DEVELOPMENT OF OUR TELECOMMUNICATIONS INFRASTRUCTURE

While government can certainly be a positive force in eradicating barriers to entry, central government planning in the selection and endorsement of technologies and investment strategies can often unintentionally stifle innovation and creativity. Nowhere is this phenomenon better illustrated than in the decision whether to fund future network construction through, on one hand, public financing and ratepayer subsidization or, on the other hand, private investment.

Today, thanks in large part to vigorous long distance telecommunications competition and the unrestrained development of technology, an infrastructure of nationwide information superhighways connecting users throughout the United States is already in place. It is the local "on and off" ramps and the metropolitan beltways that are the current bottlenecks and are in need of further expansion and modernization. Fortunately, as evidenced by MFS and other providers, this construction has been and can be accomplished through private investment—not by taxpayers or ratepayers—but it requires the removal of long-embedded roadblocks to competition in the local telecommunications market.

The promise of competition is that the cost of facilities and the task of creating employment opportunities will be borne by the private sector with an efficiency and perseverance that can not be matched by Government financing and which can be thwarted by regulatory policies that handicap technologies and industry participants. To paraphrase recent remarks by Commissioner Ervin Duggan of the Federal Communications Commission ("FCC"), the welcomed news in this time of budget austerity is that, when it comes to building new communications infrastructure, policymakers can call forth billions of dollars for investment, create immense public benefits and fuel economic growth without spending public money. As Commissioner Duggan correctly stated, this future construction can be accomplished "through intelligent regulatory behavior that inspires a private response."

Equally important to ushering in competition is that all service providers have the flexibility and market opportunities to apply technological changes without encountering and being forced to overcome artificial regulatory hurdles. Government should not seek to dictate technical standards or construction schedules. Government-mandated technology standards will almost certainly be superseded rapidly by the furious pace of technological change. The computer industry aptly illustrates the incentive for innovation generated by competition and fostered by the absence of artificial regulatory obstacles. If government had attempted to dictate or handicap technology for the computer industry, we would not have the explosion and blinding range of computing devices and applications we have today. There would be no Microsofts, Intels or Apples. Conceivably, the dire predictions of the inevitable decline of the U.S. chip-making and computer industries often voiced before Congress less than a decade ago would have occurred. Instead, the power of competition coupled with American innovation has reinvigorated our high-tech industries and made them the envy of the world. Individual corporations—some icons of American industry—have declined and lost the race. But it is clear that America and American workers are the winners.

Open entry and competition provide the very market opportunities that the private sector craves—the opportunity to invest and respond with innovative products to meet a market demand that is currently underserved and, in some instances, unserved. MFS alone has invested almost 500 million dollars in building nearly 50,000 fiber miles of local "on and off" ramps for users to access intercity and interstate information superhighways. We plan to double this number in the near term. However, our previous accomplishments are but a tiny fraction of the benefits that could be realized if outmoded, restrictive regulations gave way to government-encouraged competition. Indeed, the primary lesson we can draw from the history of competition in the long distance market, as well as from the overwhelming response of users to entrepreneurial providers of competitive services such as MFS, is that thoughtful government action can unleash the capital resources of the private sector, while preserving and enhancing affordable universal service and dispelling the persistent myth of a natural monopoly, as embodied in many hoary state and local laws.

A NATIONAL TELECOMMUNICATIONS POLICY IS CRITICAL TO ELIMINATING ARCHAIC BARRIERS TO MARKET ENTRY AND ADVANCING COMPETITION IN THE PROVISION OF ALL TELECOMMUNICATIONS SERVICES

The fruits of competition will never ripen if competitors are strangled by a thicket of inconsistent and antiquated state and local laws premised on an historically obsolete economic theory. MFS therefore strongly endorses a coordinated national policy advancing open entry and competition for all telecommunications services, as reflected in S. 1086. Indeed, a comprehensive national policy is essential to ensure that all Americans reap the benefits of full competition and universal service. Absent a coordinated national policy promoting open telecommunications entry, the deployment of a nationwide advanced infrastructure will depend on the uncoordinated and ad hoc efforts of all 50 states and the District of Columbia. Competitive progress in some jurisdictions may not occur for at least a decade. States still imbued with the natural monopoly myth and resistant to competition will effectively deny their residents, for many years, the benefits of an advanced infrastructure and unfortunately will balkanize the United States into information "haves" and "have nots."

Notwithstanding any currently valid technical or economic rationale, many states nostalgically continue to cling to the monopolistic vision of local telecommunications services advocated decades ago by the then-consolidated Bell System, but long since outdated by technology and segmentation of the telecommunications market. Even though MFS has found opportunity to compete for narrow segments of the national telecommunications market, many laws, rules and regulatory policies are still premised on a single provider model. That misguided paradigm allows exclusive or preferential treatment of the incumbent local telephone company, not-because of belief in the efficiency of monopoly, which is surely seen as oxymoronic by virtually all thoughtful observers, but simply as a vestige of its historical role as the sole provider. Unlike many historical myths, this one sadly has the effect of repressing efficient competition and of increasing costs for telecommunications users and society.

The benefits of competition have already been demonstrated in the limited markets in which competitive local carriers, such as MFS, have operated to date. Compared to the entrenched local carrier, MFS has significantly condensed the time required to initiate new services, reduced installation intervals, decreased repair time, offered a wider array of services, increased emphasis on quality, and accelerated deployment of fiber. For example, in the area of transmission quality, the BellCore objective for a standard type of high capacity circuit (a "DS-1") is one error per one million bits of information. The MFS objective is one error per one billion bits of information or one thousand times better than the BellCore standard.

As a result of competition, we are more than a little proud to note that some local exchange carriers now advertise DS-1 capacity service comparable in quality to MFS' service in select areas. Current market conditions demonstrate unequivocally that MFS and other competitors have acted as a widespread competitive spur toward local exchange carriers. Since the advent of competition, monopoly local exchange carriers have begun to offer more reliable and secure network features, including redundant electronics and diverse routing.

Similarly, the pricing of telecommunications services has changed significantly in recent years. For example, five' years ago, monopoly local exchange carriers were arguing before the FCC for the ability to set interstate DS-1 rates at "strategic" levels, i.e., levels admittedly well in excess of cost.[1] Within two years, however, after competitors began to enter limited markets, the local exchange carriers drastically and nearly uniformly reduced their interstate DS-1 rates. Significant rate reductions are also evidenced in rates for intrastate services where competition exists.

Competition is also fully compatible with the important goal of universal service—an objective which MFS fully shares. There is simply no reason to believe and no evidence to support the claim repeatedly chanted by entrenched monopoly providers that the introduction of competition into the remaining monopoly local telecommunications markets will impair universal service. While certain local exchange carriers imbued with the myth of a natural monopoly reflexively assert that competition will result in increased residential rates or the impairment of universal service, none has proven it. Moreover, whatever the hypothetical argument may be, the real world experience of states like New York and Illinois which have permitted increased levels of competition simply does not support the LEC position. In fact, MFS has consistently supported the right of all Americans to affordable local communications and has proposed an open universal service subsidy funded by all industry particpants and administered on a fair and equitable basis by a neutral entity.

INTERCONNECTION IS ESSENTIAL TO THE CREATION OF MEANINGFUL LOCAL COMPETITION

S. 1086 properly recognizes that the economical interconnection of competing networks at all feasible technical points is a prerequisite for meaningful competition in the local exchange. The critical importance of interconnection as a gateway to competition and to the elimination of the local exchange bottleneck has been embraced by the FCC and is being recognized increasingly by enlightened state regulators. As recognized in S. 1086, the public interest requires that all users of telecommunications services be able to communicate with all other users, and with all locations in the state, regardless of which carrier's network serves the customer's particular location; this ability in turn requires that all networks be interconnected on economically reasonable and rational terms. Unfortunately, the intrastate carrier-to-carrier interconnection arrangements that exist today are not generally suitable for competitive local carriers and local exchange carriers naturally have no desire or incentive to enter voluntarily into new arrangements with companies that seek to compete with them.

Forward-looking telecommunications policymakers have long recognized the central importance of equitable and cost-effective interconnection as they guided the introduction of competition in other markets. In the 1970s, the former Bell System attempted over a period of years to repeatedly deny its infant long distance competitors access to its local exchange facilities. As the District Court in the AT&T divestiture case explained, "meaningful competition in the provision of intercity services is precluded unless the non-Bell carriers are able to obtain interconnection with the Bell local distribution facilities under non-discriminatory terms and conditions.[2] Moreover, the success of the cellular telephone industry was made possible in large part by early FCC actions mandating efficient interconnection to the landline telephone network, although this process required repeated regulatory intervention.

A similar, if not more pressing, need for coordinated national policy leadership exists in the local telecommunications market. Local exchange carriers currently control ubiquitous transmission and distribution facilities that pass every residential and business premise in the nation. Competitors will have no viable opportunity to offer local services to the general public, as opposed to highly limited and select groups of customers, without cost-effective access to existing local exchange carrier networks. S. 1086 recognizes this need and provides all competitors the necessary safeguards and access to monopoly facilities and resources that will offer them the opportunity to succeed.

PROPOSED MODIFICATIONS TO S. 1086

S. 1086 contains many critical ingredients for meaningful local competition, including ensuring that all providers have access to telephone number allocations, allowing customers to change service providers while retaining their telephone numbers and mandating efficient technical and administrative network interconnections.

---

[1] See generally, Investigation of Special Access Tariffs of Local Exchange Carriers, 4 FCC Rcd. 4797 (1988).

[2] *United States* v. *AT&T*, 524 F. Supp. 1336, 1352 (D.D.C. 1981).

In addition to the competitive safeguards included in S. 1086, MFS strongly encourages this Subcommittee to consider the following modifications to the bill that will eliminate two of the most pernicious vestiges of the monopoly past that have severely discriminated against competitive local telecommunications providers. These provisions, which MFS believes are fully consistent with the overall thrust of the legislation, involve: (1) discriminatory conduit rental fees and pole attachment rates charged by telephone companies and other utilities to similarly situated telecommunications carriers; and (2) the discriminatory assessment of franchise fees and right-of-way charges by municipalities on similarly situated telecommunications carriers.

As Robert Allen, Chairman of AT&T, recently stated in specifying the requirements for effective competition:

> I am talking about such things as eliminating franchises; eliminating local telephone company control of "rights of way." I'm talking about allowing competitors to interconnect fully with the local exchange network—not just special access; and unbundling of transport, local-loop, switching and other service elements. I'm talking about cost-based rates—unburdened by unnecessary subsidies; and portable telephone numbers that give consumers real choice. * * *
>
> In short, we have to dismantle the barriers to entry. We have to provide opportunities for entrepreneurs t6 compete and for end users to have choice. And we have to set objective levels that are measurable—levels that, once met, clearly signal that the local market has become competitive.[3]

MFS respectfully submits that these additional barriers to competition inhibit entry into the local telecommunications market by subjecting competitive service providers to unreasonable and discriminatory fees not paid by their competitors. MFS suggests that S. 1086 be can expanded logically to incorporate these significant safeguards.

1. Non-Discrimination in Pole Attachment Rates Charged by Utilities to Telecommunications Carriers and Cable Television Systems

Pole attachment rates and conduit rental fees together comprise one of the largest recurring costs borne by a competitive telecommunications carrier. Competitive local carriers are often charged rates that bear no relation to the cost of supporting the attachment and that are substantially higher than the rates utilities charge cable television systems and their affiliated competitive local access providers. S. 1086 begins to ameliorate discrimination in access to pole, conduit, duct and right-of-way space in Section 229(c)(3). This section directs the FCC to prescribe regulations requiring each telecommunications carrier to provide any entity that provides telecommunications services or information services, on reasonable terms and conditions, nondiscriminatory access, where technically feasible, to poles, ducts, conduits and rights-of-way owned or controlled by the telecommunications carrier.

In order to meaningfully eliminate discrimination in access to pole attachments for competitive local telecommunications providers, MFS recommends that this provision be modified in two ways. First, the nondiscrimination requirement of Section 229(c)(3) should be amended to include poles, ducts, conduits, and rights-of-way provided by all utilities, not just telecommunications carriers. Second, in response to a 1991 decision by the FCC under the Pole Attachments Act (recently upheld by the Court of Appeals),[4] Section 229(c)(3) should be modified to explicitly bar utilities from discriminating in the price, terms and conditions for the provision of conduit space and pole attachments between independent entities providing telecommunications services and similarly situated entities affiliated with cable television systems. These two modifications would extend, in a nondiscriminatory manner, the regulatory protection against excessive conduit and pole attachment rates assured under the Pole Attachments Act to all similarly situated providers of telecommunications services.

The same discriminatory and anticompetitive practices that prompted Congress to enact the Pole Attachments Act in 1978 in an effort to protect cable television systems are now faced by independent competitive local telecommunications providers. In 1978, Congress enacted the Pole Attachments Act in direct response to concerns that telephone companies and electric utilities would use their control over poles—as well as duct and conduit—to charge cable television systems unjust and unreasonable rates for pole attachments and use of conduit. Congress was justifiably concerned that conduit rental fees and pole attachment rates charged by such utilities

---

[3] Remarks of Robert E. Allen, Federal Communications Bar Association/Practicing Law Institute Conference, Washington, D.C. (December 3, 1992).

[4] 447 U.S.C. §224; *Heritage Cablevision* v. *Texas Utilities Elec. Co.*, 6 FCC Rcd. 7099 (1991); aff'd., *Texas Utilities Electric Co.* v. *FCC*, No. 92-1032 (June 25, 1993).

would make cable services prohibitively expensive for most users or result in the complete absence of such service in the most extreme cases. Under the Pole Attachments Act, cable television system operators are entitled to seek regulatory review of pole attachment rates charged by electric companies and other utilities, including local telephone companies. The Pole Attachments Act is not limited to facilities controlled by telephone carriers.

Just last month, the U.S. Court of Appeals for the D.C. Circuit upheld a decision by the FCC asserting regulatory jurisdiction over an electric utility's pole attachment rates for a cable television system that provided not only traditional cable services but also data transmission services.[5] The FCC required the utility to charge the cable television system a single, just and reasonable rate for all pole attachments regardless of their use in the franchise area. As a result of the FCC's decision, cable television systems that use their networks to provide voice and data telecommunications services will be able to secure regulated pole attachment rates for all services in their franchise areas. This lower rate and the attendant regulatory oversight is not available to competitive telecommunications providers that do not operate as cable television systems but may compete directly with such entities in the telecommunications market.

In the case before the FCC, the electric utility charged separate pole attachment rates for traditional cable television and data transmission services. The annual rate charged for traditional cable television services was $5.00 per pole; the annual rate for non-cable television services was between $50 and $100 per pole.[6] MFS routinely has been subjected, in a clearly discriminatory manner, to the higher rates for conduit and pole attachments by utilities and telephone companies alike.

Significantly, in its decision, the FCC recognized that its regulation of all cable television system pole attachments would provide cable television system operators an advantage over other competitors who are not entitled to regulated rates under the Pole Attachments Act. The FCC, however, was evidently constrained by the limits of its statutory authority:

> We are mindful of the competitive concerns raised by TU Electric (and MFS, citations omitted). However, we cannot refrain from fulfilling our statutory mandate to ensure just and reasonable pole attachment rates, terms and conditions for cable television systems operators, simply *because our authority may not extend to every competitor in the data transmission market.*[7]

In 1978, the Pole Attachments Act was passed in response to the similar determination by the FCC that it lacked jurisdiction to adjudicate disputes between cable television system operators and utilities over conduit rental fees and pole attachment rates. MFS believes that today, as in 1978, Federal legislation is the only way to ensure just, reasonable and nondiscriminatory conduit and pole attachment rates, terms and conditions for all entities providing local telecommunications services to the public.

As the distinctions between cable companies and telecommunications carriers continue to blur, the difference in these rates will put non-cable providers of telecommunications services, such as MFS, at a significant competitive disadvantage. Accordingly, MFS encourages this Subcommittee to modify Section 229(c)(3) to prohibit utilities and telecommunications carriers from discriminating in the price, terms and conditions of attachments to poles and access to ducts, conduits and rights-of-way between different types of entities that provide local telecommunications services.

2. Nondiscrimination in Franchise Fees

The second source of recurrent discrimination against emerging competitive local telecommunications providers is the selective imposition of franchise fees and rights-of-way charges by municipalities. Once again, most of these local ordinances are rooted in the century-old theory of local natural monopolies, in which the entrenched carrier is thought to be qualitatively superior to any possible entrepreneurial upstart whose activities must be heavily conditioned and restricted. In many metropolitan areas, municipalities prohibit competitive local carriers from crossing public rights-of-way unless they first obtain a franchise from the municipality. As a condition of the franchise, MFS and other competitive telecommunications providers have been required to pay exorbitant franchise fees that are not imposed on the local telephone company for virtually identical facilities and services.

By way of illustration, the City of Pittsburgh has taken the position that MFS is required to obtain a "Private Communications System License" from the City. As

---

[5] *Texas Utilities Electric Co.* v. *FCC*, No. 92-1032 (June 25, 1993).
[6] Heritage Cablevision, 6 FCC Rcd. at 7100.
[7] Id. at 7104 (emphasis added).

a condition of the license, MFS is required to pay the City five percent of its total annual gross revenues derived from customers within Pittsburgh. significantly, the City of Pittsburgh has not required Bell of Pennsylvania—with whom MFS directly competes using virtually identical fiber facilities—to pay such a fee. In New York City, MFS is subject to a franchise fee as high as 14 percent of its total annual gross revenues. By contrast, New York Telephone is not required to pay any similar franchise fee.

The wide divergence of license and franchise fees imposed by municipalities on competitive telecommunications providers is systematic and inhibits true competition and the further deployment of fiber optic networks and other advanced technologies. If this practice of discrimination continues, it will directly undermine the threshold goal embodied in S. 1086 of providing fair competitive opportunities in the provision of all telecommunications services. Furthermore, discriminatory franchise fees will result in fragmented deployment of an advanced telecommunications infrastructure and, as with state regulatory barriers to competition, could produce a balkanized society of information "haves" and "have nots". Accordingly, MFS respectfully urges this Subcommittee to amend S. 1086 to require the uniform and nondiscriminatory application of municipal franchise fees and rights-of-way charges to all telecommunications providers.

CONCLUSION

I greatly appreciate the opportunity to testify in support of S. 1086—a bill of critical importance to the future of our nation's telecommunications infrastructure. The good news is that our national goal of universal access to a vast array of information can be promptly accomplished by private investment. Companies such as MFS stand ready to continue making the investment necessary to improve not only our nation's information highways but also the local "on and off" ramps and metropolitan beltways. S. 1086 seeks to accomplish this objective by proposing to remove many of the government-mandated, antiquated and monopolistic barriers to competition. If this bill is enacted, competition can be counted on to drive technological advancement and industrial creativity and to help ensure that America leads the world into the Information Age.

Senator INOUYE. Thank you very much, Mr. Crowe. We will be asking questions after the panel has completed its presentation. Now may I call upon Mr. Roberts.

STATEMENT OF BRIAN L. ROBERTS, PRESIDENT, COMCAST CORP.

Mr. ROBERTS. Thank you, Mr. Chairman and members of the committee. I am Brian Roberts, the president of Comcast Corp. We are the fourth largest cable operator, and we are also cellular telephone operators on the east coast. We are building cable and telephone facilities in the United Kingdom as well. And I am pleased to be here today as an officer of the national Cable Television Association and to represent NCTA at today's hearing.

I start by wanting to commend Mr. Chairman and Senator Danforth for their comprehensive approach to these issues that are truly vital to the social and economic future of our Nation. Speaking for the members of NCTA, I want you to know that we are prepared to work with you to ensure that America's telecommunications infrastructure is second to none, and to get good legislation enacted.

You see magazine covers standing behind us and more newspaper headlines every day proclaiming that this is the era of convergence in telecommunications. And there is, indeed, lots of activity and excitement in our field. We are all jockeying for position, entering new strategic alliances that cross the traditional boundaries of communications, computers, publishing, and video. There is a convergence underway, and I believe you will find cable at its center.

Cable's broadband, coaxial, and fiber optic networks are now available to 95 percent of all U.S. homes. The coaxial cable that feeds into your TV set can already carry an immense amount of information, 900 times more than the telephone company's twisted pair. And very soon, that capacity will increase dramatically as we install new fiber optic trunks and apply digital compression technology. We are also modifying the architecture of our systems to improve their ability to carry two-way communications.

All told, our industry hopes by the end of this decade to nearly double our total investment in plant and facilities. We plan to spend another $14 billion on system upgrades, and that means economic growth and up to 120,000 new jobs.

While these technologies are converging, opportunities for competition among network providers are expanding, and this will benefit every American. Your bill represents an important initial step in defining a policy framework for these changing and converging industries. We wholeheartedly agree with your core premise that the best telecommunications infrastructure is a competitive infrastructure, a network of interconnected networks, wireline and wireless, high and low capacity, fixed and mobile.

We also strongly support your goal of opening up the local loop to competition. This is the one segment of the telecommunications industry that has remained a monopoly impervious to competition. Right now, well over 99 percent of all telephone calls cross the traditional local exchange network, including almost every long distance call and cellular telephone call. Right now, for every telephone user there is no alternative to a local exchange company for purposes of connecting with the long distance telephone network or dialing across town.

Right now, all the alternate access carriers, like Metropolitan Fiber and Teleport, combined, account for one-tenth of 1 percent of the total local exchange revenue market, and are very limited, by law, in the services that they may provide. This monopoly should no longer be justified and can no longer be justified by the technology, and we believe that effective Federal policies can help change it.

Opening local phone service to competition has become a government policy priority in many other nations. We expect local exchange competition will be introduced throughout the European Community by 1998, and also in Mexico. And today Comcast is participating as a local exchange competitor in the United Kingdom. The United States should not be playing catchup. We need to get back ahead of the pack.

Mr. Chairman, we endorse your commitment to remove Federal, State, and local barriers to local exchange competition, and to promote interconnection and equal access to local exchange network functions and facilities. However, it is critically important to ensure that local networks are opened up to competition in a consistent and uniform way. Therefore, we believe the FCC and State regulators will need more detailed guidance from you on these issues.

We also understand, as you do, that competition is not a one-way street. It is inevitable, and even desirable from a public policy perspective, that everyone, including local exchange companies, should ultimately be free to compete to deliver multichannel video to the home, just as in every other aspect of the telecommunications business.

As technology advances and competition emerges, the question is no longer "whether," but "when and under what circumstances." But so long as incumbent local exchange companies continue to dominate their market, safeguards against phone company abuse of its competitors and its consumers remain essential.

We urge you to go beyond the safeguards in S. 1086 and to adopt more specific safeguards such as those contained in a bill, Mr. Chairman, that you introduced in the last Congress, S. 2112, which established detailed separate subsidiary requirements and provided for stricter rules on cost allocation and treatment of intangible assets. In addition, we believe you should establish a threshold test requiring that the local exchange telephone market must be found effectively competitive before the phone company may enter the multichannel video business.

Last year you reimposed stiff regulation on my industry, cable TV, while at the same time taking steps to ensure that cable's multichannel competitors will flourish. You insisted that the existence of real, effective competition must be demonstrated before the incumbent cable operator would be deregulated. The phone companies, we think, should be treated the same way. They should remain subject to the cross-ownership restriction until the market they dominate, local exchange service, is subject to effective competition.

Meanwhile, Congress should take steps, as you do in your bill, to ensure that the phone companies' new competitors can indeed establish themselves.

There is a fundamental linkage between permitting the local exchange companies into cable and promoting competition into local telephone exchange. The most likely providers of competitive local exchange services, I believe, are today's cable companies.

I cannot overstate the State and local barriers to entry that we face, and the problems of interconnection, equal access, number portability, and standard setting will take years to resolve. Only when these problems are solved and competition has really taken hold will the phone companies' economic incentive to shift costs to monopoly ratepayers be eliminated. Only then should those companies be permitted to deliver video within their local exchange area.

Universal service has been and remains an important social goal. We in the cable industry share this committee's commitment to it, and if we are given the chance to compete for local phone service, we will contribute our share to preserve universal service. I hope that you will give me a chance to elaborate on that point during the Q&A.

And finally, your bill appears to bar, without exception, joint ventures between local exchange companies and cable companies, or acquisitions of cable companies by local exchange companies within the service area of the local exchange carrier. In some circumstances a given market may just not sustain competing broadband networks, and it would be unfair to force companies to compete uneconomically. We urge you to establish standards to permit mergers or acquisitions under requirements that would ensure the public interest is protected.

If Congress creates the right legal environment, America can enjoy a competitive, innovative network of networks offering more choices, better prices, and more responsiveness to our diverse needs, and today's cable companies will be major participants, we hope. But this future depends upon your unswerving determination to put competition first, to break down barriers to entry and promote interconnection, and to avoid premature elimination of rules intended to protect consumers and competitors against abuses by local telephone monopolies.

We think this bill is on the right track. We offer you our help as you develop the policies that will make this future a reality.

Thank you.

[The prepared statement of Mr. Roberts follows:]

PREPARED STATEMENT OF BRIAN L. ROBERTS

My name is Brian Roberts, and I am president of Comcast Corporation, headquartered in Philadelphia, Pennsylvania. I am pleased to represent the National Cable Television Association (NCTA), of which I am an elected officer, at this morning's hearing on S. 1086, the Telecommunications Infrastructure Act of 1993.

On behalf of all the members of NCTA, I want to commend you, Mr. Chairman, and Senator Danforth for your effort to deal in a comprehensive way with telecommunications policy issues that are critically important to the social and economic future of our nation. We support the basic thrust of your proposed legislation, and in particular its strong emphasis on promoting competition in the local exchange telephone marketplace, the last remaining bottleneck in our telecommunications infrastructure. My company, and all of us in the cable industry, are eager to work with you to make sure that the policies Congress adopts will promote effective competition and permit American companies to expand their leadership in these fields into the next century.

There is a single objective that every member of this Committee, every witness before you today, and every member of the cable television industry share: we all want the United States to have a telecommunications infrastructure that is second to none, anywhere in the world. There may, however, be differences of opinion on the best way to reach this goal.

We in the cable industry agree with you, Mr. Chairman, and you, Senator Danforth: our nation should rely on a competitive marketplace to deliver to all Americans the widest range of telecommunications services at the best prices. The infrastructure model that will best serve America's needs is not a government monopoly, nor a private monopoly, but rather a network of competitive, interconnected networks—wireline and wireless, high- and low-capacity, fixed and mobile. The result should be a vast telecommunications quilt, composed of many pieces but making up an integrated whole, blanketing our nation.

The essential role of government in deploying our telecommunications infrastructure is to make competition work, to create the environment in which the entrepreneurial spirit can flourish and innovation can thrive, and in which companies like Comcast can make an important contribution to an improved quality of life for all Americans. Today, I will offer some of the cable industry's ideas on how to create such an environment.

ABOUT COMCAST CORPORATION

Let me briefly introduce you to Comcast Corporation. Our company was founded some thirty years ago by my father, Ralph Roberts, who has built Comcast into one of the nation's truly great cable communications companies, now serving some 2.7 million subscribers in 19 states. We are also an important "non-wireline" cellular telephone provider, serving a contiguous region running from the Newark (N.J.) Airport to the outskirts of Baltimore, including most of New Jersey, Philadelphia, Wilmington, and portions of Maryland. We have also invested in two alternative access companies that provide a link-up with long distance services for high-volume business telecommunications users. We also operate cable and telephone systems abroad, including combined cable/telephone networks in the United Kingdom that provide British consumers a telephone service alternative to British Telecom. We are also making important new commitments to wireless communications services in the U.S. through our investment in Nextel, which will offer new digital private radio services in major urban areas starting later this year, and through our pioneering experiments with "personal communications service" or PCS. Finally, we are partners with Barry Diller and Liberty Media in QVC, a cable home shopping service, that is an important platform for the development of new interactive video services.

Comcast, like many other companies whose core business has been cable television, is in the process of restructuring itself for the Twenty-First Century. Our goal is to integrate these myriad networks and services in order to provide consumers with unparalleled choice and convenience. We are putting in place the digital platform that will meet the telecommunications needs of generations to come.

CABLE IS THE BROADBAND SUPERHIGHWAY * * * TODAY

If magazine covers and newspaper column inches are a fair measure of the importance of an issue to the public, the convergence of telecommunications and related technologies certainly ranks high.

This is a time of great activity and excitement in the telecommunications field, as scores of companies jockey for position and enter new strategic alliances that cross the traditional boundaries of communications, computers, publishing and video.

America's cable television companies have an important role in that future. We already have in place coaxial and fiber optic networks of immense capacity, available to 95 percent of all U.S. homes. The coaxial cable that feeds into your TV set can carry 900 times as much information as the telephone company's twisted pair. That is a powerful connection. As we install new fiber optic trunks over the next five years, cable's already great capacity will double, and cable system reliability will increase dramatically. Meanwhile, as we apply digital compression technology and install new converters in cable homes, channel capacity will increase tenfold—to 500 channels or more. And as we modify our cable system architectures to improve their two-way capability dramatically, we expect an explosion of interactive media, giving Americans new ways to communicate, learn, complete commercial transactions, and just enjoy themselves.

By the end of this decade, the cable industry as a whole plans to invest another $14 billion in system upgrades. That represents major technological advancement. That also represents economic growth and job creation for America.

We have provided for the record of this hearing two recent reports by the National Cable Television Association that describe cable's role in the information age and in the advancement of America's telecommunications infrastructure. As these reports show, cable television systems, standing alone or interconnected with other wireline and wireless networks, are an important part of the technological mix that will meet a wide spectrum of residential and commercial telecommunications needs. Cable TV systems are providing a ubiquitous broadband superhighway today.

THE "NETWORK OF NETWORKS" WILL BENEFIT ALL AMERICANS

Simultaneous with the convergence of technologies is growing competition among network providers. The growth of competition in the telecommunications marketplace will pay dividends for every American.

Consider the results of the break-up of AT&T almost ten years ago, which produced new competition in long distance, telecommunications equipment, and information services. Long distance rates have dropped dramatically, while quality and innovation have improved. The black rotary telephone, still commonplace a decade ago, is a distant memory, and we enjoy a previously unimaginable proliferation of devices—cordless phones, answering machines, fax machines, sophisticated multiline phones, and more. The wisdom of relying on competition is confirmed by the steady stream of other countries liberalizing their own telecommunications networks. For instance, the European Community has resolved to open up all local exchange services to competition by 1998, and Mexico plans to do the same by 1996.

As a cable company, we face growing and real competition in our core business, as direct broadcast satellites, wireless cable and other multichannel video competitors enter the marketplace. Government policies are encouraging, supporting and nourishing these new entrants. This competition is genuine, it is inevitable, and we must all respond to it.

Now is the time for comparable policies to open up the telecommunications bottleneck that has remained impervious to meaningful competition: the local telephone exchange.

Looking ahead, we believe that technology may eventually make possible new entry in the local exchange telephone market, which currently generates $100 billion a year in revenues. Americans will get the benefits of new competition in this marketplace before most other nations will. And the growth of competition in this market will promote, rather than endanger, the important social goal of universal service, as I will discuss later.

Before Congress can make appropriate public policy on these matters, it is essential to have a realistic grasp of the current state of competition in the local exchange market.

The local exchange network is distinguishable from every other telecommunications facility in America. It provides essential switching and distribution services upon which every other provider of telecommunication services depends—in short, a monopoly. These same companies control the poles and conduits through which every other wireline provider of telecommunications services must pass. And this will not change overnight.

Right now, well over 99 percent of all telephone calls cross the traditional local exchange network, including almost every long distance call and almost every cellular call. Right now, for virtually every telephone user, there is no alternative to the local exchange company for purposes of connecting with the long distance telephone network or dialing across town. Right now, all the alternative access carriers combined account for about 0.1 percent of the total local exchange revenue market.

This is a simple, numerical refutation of the argument that new entry by cable, cellular, or any other technology threatens the revenue base of local telephone companies now or in the immediate future. Quite the contrary—new entrants will mean market growth and increased demand, and this will drive the local phone companies to improve service, reduce costs and respond to the market. Every increase in traffic on alternative carriers means more incremental revenues for the local exchange company whose facilities are still essential to the completion of every call.

By the same token, regulations premised on the existence of a local telephone monopoly are still entirely appropriate. Local exchange companies will retain the ability to abuse consumers and competitors until such time as their bottleneck is eliminated. The facts are plain—that bottleneck remains secure, and will for some time.

OBSERVATIONS ON S. 1086

Government has an essential role in making competition work. And given the right ground rules, the benefits of competition should ultimately be within the reach of every consumer. Your proposed legislation, S. 1086, "The Telecommunications Infrastructure Act of 1993," represents an important initial step in defining a policy framework for these changing and converging industries.

Your bill embraces a number of key principles that are essential if America is to have the world's best telecommunications infrastructure.

S. 1086 recognizes that the best telecommunications infrastructure is a competitive infrastructure and places maximum reliance on private investment to build out the infrastructure. In order to promote competition, your bill would begin to break down unnecessary regulatory barriers, particularly those that preclude entry in the local telephone exchange market. S. 1086 recognizes that equal access, number portability, and interconnection of networks on fair and reasonable terms and conditions are a sine qua non to competitive entry. Your bill also calls for a review of state regulatory limits on competitive entry to ensure that they do not thwart procompetitive federal policies.

S. 1086 recognizes that competition in virtually all aspects of telecommunications is both inevitable and desirable. It also recognizes that government cannot merely declare the marketplace open and let Darwin's law prevail. We have not yet even reached the threshold of meaningful competitive entry in local exchange telephone service, the last remaining bottleneck in the terrestrial communications network. Federal policies that presume the existence of competition before it actually exists, and premise the elimination of restrictions on dominant local exchange carriers on that presumption, will have the perverse effect of killing off competition before it has even germinated. You must ensure that the policies you adopt, taken as a whole, provide a nurturing seedbed for emerging competitors.

With that preface, let me address several of the salient elements of your proposed legislation.

Competition in the local telephone exchange: We strongly support your goal of opening up the local telephone loop to competition. Every American will benefit from the new choice and innovation in local exchange services that only competitive entry can bring. But federal policy must ensure that this competition is real and sustainable, not illusory.

We applaud your commitment to remove federal, state and local barriers to entry in telephony and to promote interconnection and equal access to local exchange network functions and facilities on a nondiscriminatory basis. Your overall policy direction is on the right track, and a bill of this kind can help to ensure consistency and uniformity in the opening up of local networks by the Regional Bell Operating Companies and other local exchange companies—which will be essential if competitive entry is to succeed. However, we hope that any final legislation will provide more detail and guidance to the FCC and state regulators to ensure that Congressional intent is fully carried out.

Telephone entry into the video programming delivery business: We recognize the inevitability, and even the desirability, from a public policy perspective, of ultimately permitting everyone to compete freely in the multichannel video delivery business as in every other aspect of the telecommunications business. As technology advances and competition emerges, the question is no longer "whether," but "when and under what circumstances."

Your bill concedes that the dominant position of incumbent local exchange companies requires that certain pro-competitive safeguards be imposed on their entry into the delivery of video services. Separate subsidiary requirements and proctions against inappropriate cross-subsidies are essential elements of any such legislation, and S. 1086 provides for some of these. We believe that the more specific safeguards contained in the bill that you introduced in the last Congress, Mr. Chairman—S. 2112—should be incorporated into S. 1086. Among other things, that earlier bill established detailed separate subsidiary requirements and provided for stricter rules on cost allocation and treatment of intangible assets.

In addition, we believe there is a critical threshold test that must be a part of any legislation: the requirement that the local exchange telephone market be found "effectively competitive" as a precondition to phone company entry.

Just last year Congress reimposed stiff regulation on the cable television industry, while at the same time taking steps to ensure that cable's multichannel competitors will flourish. Congress declared that much of its regulation of cable will be lifted when the multichannel marketplace is subject to "effective competition," which Congress defined in relevant part as the availability of one or more alternative video providers to at least 50 percent of homes in a service area, and subscription to such alternative providers by at least 15 percent of all homes taking multichannel services.

Today, in an ever-growing number of communities across America, consumers will soon have a real choice among multichannel video providers, including competing cable systems, wireless cable operators, and home satellite dishes. By the end of 1994, Hughes Space Communications will begin beaming 150 video channels to small dishes at homes across America, and U.S. Satellite Broadcasting will also enter the market.

In establishing these new restriction on cable, Congress did not presume that the multichannel video marketplace would be competitive by a date certain, or that by the mere enactment of legislation that marketplace could be deemed competitive—it insisted that the existence of real, effective competition be demonstrated before the incumbent cable operator would be deregulated.

The telephone companies should be treated the same way. Any relief from the restriction on telephone company provision of video programming services should be similarly conditioned on the existence of "effective competition" in the market currently dominated by the local exchange companies. The harm you are trying to curb through regulatory safeguards and other requirements is to make it uneconomic for local exchange companies to shift costs, harming consumers and competitors. Therefore, the best safeguard of all is not to rely solely upon regulation, but to eliminate the local telephone company's economic incentive to shift costs by subjecting it to competition and ensuring that competition can be sustained.

There is a fundamental linkage between letting local exchange companies into cable and promoting competition in the local exchange. The most likely providers of competitive local exchange services are today's local cable companies. Our companies face immense barriers to entering that market, not the least of which is the resistance of many state regulators (fueled by the telephone companies) to grant operating certificates to new entrants, no matter how specialized the market they may seek. The problems of interconnection, equal access, number portability, standard-setting, and others will likely take years to resolve through regulatory and judicial means. Letting the local exchange companies into the video business before their core business is subject to effective competition will simply wipe out economic incentives for competitive entry.

Certainly, conditioning phone company entry into cable merely on the existence of an FCC-approved interconnection tariff, as S. 1086 in its current form would do, is no guarantee that competitive entry will ever occur. On the contrary, freeing the local exchange company to invade the cable business from the moment that such tariff is approved means that the game will be over before it has begun. We have precedent that speaks clearly to the shortcomings of this approach—the FCC's efforts to implement "open network architecture," or ONA, as a means of promoting competition in certain network and information services have been highly unsatisfactory. The local exchange companies cannot be held in check by the mere existence of an open-entry tariff. Congress must ensure that real and sustainable competition has taken hold in the local exchange before LEC entry into the video marketplace is allowed. This is a matter of fundamental fairness, for consumers and competitors alike.

Universal service: Government must ensure that universal telephone service is preserved. This is a critically important social goal, but the way in which we achieve it must be reexamined. Regulations that maintain a monopoly network and allow cross-subsidies to promote access to that network are simply not a viable policy for the next century. Now is the time to explore new approaches.

Your proposed legislation recognizes this fact by stressing that the states and the FCC "shall have as their goal directly assisting individuals or entities that cannot afford the cost of their telecommunications service or equipment." As government policy promotes competitive entry, the focus must shift away from intra-network cost-shifting and toward more targeted assistance to those who truly need it.

We agree that all companies that participate in the local exchange marketplace—incumbents and new entrants alike, including cable operators—must bear some responsibility to contribute to universal service. But the structure of the universal service program, and the flow of funds, must be changed.

There currently are programs in place to ensure universal service which involve government-administered transfers from telecommunications users directly to local telephone companies. In the future, the better approach may be for government to give anyone who can demonstrate that they are the least-cost provider—whether the current local exchange company, a cable operator, an alternative access provider, a wireless operator—a chance to bid to become the "universal service provider." In other words, let anyone who can do the job, do the job. Another approach could be to allow competitive provision of universal service, and allocate these universal service funds to service providers in accordance with their share of the subsidized market; this could have the added advantage of expanding incentives for competitive entry into markets that would otherwise be economically unattractive.

We encourage you to weigh these or other effective ways to accomplish the important social goal of universal service, and we are prepared to work with you on this issue.

Joint ventures. mergers and acquisitions by and between local exchange companies and cable systems: S. 1086 appears to bar joint ventures between local exchange companies and cable companies, or acquisitions of cable companies by local exchange companies, without exception. We believe there must be flexibility in these provisions. In any market where competition is simply not sustainable, it would be extremely unfair and counterproductive to force one or both contenders to continue to provide service. And this may not simply be limited to rural markets. Therefore, any legislation should include appropriate policies to anticipate such circumstances and to permit mergers or acquisitions under requirements that ensure the public interest is protected, including the continued application of detailed pro-competitive and pro-consumer safeguards that would generally apply to their provision of multichannel video services as we have detailed above.

CONCLUSION

Mr. Chairman, you have shown your determination to adopt policies that will bring Americans a Twenty-First Century telecommunications infrastructure, second to none, and faster than the rest of the world. Given the right legal environment, America can have a choice among multiple wireline and wireless providers, offering competitive prices, innovative service options, and myriad combinations of capacity, capability and mobility that will meet our nation's diverse information and entertainment needs.

But the ability of Comcast, other cable companies, and other telecommunications entrepreneurs to contribute to this infrastructure depends upon your unswerving determination to put competition first, to eliminate federal, state and local barriers to new entrants, and to promote compatibility and interconnection of networks. Equally important, we hope you will avoid policies that permit local exchange companies to enter the multichannel video programming marketplace prematurely, or under regulatory safeguards that won't protect consumers or competitors against monopoly abuses. And we also encourage you to permit flexibility in joint ventures, mergers and acquisitions to account for the likelihood that not every community will be able to sustain competing broadband networks, in order to ensure that no community will be deprived of the best possible network.

On behalf of the members of NCTA, thank you again for the invitation to participate in this vitally important dialogue. We are very eager to work with you to ensure that our vision of a competitive telecommunications future can become a reality.

Thank you.

["Cable Television and America's Telecommunications Infrastructure," and "Twenty First Century Television—Cable Television in the Information Age," by the National Cable Television Association may be found in the committee's files.]

Senator INOUYE. Thank you very much, Mr. Roberts. Mr. Ware.

STATEMENT OF LAWRENCE C. WARE, MANAGER, GARDEN VALLEY TELEPHONE CO.

Mr. WARE. Good morning, Mr. Chairman and members of the subcommittee. My name is Lawrence Ware, and I am the manager of Garden Valley Telephone Co. in Erskine, MN. My company is a cooperative that serves approximately 13,000 subscribers in 24 exchanges. Our density is 2.6 subscribers per mile of line. With the help of Federal and State programs, we provide single-party telephone service with 23 digital offices. We also have over 300 miles of fiber optic cable in service and provide a fully interactive television service connecting six area schools.

I am appearing for the Rural Telephone Coalition to explain our grave concerns about S. 1086. The coalition is an alliance of NRTA, NTCA, and OPASTCO. The three associations represent more than 850 small rural telephone systems.

Mr. Chairman, we support universal affordable service and infrastructure development throughout the United States. We understand that is the purpose of S. 1086. However, the coalition believes that S. 1086 is far more likely to raise rural rates, stall rural infrastructure improvements, and create new barriers to expanded service choices for most rural consumers.

S. 1086 would preempt State authority and open local telephone service to competition. Economic theory generally says competition will increase efficiency, lower prices, and improve service, but rural systems typically serve thinly populated areas that support a single network only with the help of existing support structures. Indeed, small rural companies came into existence because marketplace forces chose not to serve high-cost, low-density areas.

Encouraging duplication in rural areas is almost certain to erode our ability to provide high-quality service at affordable rates and upgrade our networks. Selective duplication and massive unbundling on request will erode the limited economies of scale available in low-density markets. Establishing competitive pricing for high-volume customers in rural areas will raise rates for residential and small business customers. Add in the huge cost of local number portability, and you have a recipe for rural disaster.

Now, I want to be clear. We are only asking you not to mandate rural competition. We ask that you continue to let regulators, State regulators, decide when the benefits of rural competition will outweigh the costs. They understand the rural needs and are responsible to the local customers.

The bill does contain some universal service and rural infrastructure provisions, so it is clear that the subcommittee wants to help. In fact, the coalition agrees wholeheartedly with the provision requiring universal service contributions from all carriers. The rest of the universal service and infrastructure provisions will not do the job. The bill deprives the States of most of their time-tested tools to aid and ensure universal service and then charges them with protecting universal service.

Moreover, the bill endorses direct support flow to low-income users rather than support for high-cost areas, and it calls for cost-based rural rates. The existing support flows are efficient and they work well. The proof of that is in our affordable and world-class public switch network. However, provisions of this bill seem to conflict with the existing support measures. We ask the subcommittee not to change the market structure for basic local service until you are sure that rural areas will have adequate support to prevent huge rate increases and a growing information gap between rural and urban areas.

We need your help. Rural areas will need major investments to replacing aging plant, maintain adequate services, and bring needed information services to rural residents. The Coalition urges Congress to enact the provisions in S. 570 that require joint telco planning and give small telcos the option to share network capabilities with their larger neighbors. Those policies will do much more for rural service, rates, and infrastructure development than S. 1086 in its present form.

Mr. Chairman, thank you for this opportunity to speak for the Rural Telephone Coalition. We look forward to working with the subcommittee to craft legislation that effectively promotes public network improvements without making rural consumers the victims of a risky experiment with the national public switch network.

Thank you, sir.

[The prepared statement of Mr. Ware follows:]

PREPARED STATEMENT OF LAWRENCE C. WARE

Good morning, Mr. Chairman and members of the subcommittee. My name is Lawrence C. Ware and I am the Manager of Garden Valley Telephone Company in Erskine, Minnesota. Garden Valley is an independent telephone cooperative that serves approximately 13,500 telephone subscribers in 24 exchanges spanning over 3700 square miles in eight counties in northwest Minnesota. Our density is 2.6 subscribers per mile of line. We provide all one party telephone service with 23 digital central offices and have over 300 miles of fiber optic cable in service. Garden Valley provides a fully interactive television service connecting six area schools. We provide cable television service to approximately 2000 subscribers in eleven communities, and are also involved in providing cellular and paging services. Garden Valley has served rural Minnesota for 87 years. We are an REA borrower.

I appreciate the opportunity to present the serious concerns of the Rural Telephone Coalition (RTC) with S. 1086, the Telecommunications Infrastructure Act of 1993.

The Rural Telephone Coalition is an alliance of the National Rural Telecom Association (NRTA), National Telephone Cooperative Association (NTCA), and the Organization for the Protection and Advancement of Small Telephone Companies (OPASTCO). More than 850 small and rural telephone systems scattered through 46 states are members of the three associations.

I. THE RURAL TELEPHONE COALITION SUPPORTS THE ULTIMATE POLICY GOALS OF S. 1086, AS IT UNDERSTANDS THEM

According to the findings and purposes of S. 1086, the ultimate goals of the bill are to develop the nation's telecommunications infrastructure; promote universal service; and provide consumers with advanced and reliable capabilities and services, reasonable prices, and a wide variety of service options. Mr. Chairman, the Rural

Telephone Coalition (RTC) strongly supports these ultimate goals of the bill.[1] They should be mandated as federal policy goals for the FCC to pursue, not just identified as non-substantive purposes of the bill. Indeed, small and rural telephone companies and existing rural telecommunications policies were created in order to realize in rural areas the exact results S. 1086 envisions.

The RTC has long advocated the development of a nationwide advanced telecommunications infrastructure. Much has been written on this subject in recent years, which the RTC will not repeat in detail here. Telecommunications services play an important role in overcoming some of the problems associated with the long distances and isolation of rural life. For example, distance learning programs make it easier for rural communities to combine teaching and facilities resources, giving their students access to classes that the communities would not have been able to support individually. Similarly, medical imaging can allow physicians in small towns to tap into the consultation and diagnostic resources of larger urban hospitals without leaving their communities. Advanced services provided over state-of-the- art telecommunications facilities will allow more communities to benefit from these and other telecommunications services.

The RTC has also always been fully committed to the federal and state universal telephone service policies. Small and rural telephone companies have striven to extend service to those customers most at risk of marketplace failure: low-income subscribers and subscribers located in areas that are more costly to serve. The RTC has also supported the various state and federal programs designed to help achieve and sustain the national goal of universal service, such as the Lifeline and Link-Up programs, the Universal Service Fund and Rural Electrification Administration financing. An important aspect of universal service is the availability of services at reasonable rates. The RTC believes that affordable local rates and nationwide averaged interstate toll rates are the hallmarks of an important and successful national policy, which contributes to rural economic development by decreasing the difference between the cost of living and doing business in rural and urban areas.

Small and rural telephone companies also see the benefit of maximizing consumer choices, of technology and services. Many rural LECs have deployed, or are planning to deploy, digital switching, fiber optic transmission, and signalling system 7 (SS7) in order to provide their rural customers with the same array of technology and service options that urban customers have. An advanced affordable network will allow multiple choices of information and other services to be available to rural residents and businesses. The RTC believes that rural areas must keep pace with urban areas in development and affordability if rural areas are to succeed in their efforts at economic development.

In the following testimony, the RTC will discuss (1) why rural areas have needed strong government policies and programs to achieve the telecommunications networks that serve them well today at reasonable prices; (2) why competition will not provide affordable, high quality service; (3) why rural differences require the development of policies and mechanisms that will effectively ensure universal rural service at reasonable rates and rural infrastructure developments Congress before makes radical changes in the nation's telecommunications market structure and (4) some policy initiatives that will serve rural customer interests better than those proposed in S. 1086.

II. FEDERAL POLICIES AND MECHANISMS, NOT THE MARKETPLACE, HAVE BROUGHT HIGH QUALITY TELECOMMUNICATIONS SERVICE TO RURAL CONSUMERS

A. Rural LECs Grew Up to Serve Unserved Areas

Small and rural independent telephone companies have a long history of serving those areas that the Bell companies did not choose to serve. But the marketplace alone did not bring about today's ubiquitous, affordable telephone network.

Commercial telephone service in the United States began in 1877. Initially the Bell companies held the necessary patents and essentially enjoyed a complete monopoly. However, spurred only by the marketplace, the Bell companies did not extend service to many rural and residential areas. Consequently, after 15 years of Bell infrastructure development, there was only approximately one telephone for every 240 people in the United States.

Independent telephone companies began to appear to meet the unmet need in their communities in 1894 when the essential Bell patents had either expired or were narrowly construed by the courts. In addition, competition by independent

[1] We discuss, in Section III, our concern that the means chosen to achieve these laudable goals will not do so for rural areas and, in Section IV, the measures we believe will better serve rural consumers.

local exchange carriers (LECs) appeared in only the most profitable locations. Less populous areas were generally served by systems under cooperative ownerships, or by often rudimentary "farmers lines," if they were served at all.

Telephone service was necessary for those living in the rural and remote areas. Moreover, it allowed them to communicate quickly without the necessity of an expensive and time-consuming trip. In fact, rural residents still tend to rely heavily on telecommunications because of their isolation in comparison with metropolitan area residents.[2]

B. Rural LECs Have Shown a Commitment to Rural Infrastructure Development

The pioneering spirit of the owners of the cooperatives and those who strung the "farmers lines" began the process that brought service to rural and remote America. Small and rural telephone companies have continued to upgrade their networks and bring new services and technologies to their customers, often at a faster pace than the larger companies. The National Exchange Carrier Association (NECA) compiled a report regarding the technological standing of the more than 1000 small telephone companies electing to participate in the NECA interstate access service tariff. These companies serve many of the rural and remote areas of the U.S. Over 90 percent of these companies serve less than 10,000 access lines—82 percent of these small companies have fewer than 6500 access lines. On an average they serve about 5600 access lines per company. By the end of 1992, 93 percent of their central office switches were expected to be digital. By the end of 1990, NECA companies had increased their digital switches to 76 percent of their total switches. This compares to 59 percent for the RBOCs. These NECA member companies continue to be leaders in the deployment of digital switches.

More than 800 of these companies' central offices, serving over a million access lines, will be upgraded to new software programs by the end of this year. Approximately 35 percent of these companies have installed some fiber optic interoffice facilities. Ten percent of the offices will have at least some fiber in their loop facilities by year-end 1994.

C. Rural LECs Have Relied on Government Programs and LEC Network Integration to Promote High Quality, Affordable Service

Independent LECs have been a critical factor in the equation that makes available "so far as possible, to all the people of the United States a rapid, efficient, nation-wide, and world-wide wire and radio communication service with adequate facilities at reasonable charges. But rural LECs could not have done it alone." The other factors in the equation necessary to achieve universal service and infrastructure development are federal and state support programs.

The Rural Electrification Administration (REA) telephone loan program has been instrumental in ensuring that modern telecommunications technology and services are available in rural areas in the United States. Since the inception of the telephone lending program in 1949, REA financing has provided telephone service to more than 5 million subscribers. REA borrowers have grown from less than 20 to 960 since the program began. All of the REA telephone loan funds have been used to further the purpose of the law—"to furnish and improve telecommunications services in rural areas." This federal program has been indispensable in ensuring that the nation's universal service objectives are met. Indeed, the statutory purpose is to provide and improve service for "the widest practicable number of rural users * * *"[3] The REA plays a pivotal role in the development of the infrastructure in rural areas. Moreover, the continued efficient operation of the Rural Telephone Bank and the REA loan programs is essential to the continued development of the telecommunications infrastructure in the rural areas of the United States.

The Universal Service Fund (USF) is a Federal/State program which helps offset the high operating costs of some telephone companies, enabling them to keep local rates at reasonable levels in high cost areas—most often rural and remote areas. Since the establishment of the USF, many of the small and rural LECs have extended service to remote areas and upgraded their facilities without significant impacts on their customers' local rates. Thus, the USF significantly contributes to universal service and infrastructure development.

The federal commitment to geographic averaging of interstate toll rates is an essential component of the Federal Communications Commission's (FCC's) universal service policy. Through interstate averaging, price reductions for high volume routes also benefit customers making calls on low volume routes (mostly to or from rural

[2] Rural America in the Information Age, Edwin Parker, Heather Hudson, Don A. Dillman, Andrew Roscoe, p. 34 (1989). (Parker).

[3] 7 U.S.C., Sec. 721.

and remote areas). Geographically averaged toll rates are very important to customers in rural areas. Rural residents use telecommunications more and spend more of their disposable income on telephone calls than their urban counterparts. For example, Native people in northern Canada and Alaska spend more than three times as much as their urban counterparts on long-distance calls, even though their average income is generally lower.[4]

Members of the rural community make many more long distance calls because of their location. For instance, a rural resident's call to a physician is likely to be a long distance call because of the limited geographic scope of the local service in these remote and rural areas and the relative lack of doctors. Consequently, without geographic rate averaging, people in some rural communities would pay substantially more for calls that metropolitan area residents take for granted, since they are within the scope of a "local call." Higher rates would particularly disadvantage those on fixed incomes and most in need of communications services, such as the elderly. High cost access to information services in rural areas could stunt economic development and deprive consumers of the benefits of information sources available at reasonable prices in cities and suburbs. Geographic rate averaging thus remains essential to universal service and infrastructure development. Losing the benefit of nationwide toll rate averaging would increase the economic disadvantages and isolation of rural and remote areas.

The local franchise has provided a successful vehicle to achieve the nation's paramount public interest goal—universal, affordable, efficient, high quality communications service. In fact, the local franchise was developed specifically because it could satisfy public objectives where competition failed.[5] The franchise allows each state to promote efficiency, service to remote areas, and affordable local and long distance services by viable LECs.

Regulators concluded that competition interfered with economies of scale because of the large investment required to develop and maintain telephone operations and the high levels of fixed costs. Concern with the importance of telephone service and the belief that it was a natural monopoly, most efficiently provided by a single company in a geographical area, led state legislators in a growing number of states to apply "public utility" regulation to telephone communications as industries "affected with the public interest."

The local franchise concept has worked, in part, because it gives each state latitude to tailor regulation to satisfy public goals to conditions within the jurisdiction. State jurisdiction over local exchange service produces a package of rights and obligations fine-tuned by each state to obtain the desired public benefits. The development of universal service and the advanced infrastructure has been substantially helped by the co-provision of services by small and large LECs. LEC-to-LEC interconnection and joint service arrangements increase small LEC efficiency, in spite of the inherent limitations on economies of scale in low density rural markets. This traditional form of infrastructure sharing has allowed the small LECs to establish interconnection arrangements with their larger Bell neighbors that afford them a greater opportunity to bring new and advanced services to their customers in rural and remote areas. Such co-provision of services establishes a publicly available core distribution network that serves as a foundation for the advanced services and information opportunities that are becoming available.

Thus, one indispensable ingredient that has allowed LECs to provide rural residents and businesses with high quality, affordable telecommunications and technologically advancing network capabilities has been the federal and state programs aimed at serving high cost rural areas. Preserving the hard-won benefits of these programs must be the backdrop for this Subcommittee's consideration of a new public policy framework. These programs—the REA telephone loan programs, the Universal Service Fund, geographic toll rate averaging, the local franchise and LEC public network integration—and the service orientation of the small and rural LECs combine to drive the development of the rural infrastructure and help meet the nation's universal service objectives.

---

[4] Parker, Supra. 2 p. 34.

[5] See, W.G. Lavey, The Public Policies Which Changed the Telephone Industry Into Regulated Monopolies: Lessons from Around 1915 (1987).

III. ENACTING S. 1086 IN ITS PRESENT FORM, WOULD UNDERMINE UNIVERSAL SERVICE, INFRASTRUCTURE DEVELOPMENT AND REASONABLY PRICED CONSUMER CHOICES FOR RURAL AMERICANS

A. Preemptively Mandating Competition Cannot Be Assumed To Benefit Rural Consumers.

Mr. Chairman, the RTC recognizes the good intentions underlying S. 1086. However, the RTC is concerned that the bill's combination of regulatory and marketplace tools will not achieve the legislation's goals in rural areas.

> A basic premise of S. 1086 (p. 2, ll. 18-22) is that fostering competition will encourage infrastructure development and have beneficial effects on the price, universal availability, variety, and quality of telecommunications services.

Based on this belief, S. 1086 proposes heavy reliance on competition and marketplace forces to achieve the paramount national telecommunications goals of universal service, infrastructure development and increasing consumer choice. To stimulate competitive entry, S. 1086 imposes extensive interconnection and unbundling obligations unparalleled in competitive industries.

This proposed national policy would initiate a large-scale test of telecommunications economics: Alfred Kahn, a strong advocate of competition, has testified in the House of Representatives that only experimentation can demonstrate whether local exchange operations are a natural monopoly[6]; another noted economist, Lewis Perl, has testified that the feasibility and impact of intraLATA competition can also only be determined by experimentation.[7]

The RTC cannot comment on how this proposed experiment in revolutionizing the nation's local exchange and access market structure and intercarrier obligations will affect metropolitan markets. It may be time to experiment with local and intraLATA competition in metropolitan areas. That judgment call is for Congress, after it has identified and dealt with the likely side effects discussed below. However, the risk of such an experiment in rural areas is not one that members of Congress should impose on the residents of these high cost, economically fragile portions of the nation.

Northwestern University Economics Professor John C. Panzar has warned that local and intraLATA competition in rural areas can decrease efficiency and erode the support flows underlying economically and socially desirable universal service.[8] Moreover, small and rural LECs have long experience in serving areas with lower population density and consequently higher costs of service per customer. They understand, as the National Telecommunications and Information Administration (NTIA) recognized in NTIA Telecom 2000, the pressing need to:

> assess each competitive policy as to its potential to be counterproductive in sparsely populated areas where the economic base may not be sufficient to support a single firm, much less competing firms.[9]

S. 1086 does not follow this principle of cautiously assessing the feasibility and effects before mandating competition in rural areas. Instead, the policy in S. 1086 would encourage new entrants to build duplicate facilities in competition with established rural universal service providers to take advantage of any perceived opportunity for profit. The consequences for rural consumers and the LECs that serve as rural carriers of last resort could be devastating.

1. The Costs of Rural Competition Will Likely Exceed the Benefits

To the extent that competitors come to rural areas at all, the costs often will exceed any local benefits. The generally low volume of traffic in rural areas already makes it difficult to achieve economies of scale in trunking, switching and other network functions. Long local loops further add to the cost to serve rural customers. Duplication of selected parts of the network by competitors would reduce traffic loadings, further diminishing efficiency.

New entrants seldom serve the places and customers that are most costly and difficult to serve; they choose to compete for high volume, low cost customers and places. The higher cost customers and places are left behind for the legally obligated "carrier of last resort" to serve. Thus, the existing provider becomes saddled with more costs to recover from a diminishing customer base.

---

[6] Testimony of Alfred E. Kahn before the House Subcommittee on Telecommunications, p. 5 (July 15, 1987).

[7] Testimony of Lewis J. Perl before the Kentucky Public Service Commission, pp. 3-4 (September 8,1989).

[8] J.C. Panzar, The Continuing Role for Franchise Monopoly in Rural Telephony (1987).

[9] U.S. Department of Commerce, (NTIA) NTIA Telecom 2000: Charting the Course for a New Century, p. 97 (1988) (NTIA Telecom 2000).

Cost and rate averaging between lower cost areas and customers and higher cost areas and customers is increasingly unsustainable for a rural LEC faced with competition. Great Plains Communications in Nebraska has stated that it costs 10 times more to serve the outlying parts of its service area than to serve a core rural community. Deaveraging at any level means higher rates for the customers it costs more to serve.

There is little incentive for the universal service provider to upgrade the publicly available network infrastructure when cost recovery is in doubt. Accordingly, to the extent that the bill encourages rural duplications the result is likely to be higher rates and less network modernization for typical rural residential and small business customers. The benefits to larger business customers and other "winners" from imposing a federal local exchange and intrastate competition policy will be paid for by the rural and small business "losers." These ill-fated consumers will also likely form a growing rural class of information have-nots, as their infrastructure falls behind, fulfilling the Office of Technology Assessment's warning of an increasing rural-urban infrastructure gap.[10]

2. Many Rural Consumers Are Unlikely to Have Reasonably Priced Choices Among Providers

Rural areas are not immune from competition. Rural cellular systems are in operation providing service at premium prices, and PCS systems will be authorized before long. The dense cores of many rural areas have cable service, although the system often does not "pass" residents in the surrounding areas. But it is a serious mistake to assume that marketplace forces will offer the typical rural residential or small business customer competitive choices and access to advanced network capabilities, or even maintain reasonably priced basic services and expand access to new services. Entrants concentrate on high volume customers with the highest potential profit margin. Lower volume customers are then left with only one choice of provider, but almost certainly must pay higher rates. Indeed, typical rural consumers would have more competitive service choices if the LEC that provides universal service could count on recouping the cost of widely-available advanced public network capabilities. Affordable alternative services could then be offered using the public network.

3. Far-Reaching Requirements for Interconnection, Number Portability and Intercarrier Arrangements Will Overburden Rural Areas

Rural consumers that experience no benefits would nevertheless shoulder a share of the expense for S. 1086's policies to foster competition. Local number portability and selective, minutely unbundled interconnection, both required upon request wherever "technically feasible" under S. 1086, could impose enormous costs and administrative burdens on the universal service providers. For example, even small LECs would require extensive data bases and signaling capabilities to accommodate floating local numbers, even if few customers have access to alternative providers. Unbundling would be available in rural networks, on request, without regard to the effect on the loading and efficiency of existing or planned facilities, the different economic consequences for low volume rural markets or the extent of demand for unbundled network elements.

The FCC has limited its expanded interconnection requirements for interstate special access—and those proposed for switched access - to the largest Tier 1 LECs. This bill goes far beyond the expert federal agency's carefully considered application to sweep in small and rural LECs, without any analysis of their different circumstances.

S. 1086 Could Interfere with Efficient Network Development and Integration Mechanisms

Rural Electrification Administration (REA) programs that provide capital to improve the infrastructure in rural areas are unavailable where duplicative rural telecommunications facilities may be involved. To protect the U.S. government's loan security, the Rural Electrification Act restricts loans to situations where a state commission issues a certificate of convenience or the Administrator makes an equivalent ruling that "no duplication of lines, facilities or systems, providing reasonably adequate services will result therefrom." 7 U.S.C. § 922. By preempting state certification and making duplicative entry a national policy, the bill may obstruct needed loans for rural infrastructure development and compromise the security of existing REA loans. It would be ironic if legislation purporting to stimulate infrastructure

[10] U.S. Congress, Office of Technology Assessment, Rural America at the Crossroads: Networking for the Future, p. 79 (1991).

development instead interfered with the highly effective REA rural infrastructure development programs.

In fact, S. 1086 could also limit the ability of rural LECs to buy rural cable television systems or engage in joint ventures to increase the efficiency of the rural infrastructure. Congress allowed the acquisition of rural cable systems in the 1984 cable legislation and left the definition of "rural" up to the FCC. However, S. 1086 (pp. 17-19) would amend the 1984 Cable Act cross-ownership provision to limit LEC acquisitions and most joint arrangements to cable television systems within "rural" service areas as currently defined. The FCC has before it compelling evidence that the definition of rural areas should be expanded to remedy gaps in rural cable television availability. In rural areas, the ability to integrate telephone and cable operations could spell the difference between broadband telecommunications availability and arrested network development. Thus, there is no reason to change the 1984 rural acquisition policy or restrict the FCC's discretion to define "rural" for purposes of the rural exemption.

The interconnection and mutual compensation sections might even be interpreted to require termination of the long-established co-provider relationships among LECs. In that case, rural and residential customers could suffer further setbacks in the price and availability of services and network capabilities.[11]

5. Changes in the Interexchange Provisions of the MFJ Must Also Be Carefully Evaluated

S. 1086 also proposes major changes in the interexchange restrictions currently applicable to the Regional Bell Operating Companies under the Modification of Final Judgment that implemented AT&T's divestiture of its local exchange and access operations in the government's antitrust case.[12] The RTC has generally opposed removal of the interexchange restrictions, except for narrowly limited relief to rectify specific problems.

The interexchange relief proposed here includes an oddly-assorted menu of changes. The proposal to remove the restriction with regard to cable television operations presents new questions which the RTC is still studying. The requirement of equal access for cellular systems would inexplicably extend a requirement now applied to the RBOCs to the rest of the cellular industry, with no recognition of the significant differences involved. The other cellular relief seems to duplicate relief already granted through court-approved waivers. In contrast, the RTC has supported relief to allow interLEC signalling needed for network operations. This beneficial and necessary change is not proposed by S. 1086.

Clearly, the smorgasbord of interexchange relief proposed here must be evaluated carefully. One criterion for all such proposals should be protecting adequately against further geographic deaveraging of interstate toll rates.

6. Limiting the New National Policy Mandate and Requirements to Metropolitan Areas or Large LECs Would Provide Partial Relief for Rural Areas

Many of the problems raised for rural areas by S. 1086 could be alleviated by restricting the bill's competitive mandate and implementing requirements to metropolitan areas or (as the FCC has done for special access interconnection) to the largest (Tier 1) LECs.

This approach would not preclude competition in rural areas. It would simply leave the authority to make determinations about rural competition where it is today, with the States. States would remain free to allow competition whenever they decide that the benefits of competition in rural areas outweigh the costs. A Tier 1 limitation would allow competition with the LECs serving most of the nation's customers. Tier 1 carriers serve more than 92 percent of the nation's access lines. Even excluding the large LECs' rural areas would leave the majority of U.S. telecommunications customers within the local competition experiment. Thus, a more cautious policy that recognized rural differences would still represent a massive change in national telecommunications policy, while decreasing the unacceptably high risks S. 1086 poses for rural consumers.

B. S. 1086 Undermines Crucial Universal Support Mechanisms and Jeopardizes Toll Rate Averaging

Mr. Chairman, S. 1086 recognizes (P. 4, LL 21-24) that "additional regulatory measures are needed to allow consumers in rural markets and non-competitive markets the opportunity to benefit from high-quality telecommunications capabilities."

---

[11] Section IV, below, discusses the vital importance of infrastructure sharing to rural infrastructure advances.

[12] *United States* v. *Western Electric*, Civil Action No. 82-1092 (U.S. District Court, District of Columbia).

The bill even seeks, in Sections 229(d) and 230, to address the inherent problem of maintaining universal service and promoting infrastructure development throughout the U.S. in a competitive environment. The RTC appreciates the effort to accommodate rural needs.

Unfortunately, the curative efforts are vague and uncertain. They boil down to empty assurances. Thus, even if S. 1086 is amended to limit its pro-competitive mandates to metropolitan areas or Tier 1 LECs' service areas, further changes are necessary to protect universal service, promote rural infrastructure advances and keep rates reasonable for rural and residential consumers.

1. Competition (Without New Mechanisms) Is Incompatible with Toll Rate Averaging and Existing Support Flows

Economic theory holds that competition will bring prices down to incremental cost. Competition in metropolitan areas is already creating pressure to lower urban and large customer rates that are above LECs' costs, to permit deaveraging and to reduce or eliminate support flows, whether direct or achieved through public policy-driven cost allocations. These have been the traditional sources of support flows to keep the rates in higher cost areas and rates for low volume customers lower than they would otherwise be. The same kind of cost allocation and pricing policies have traditionally kept business rates above their average costs so that residential rates could be kept low and have supported interstate toll rate averaging.

Competition will inevitably eviscerate support mechanisms that require some competitors (i.e., LECs) to charge above-cost rates for service subject to competition, while new entrants can charge lower, cost-based rates and do not bear equivalent regulatory costs and burdens. Section 229(d)(1) proposes a major step in the right direction by requiring all competitors to contribute to universal service. The RTC supports requiring contribution for universal service from all providers of communications service. The contribution scheme should recognize that LECs already contribute significantly by actually providing service to all in their service areas upon reasonable demand, and serving as the de facto backup carrier for other providers and as the carrier of last resort. They should receive "credit" in any contribution mechanisms for continuing to perform these duties from the era of public utility regulation.

2. S. 1086 Gives States the Responsibility but not the Means to Protect Universal Service and Averaged Toll Rates

The bill places upon the States the primary responsibility for "ensur[ing] the preservation and advancement of universal service," (§ 229(d)(2)) and encourages them to use "regulatory incentives" or, failing that, unspecified "other methods" to provide high quality facilities and capabilities to rural and non-competitive areas. if the states do not successfully ensure rural infrastructure and service advances, the FCC "may take any action necessary to achieve that goal." Of course, neither the States nor the FCC would be free to adopt universal service or infrastructure development approaches unless they were consistent with S. 1086.

Despite their greater familiarity with rural conditions and their greater accountability to their constituents, State regulators have been preempted from their traditional full authority over local and intrastate entry, interconnection and even rates. For example, the "local franchise" package of public utility rights and responsibilities—long a valued tool in achieving universal service[13]—would become unlawful a year after enactment of S. 1086.

To illustrate further, section 230(a) requires that "consumers in rural and non-competitive markets" have network capabilities and facilities at "nondiscriminatory rates that are based on reasonably identifiable costs of providing such services." Local rates and access charges based on the "reasonably identifiable costs" of serving low-density rural and non-competitive areas will be higher than the averaged rates available under existing support mechanisms. The FCC has already allowed Tier 1 LECs to begin the process of deaveraging their access charges into density-based zones to meet special access competition. Such access deaveraging increases the pressure to deaverage interstate toll rates, raising rates for service on low density routes. S. 1086 thus seems to embrace the concept of higher rates for rural or remote communities in Alaska, Hawaii, Montana and other distant or sparsely populated states as the price of competitive policies that will help big businesses and high volume urban customers.[14]

---

[13] See Panzar, supra, n. 3.

[14] The requirement of rates based on "reasonably identifiable costs" of rural service might even be used to abolish traditional support through public policy-driven cost allocations.

3. *The Bill's Universal Service Proposal Conflicts With Existing Support Mechanisms*

By settling the responsibility for universal service on the states, aided solely by FCC "coordination," S. 1086 seems to undermine support flows that cross state lines. Beyond that, it is even more difficult to see how nationwide toll rate averaging and support for high cost areas can survive the enactment of this bill: S. 1086 states the national policy goal of support that "directly assist[s] individuals or entities that cannot afford the cost of their telecommunications service or equipment." (Section 229(d)(3)). If the intent is to retain existing support to protect against rate increases and greater disparities, amendments are necessary.

4. *S. 1086 Also Fails to Resolve the Inherent Conflict Between Rural Infrastructure Development and Marketplace Reliance*

S. 1086 also delegates to the States the primary responsibility for rural infrastructure development. Here, too, preemption of the States' current authority over market structure, interconnection and at least some elements of rate-making and intercarrier compensation arrangements raises serious questions about States' ability to assure infrastructure development. Under the mutual compensation requirement, can the states continue to require large LECs to share infrastructure and provide joint-through service with smaller LECs? What "regulatory incentives" are available if selective cream-skimming is encouraged, rural rates must be cost-based and support flows are, at best, extremely uncertain and limited to low income assistance? Low income individuals may not even use their direct assistance to connect to the universal service provider's network. The same questions would be pertinent if the FCC tried to rectify marketplace failure in rural infrastructure development under the default provision of Section 230(a).

5. *The Rural Rates and Infrastructure Dilemma Must Be Resolved Before Local and Intrastate Competition Are Unleashed*

Rural infrastructure development is a real and immediate need. Analysis of S. 1086 makes it clear that the bill lacks adequate universal service and rural infrastructure development safeguards; and real competition must inevitably erode the current mechanisms that unfairly disadvantage incumbent universal service providers. Thus, this Subcommittee should again show the keen sensitivity to the needs of rural consumers it has demonstrated in the past: Before moving forward on a proposal to preempt the States and mandate a nationwide competitive market structure for local and intrastate services, the Subcommittee should find effective, competitively-neutral support mechanisms that can be sustained in a competitive environment.

IV. CONGRESS SHOULD ENACT POLICY MANDATES AND LEC INFRASTRUCTURE SHARING AND JOINT PLANNING LEGISLATION TO ATTAIN THE GOALS OF INFRASTRUCTURE DEVELOPMENT AND UNIVERSAL SERVICE

The RTC cannot yet suggest a comprehensive package of safeguards and universal service mechanisms that will enable Congress to mandate local and intrastate competition without endangering universal service and rural infrastructure development or threatening rural residents and businesses with skyrocketing rates. However, the RTC believes that some modest measures are necessary to protect and develop the national universal service network while Congress develops the solutions it needs before drastically changing the nation's telecommunications ground rules.

The RTC has for the last several years advocated legislation that would greatly assist small and rural LECs and their subscribers in obtaining the benefits of a nationwide, reasonably priced infrastructure. We believe that to meet the goals of infrastructure development, universal service and consumer choice, Congress should reaffirm that national telecommunications policy includes nationwide access to information services via an advanced public network infrastructure. Congress should also provide for joint planning and infrastructure sharing by the nation's LECs which together provide the nation's core public distribution network. The RTC supports the approach taken in S. 570 to ensure the continued development of the nation's public switched network. Specifically, we believe that S. 570's provisions pertaining to joint LEC planning and infrastructure sharing are essential to the availability of new technologies and services in the rural U.S.

Before its breakup, AT&T served as the telephone service network manager and standards setter for the public switched network. When AT&T was divested of its local exchange companies, a void was created. At present, the nation's local telephone companies participate on a voluntary basis in some aspects of network planning. However, there is no comprehensive or formal policy or mechanism which covers all aspects of network management and planning needed to ensure the continued interconnectivity and interoperability of the LEC public switched network. Nor is there any policy or mechanism in place to ensure that new services and technologies reach the rural areas of our nation. As competition develops, preserving the national public network as a universally available resource will become more difficult, but no less important.

As a result of the AT&T divestiture and the ensuing deregulatory trend in telecommunications policymaking, the public switched network has become increasingly fragmented. Prior to the breakup, expanded technological capabilities and services were traditionally spread throughout the nation's local exchange service areas by aggregated pricing and cost sharing mechanisms. This process is jeopardized when an unrestricted marketplace drives prices to costs on low cost parts of the network, the heavily populated urban areas. LECs serving metropolitan areas become increasingly unwilling to share with their rural neighbors. Rural America needs help in dealing with the effects of policies implemented primarily for a competitive urban environment.

The fragmentation of the predivestiture nationwide partnership has made an integrated network more difficult to achieve by fundamentally changing the relationships among carriers. For example, if 557 had been a truly joint effort, with all the participants' technical needs and financial situations taken into consideration, the RTC believes that independent LECs probably could have participated more fully and cost-effectively in this technology. As it turned out, independent telephone systems had to play a fast game of catch-up in order to bring this signaling technology and the services it supports to their subscribers. Thus, there must be affirmative public policy requirements and safeguards to ensure that the largest telephone systems work with other telephone systems to develop a nationwide advanced public communications infrastructure. Congress should not leave the nation's communications network to be shaped by large business needs and the largest LECs' perceptions of the marketplace.

Mr. Chairman, communications and information will play a major role in the life of this country and its citizens. The benefits of telecommunications and access information should not be limited to our cities. In fact, the availability of modern telecommunications networks in rural areas is the key to revitalizing rural America.

Along with joint planning requirements, small and rural LECs need infrastructure-sharing in order to provide new services and technologies to their subscribers. Through infrastructure sharing, small and rural LECs would be able to share information and technology with larger telephone companies, making it far more economical to offer a full array of advanced telecommunications services to their customers.

To participate in the Information Age small and rural LECs must be able to choose the most economical way to install or obtain access to the complex and expensive infrastructure required to deliver new services. They should be able to work with larger LECs until it becomes cost effective to install their own network capabilities. Joint network planning and infrastructure sharing policies would give smaller LECs the ability to upgrade their networks sooner. These sharing arrangements are also key to maintaining a fully functional nationwide public telephone network.

Infrastructure sharing is not a new concept. For example, some small LECs use a larger LEC's tandem switch to deliver traffic to interexchange carriers. However, as new network technologies are introduced and competition develops, it is essential for public policy to ensure that necessary infrastructure sharing continues to protect customers in areas where competitive choices will become available slowly, if ever. It is not economically feasible to install vast databases and advanced switching equipment in some rural areas. This does not mean, however, that rural communities must go without advanced services. Using fiber optics and other transmission media, LECs serving rural areas can access the databases and advanced switches that larger LECs install in more populous areas.

Infrastructure sharing is a vital ingredient for the infrastructure development goal. Public policy requiring infrastructure sharing at the request of small LECs would help ensure that all Americans, urban and rural alike, would have access to a nationwide information network.

Mandating planning and infrastructure sharing by neighboring LECs that collectively make up the only nationwide regulated network which provides universal service, is different from requiring sharing among regulated LECs and their largely unregulated competitors, which serve only the locations and customers they choose. Treating carriers with different obligations and motives as if they are the same is just as unreasonable as treating identical carriers differently.

To attain the goal of infrastructure development and nationwide access to new services, Congress should adopt a national telecommunications policy statement that promotes an integrated nationwide LEC network, joint LEC network planning and LEC infrastructure sharing.

V. CONCLUSION

The RTC has long stated that Congress should set national telecommunications policy to promote universal service and nationwide enhancement of the public telecommunications network. Legislation, if it is to protect the interest of rural America, must accommodate the major differences in cost of service and conditions that confront rural providers of universal service.

Congress cannot assume that the marketplace will suffice to protect rural areas. Different policies and specific, effective, sustainable support mechanisms must be developed to prevent high rural rates, a second class rural network and new barriers to expanded rural service choices.

As S. 1086 is currently drafted, it would jeopardize rural service rates and infrastructure development. The RTC looks forward to working with the Subcommittee to craft legislation that will meet the needs of America's rural areas.

Senator INOUYE. Thank you very much, Mr. Ware. You may be assured that this subcommittee will be working with you. As you can look at the subcommittee, they represent rural America. May I now call on Mr. Coe.

STATEMENT OF KERMIT COE, DIRECTOR OF MERCHANT ACCEPTANCE STRATEGY, VISA INTERNATIONAL

Mr. COE. Mr. Chairman and distinguished members of the committee on communications, my name is Kermit Coe and I am a director with the VisaNet Market Development Group of VISA International, and I am pleased to have this opportunity to testify in support of S. 1086, the Telecommunications and Infrastructure Act of 1993.

VISA International is an association of over 21,000 financial institutions throughout the world that are licensed to use its service marks in connection with payment systems, notably credit and debit cards; check authorizations; automated teller machines; and related services. VISA is the world's most widely used and accepted credit card. The approximately 300 million VISA cards now in circulation are accepted by 10 million merchants around the world. Last year alone, nearly 6 billion transactions were completed using a VISA card.

Communications is the lifeblood of VISA. It has been said that modern banking consists largely of computers talking to each other over telephone lines about money, and nowhere is that more true than in credit card authorizations and related transactions. VISA now is receiving close to 5 billion inquiries per year over switched and dedicated telephone lines, and that number is growing by more than 500 million each year.

In addition to being a major telecommunications user, VISA is also a substantial provider of electronic publishing services. Transaction processing is by far the largest segment of the information services market.

Because of the importance of telecommunications to VISA, we have been an active participant in various FCC and State public utility commission proceedings. That participation and our activity in the marketplace have taught us a few things about telecommunications regulation.

First and foremost, competition, competition, and competition are the three most important factors in improving telecommunications service quality, bringing new and innovative services to the market quickly, and reducing prices. Enlightened regulation is a fourth important factor because it encourages the development of competition.

The appearance and growth of competition in the long distance and equipment markets has enormously benefited both large users like VISA and consumers. Competition has, for example, improved the speed of credit card authorizations, driven down the cost of queries, and provided a cost effective means of requiring authorization of nearly every credit card transaction, which substantially reduces credit card fraud.

We are absolutely certain the introduction of competition in the local exchange market, the last mile and the last bastion of monopoly, will have similar effects. S. 1086 would bring us closer to that goal and VISA therefore supports it.

Second, the success of new legislation and the regulations that implement it will be measured by its ability to produce substantial results in a timely fashion. If efforts to introduce competition into the local communications market are to succeed, the Congress must assure that straightforward, enforceable standards govern access to essential network services and ensure a level playing field for users and competitive service providers.

The standards supported by VISA are both simple and comprehensive, and all of them are found in one form or another in S. 1086.

First, we urge the committee to require the RBOC's to provide all information services through fully separate subsidiaries. Because when all is said and done, structural separation remains the only effective means of protecting competition and captive ratepayers from discrimination and potentially unlawful cross-subsidization. Separation will also facilitate FCC enforcement of other safeguards such as rules governing nondiscriminatory access.

Section 233 of the bill embodies much of what we seek in this regard, although it creates considerable confusion by referring to "information services" in section 233(a) and then using the more restrictive phrase "electronic publishing" in section 233(b)(1), only four lines later. VISA disagrees with efforts to impose different standards on content based, as opposed to noncontent based information services. Indeed, based on our own experience, we doubt that the two can even be distinguished.

We urge the committee to use the phrase "information services" throughout the section. I understand that this was the approach taken in legislation introduced by Senator Inouye last year. It was the right solution then, and it is the right solution now.

Second, we recommend adoption of a strong, clear nondiscrimination requirement to govern the provision of essential services to independent entities that seek to compete with local telephone companies. What is needed is something akin to the MFJ's successful equal access requirement. The owners of the local exchange companies should be prohibited from offering themselves or their affiliates' services, including features, functions, and interconnections, that they do not offer on at least equivalent terms to independent providers.

Third, we support efforts to strengthen existing FCC enforcement mechanisms so that violations of the above requirements can be dealt with swiftly and effectively.

Finally, we note with approval your efforts to accelerate standards development and mandate coordinated network planning. Increased coordination is essential to a compatible telecommunications infrastructure. While some carriers solicit input as new products and services are deployed, such efforts are the exception and not the rule. Section 230 of the bill seeks to cure this problem, but as presently written section 230 neglects to include users among those tasked with developing standards to ensure network interconnection. We urge the committee to correct this oversight by including users, the intended beneficiaries of network innovation, in such efforts.

S. 1086 goes a long way to meeting our concerns and VISA is therefore pleased to support it. Beyond the suggestions made above, we have three technical suggestions which are detailed in my written statement.

Thank you very much for the opportunity to appear here before you today to present VISA International's view on this important legislation.

[The prepared statement of Mr. Coe follows:]

Prepared Statement of Kermit Coe

Mr. Chairman, and distinguished Members of the Subcommittee on Communications, my name is Kermit Coe. I am the Director of the VisaNet Market Development Group of VISA International, and I am pleased to have this opportunity to testify in support of 5.1086, the Telecommunications Infrastructure Act of 1993.

INTRODUCTION AND BACKGROUND

VISA International is an association of over 21,000 financial institutions throughout the world that are licensed to use its service marks in connection with payment systems (notably credit and debit cards), check authorizations, automated teller machines and related services. VISA is the world's most widely used and accepted credit card. The approximately 300,000,000 VISA cards now in circulation are accepted by 10 million merchants around the world. Last year alone, nearly 6 billion transactions were completed using a VISA card.

Communications is the lifeblood of VISA. It's been said that modern banking consists largely of computers talking to each other over telephone lines about money, and nowhere is that more true than in credit card authorizations and related transactions. VISA is now receiving close to 5 billion inquiries per year over switched and dedicated telephone lines, and that number is growing by more than 500 million each year.

In addition to being a major telecommunications user, VISA is also a substantial provider of information and electronic publishing services. Although few people realize it, as the numbers above indicate transaction processing is by far the largest segment of the information services market.

Because of the importance of telecommunications to VISA, we have been an active participant in various FCC and state public utility commission proceedings. That participation, and our activities in the marketplace, have taught us a few things about telecommunications regulation that I'd like to share with you today.

THE IMPORTANCE OF COMPETITION

First and foremost, competition, competition, and competition are the three most important factors in improving telecommunications service quality, bringing new and innovative services to the market quickly, and reducing prices. Enlightened regulation is a fourth important factor, because it encourages the development of competition.

The appearance and growth of competition in the long distance and equipment markets has enormously benefited both large users like VISA and consumers. Competition has, for example,improved the speed of credit card authorizations, driven down the cost of queries and provided a costffective means of requiring authorization of every credit card transaction, which substantially reduces credit card fraud. We are absolutely certain that the introduction of competition in the local exchange market—the "last mile" and last bastion of monopoly—will have similar effects. S. 1086 would bring us closer to that goal, and VISA therefore supports it.

Second, the success of new legislation and the regulations that implement it will be measured by its ability to produce substantial results in a timely fashion. Here, the FCC's Open Network Architecture ("ONA") initiative is a model of how not to proceed. ONA was to be the mechanism for unbundling local network services so that information service providers could purchase only those features and functions that they needed to build their own services. VISA, in coalition with other financial service providers, devoted substantial time and resources to ONA. But good intentions proved insufficient. Almost no one buys ONA services today if they have a choice; indeed, information service providers spend a fair amount of time fighting off efforts to compel them to take services under ONA tariffs. The FCC has not given up, but when it recently took important steps to open up local exchange networks, it did so through an entirely new proceeding.

The lessons of ONA should not be lost on this Committee. If efforts to introduce competition into the local communications market are to succeed, the Congress must assure that straightforward, enforceable standards govern access to essential network services and ensure a level playing field for users and competitive service providers.

CREATING AN ENVIRONMENT THAT FAVORS COMPETITION

The standards supported by VISA are both simple and comprehensive, and all of them are found in one form or another in S. 1086.

First, we urge the Committee to require the RBOCs to provide all information services through fully separate subsidiaries. When all is said and done, structural separation remains the only effective means of protecting competition and captive ratepayers from discrimination and potentially unlawful cross-subsidization. Separation will also facilitate FCC enforcement of other safeguards, such as the rules governing nondiscriminatory access.

Section 233 of the Bill embodies much of what we seek in this regard, although it creates considerable confusion by referring to "information services" in Section 233(a) (page 24, line 13), and then using the more restrictive phrase "electronic publishing" in Section 233(b)(1), four lines later. VISA disagrees with efforts to impose different standards on "content-based" as opposed to "non-content based" information services—indeed, based on our own experience, we doubt that the two can even be distinguished. We urge the Committee to use the phrase "information services" throughout the Section. understand that this was the approach taken in legislation introduced by Sen. Inouye last year. It was the right solution then, and it is the right solution now.

Second, we recommend adoption of a strong, clear nondiscrimination requirement to govern the provision of essential services to independent entities that seek to compete with local telephone companies. While the Communications Act's prohibition on "unreasonable discrimination" is appropriate in many circumstances, it is inadequate to protect against self-dealing by entities which are both monopoly providers and competitive consumers of essential services. What is needed in such situations is something akin to the MFJ's successful "equal access" requirement. Section 229(c) of the Bill should accordingly be amended to require telcos to furnish features and functions to information service providers and other large users that are "economically and technically equivalent" to those that they provide to themselves or their affiliates. As a corollary, the owners of local exchange companies should be prohibited from offering themselves or their affiliates services (including, features, functions and interconnections) that they do not offer on at least equivalent terms to independent providers.

Third, we support efforts to strengthen existing FCC enforcement mechanisms so that violations of the above requirements can be dealt with swiftly and effectively. Thus, we urge the Committee to consider modifying the "patently unlawful" standard of tariff review, which permits almost all tariffs to go into effect pending investigation and allows a carrier to exploit an unfair competitive advantage for months or years until an investigation is completed or a formal complaint can be resolved.

Finally, we note with approval your efforts to accelerate standards development and mandate coordinated network planning. Increased coordination is essential to a compatible telecommunications infrastructure. While some carriers solicit input as new products and services are developed, such efforts are the exception, not the rule. Section 230 of the Bill seeks to cure this problem, but as presently written Section 230(d)(1)(A) (p. 15, lines 10-16) neglects to include users among those tasked with developing standards to ensure network interconnection. We urge the Committee to correct this oversight by including users—the intended beneficiaries of network innovation—in such efforts.

S. 1086 goes a long way to meeting our concerns, and VISA is therefore pleased to support it. Beyond the suggestions made above, we suggest three technical changes:

- The definitions should be amended to make it clear that electronic publishing services are a subset of information services.
- The phrase "network functions" as used in Section 229(c)(4) (page 11, lines 4-5) is undefined and has no clear meaning in the industry. We ask that you consider substituting "operations, support, signaling and control systems", or adding a definition to the same effect.
- The words "other than the customer to which it relates" should be inserted after the word "person" in Section 234(a)(1) (page 28, line 6) in order to make it clear that a customer is always entitled to his or her own Customer Proprietary Network Information.

Thank you very much for the opportunity to appear before you today to present VISA International's views on this important legislation. I would be pleased to answer any questions.

Senator INOUYE. Thank you very much, Mr. Coe. Now, may I call on Mr. Niggli.

STATEMENT OF MICHAEL NIGGLI, SENIOR VICE PRESIDENT, ENTERGY CORP.

Mr. NIGGLI. Thank you, Mr. Chairman, good morning. Good morning to members of the committee. I would like to thank you for the opportunity to speak with you today. We are here today because members of the committee staff have heard about PowerView and are intrigued by its potential applications. And frankly, we are excited about its prospects. We hope to bring to you a new perspective on how advanced communications technology can be used in my industry, the electric power industry.

Entergy serves about 1.7 million electric utility customers in the primarily rural States of Arkansas, Mississippi, and Louisiana. We are continually looking for ways to better serve our customers, reduce bills, and concurrently reward our shareholders. The PowerView system, we believe, will help us do all three. Today, I would like to briefly tell you about what the PowerView system is and how we plan to use it.

PowerView is a high-capacity bidirectional communications system that connects us directly to our customers. PowerView allows us to link up with our customers in an effort to improve the efficient use of electric energy. The result will permit us to defer or displace the need for certain new powerplants in the future, thus saving our customers money.

PowerView allows the customers to make active choices on how he or she will participate in energy conservation and efficiency measures. PowerView provides a continuous automated feedback system that permits verification of actions taken by the utility or the customer to control energy consumption. In short, it provides us with an educated, active consumer and an automated feedback system that helps us reduce our cost of service.

How do we plan to use PowerView and its related system? A couple of examples might best explain how we plan to use this system for energy efficiency, conservation, and customer service improvements. First, we intend to send continuous and current pricing signals to our customers, a spending rate, if you will, the price of electricity. This price will vary over time. It can vary over time of day, it can vary over the time of the week, or on a seasonal basis.

Customers react to spending rates, and especially if they have current information. They will change their consumption patterns. They will actually move energy from onpeak to offpeak periods, similar as to the situation that many other industries have used to price their products. We do not use that concept substantially today, but we have an opportunity to now.

Have you ever heard of the concept of an energy sale from your electric utility—one-half price sale? Probably not. We have the opportunity, we believe, with this type of communications system, on Wednesday to give you a signal that says Saturday and Sunday you have one-half or three-quarter price energy sale.

Why would an electric company do that? Primarily for two reasons. One, to get you to shift your consumption to an offpeak period, and second, because I have personally 3 to 4 billion dollars' worth of plant sitting idle on Saturdays and Sundays.

We intend to give our customers choices on appliance control. An example might help here. We may give a customer an option A—option A is a discount, possibly 3 percent or some number like that, if he allows us to control an air conditioner in their home. That control would be in 15-minute increments every couple of hours during the peak periods. Another customer might choose option B. Option B is the air conditioning control and a water heater control. Option C could be air conditioning, water heating, and another appliance.

The whole point I am making here is that customers would have the option, they would make the selection on these types of control devices, and we would then have the opportunity to control the consumption from a remote location to help our system achieve a lower peak demand and possibly to better utilize our facilities.

We also intend to provide itemized billing. We believe that an educated customer is essentially our best customer. In terms of itemized billing, this has been done by the telephone companies for decades. What do you get on your electric bill? You get two lines. One line says your energy consumption, the second line says your bill. Well, what we are hoping to do with this type of a system is to tell you that out of that $120 bill you may have had last month, $77 was for air conditioning, $34 is for lighting, $12 is for your refrigerator, that type of information. What does that give to you? That gives you, the consumer, the opportunity to make a good choice as to how you will use the energy in the future.

We have other applications for the PowerView system, including remote on/off devices. When people move in and out of apartments or homes we would have the ability to turn them off remotely rather than dispatching someone in a hardhat and a truck to go out and turn on or turn off the service.

We also would have the opportunity for poweroff indication where we would know if your service is off. Right now, when your service is off, you have got to call us. We would have the opportunity for automatic indication. Now, we have several other opportunities with this type of a system that can be applied to the electric utility industry.

In summary, Entenergy is rapidly pursuing the development and deployment of PowerView. We will be using our existing fiber optic backboned system which extends the length of our service territory to interconnect with our customers. Our backbone system goes through towns such as Brookhaven, Tylertown, Onward, MS; Bogalusa, Bastrop, and Slidell, LA.

The electric utilities have communicated with their customers and own facilities in many ways for decades. We are excited about bringing this advanced technology to bear on the issues of energy conservation and efficiency, and for the benefit of our customers.

We thank you for the opportunity to speak today.

[The prepared statement of Mr. Niggli follows:]

PREPARED STATEMENT OF MICHAEL R. NIGGLI

INTRODUCTION

Good morning. My name is Michael R. Niggli. I am the Sr. Vice President for Marketing with Entergy Corporation. Entergy Corporation is an electric utility holding company that serves approximately 1.7 million customers in the states of Arkansas, Mississippi, and Louisiana. This service is provided by four subsidiaries of Entergy Corporation: Arkansas Power & Light Company, Louisiana Power & Light Company, Mississippi Power & Light Company, and New Orleans Public Service, Inc. We currently have pending before various regulatory agencies a merger with Gulf States Utilities Company. Approval of the merger will increase the number of customers served to 2.3 million, and will add service area in southeast Texas and western Louisiana. I would like to thank the subcommittee for the opportunity to provide testimony at today's hearing.

SUMMARY

The purpose of my testimony is to identity the role that electric utilities can play in establishing the "information superhighway" to the residential customer. Further, I will briefly describe a project that Entergy currently has underway in Little Rock, Arkansas, which is bringing a high capacity fiber optic/coaxial cable infrastructure to the residential customer. This project utilizes the "PowerView" technology to deploy cost-effective energy efficiency and conservation programs. The PowerView technology has been jointly developed by Entergy and First Pacific Networks (a technology supplier headquartered in Sunnyvale, California.)

Entergy Corporation believes that electric utilities have a definitive role to play in establishing a high capacity information system to the residential customer. This can happen in a time frame that otherwise may not be achievable without the economic justification brought to bear by our industry. This economic justification is driven by energy efficiency and conservation in the form of "demand-side management" programs and concepts. These demand-side management programs can help defer or displace the need for new generating facilities, thus economically-justifying the installation of PowerView. Implementation of PowerView on a commercial basis will permit the deferral or displacement of new electric capacity at least equivalent to a large coal-fired generating unit on the Entergy system.

Exhibit 1 depicts the economic justification (in terms of benefit/cost ratio) and the potential impact that PowerView can have on the Entergy system. The potential impact of PowerView is significant because the system (1) effectively combines a number of demand-side management applications (as discussed later), and (2) is flexible to adapt to future applications. Exhibit I also depicts the impacts of numerous other demand-side applications that Entergy is pursuing. The graph illustrates that PowerView can be one of the more dominant mechanisms for implementing effective demand-side management programs.

PROGRAM DEPLOYMENT

Entergy is pursuing rapid deployment of the PowerView technology. Entergy's involvement with PowerView and First Pacific Networks has been approved by the Securities and Exchange Commission, without opposition or intervention by any private party or regulatory agency.

70-334 O - 94 - 3

On December 1, 1992 Entergy submitted a filing simultaneously to its four retail regulators regarding the Company's proposed Least Cost Plan. The Least Cost Plan (LCP) includes Entenergy's 20 year outlook for new resources required to meet customer needs. The LCP filing identifies PowerView as the focal demand-side management program in our service area. For that reason, the PowerView application has been reviewed, along with other proposed programs, during a series of three public conferences held in Entergy's service area. These conferences have been attended by representatives of the respective public service commission staffs and/or their consultants, major industrial customers, contractors, environmental organizations, various state agencies, senior citizen groups, and other special interest groups. The Arkansas Public Service Commission has recently noted that PowerView is the most innovative program to be included in Entergy's demand-side management filing.

The LCP filing notes that PowerView is expected to penetrate at least 442,000 homes in the current Entergy service area. This includes homes in urban, suburban and rural locations in Arkansas, Louisiana, and Mississippi. Entergy's fiber optic back-bone system—which extends over 700 miles—will bring the opportunity for PowerView to be deployed to a wide range of customers throughout the company's service area. Further penetration is anticipated as Entergy's infrastructure costs are reduced over time. Entergy plans to deploy PowerView to 10,000 homes in the first fill' year of commercial operation (now scheduled to begin in 1994).

Entergy and First Pacific Networks are currently deploying the technology as a pilot research and development program in a suburb of Little Rock, Arkansas. The pilot project (known as Customer Choice 2000) has completed installation of a fiber optic/coaxial cable infrastructure to 50 homes in the test area. The PowerView technology, with fully functional intelligent utility units, will be connected to these homes during the 3rd and 4th quarters of 1993. The initial work on Customer Choice 2000 has been enthusiastically received by our participating customers. We intend to follow through with a series of customer interviews to obtain feedback on customer satisfaction with this advanced technology. The results of this pilot program will be used to further enhance the scheduled commercial deployment of the technology in 1994.

THE NEED FOR POWERVIEW

The electric utility industry requires large sums of capital to fund the installation of major generating facilities to accommodate new growth in our society. We are always seeking new and more efficient ways to utilize that capital in meeting requirements that our customers place on our service. The focus on energy efficiency and conservation is one that allows the electric utility industry to defer or possibly displace the need for large generating facilities, thereby deferring the need for large amounts of capital expenditures. In many cases the deferral of these large generating facilities provides the economic justification for installation of a fiber optic/coaxial cable network. The establishment of a bi-directional, high capacity communications network with our customers will permit the implementation of cost effective demand-side management applications and potential operating cost reductions.

Exhibit 2 identifies the problem that the electric utilities are attempting to address. Essentially, our industry constructs facilities with sufficient capacity to serve the needs of our customers at peak load periods. During off-peak hours, the systems have excess capacity available to serve additional load. Given that we generally construct new facilities to meet peak load requirements, it can be economically beneficial to shift demand requirements of our customers from the peak load period into the off-peak period where additional capability is available. Certain energy conservation and efficiency programs can assist us in moving our peak load to an off-peak period, thus deferring the need for new capacity. These programs include (1) direct control of customer appliances (as selected by the customer), (2) flexible time-of-use pricing, and (3) energy efficiency education. Entergy believes that utilization of high-capacity information technology can assist in cost effectively shifting this load, as explained later.

Many industries are faced with the same problem of constructing or acquiring facilities to meet peak load demands. Several industries have attempted to address the issue through establishment of advanced information systems. For instance, telephone companies, overnight delivery services, movie theaters, and other industries have addressed this problem through peak load pricing, wherein they price the utilization of their service during on peak periods at rates higher than during off-peak time periods. This allows the customer to determine the value they place on the transaction during peak periods, and possibly shift demand to a period when the service is less expensive. This permits these industries to better utilize their infrastructure, construct fewer facilities, and be more efficient serving customer requirements.

ELECTRIC UTILITY APPLICATIONS

As noted in Exhibit 3, Entergy anticipates utilizing a high-capacity information system to the residence in order to apply certain demand-side management and operational applications to the utility customer. Current technologies allow the transmittal of voice, video and data on a single infrastructure via various communications media, including coaxial cable and fiber optic facilities. The capacity of such a system is generally maximized when utilizing fiber optic communications capabilities and is quite cost effective when the installation includes a fiber optic/coaxial cable infrastructure. The electric power industry could possibly use this technology in a cost-effective manner for the following applications: (1) demand-side management with continuous, automated feedback, (2) accumulation of customer usage information (i.e. load research), (3) remote reading of meters which could be applicable to electric, gas, water, and other devices, (4) automated billing for consumers, (5) remote turn on/turn off capabilities, (6) automated "power-off" indication, (7) automated current diversion, (8) distribution automation, and (9) enhanced customer service.

Exhibit 4 identifies some of the specific opportunities that can be implemented with demand-side management via a high-capacity communication system. First, the utility has the opportunity for verifiable load control options, including the positive identification of control applications as well as the establishment of various time of use pricing applications. Demand-side management programs throughout the nation have often suffered from a perception, and sometimes reality, that the savings associated with these programs were "soft" or non-existent. This occurs because the effect of many programs cannot be directly measured, or are engineering estimates. For example, a program that provides a rebate for an energy-efficient refrigerator assists in promoting the installation of such refrigerators. However, such a program does not measure effectiveness over any time period, nor does it ensure that the refrigerator will not be relocated out of the utilities service area. The application of a high-capacity information system implements verifiable demand-side management savings and provides immediate feedback to the utility to quantity the results of our various programs and control actions. This is a substantial advance in our ability to determine—in a much more accurate manner—the cost-effectiveness of various demand-side management options that are implemented.

The immediate feedback of results via the bi-directional communication system will significantly enhance the operability—and acceptability—of demand-side applications in the electric utility industry. Such immediate verification of results provides the utility system operator with an acceptable option for meeting the real-time energy needs of his customers. This allows demand-side applications to be considered equally with supply side alternatives, and thereby enhance the penetration of energy efficiency and conservation efforts.

Next, with respect to demand side management, a high capacity information system provides flexibility to approach real-time pricing. Installation of appropriate communications infrastructure allows the electric utility flexibility in the pricing of energy and in the establishment of time periods applicable to its pricing structure. These time periods could be established with an eye toward conserving energy on peak and promoting consumption during off peak periods when capacity is available for consumers. This also defers the need for new generating facilities to be constructed. A real-time pricing scheme can be established wherein pricing signals can actually be sent, if desired, at intervals as small as every 6 seconds. Further, it will be possible to approximate the consumption of each and every major appliance in a household and provide the customer with an itemized bill. This type of itemized bill provides significant new information to the customer regarding the consumption patterns of home appliances. Moreover, it encourages energy awareness on the part of the customer resulting in a more efficient utilization of energy. With a system such as PowerView, the utility also has the ability to provide energy efficiency education materials in a video mode. Finally, as an adjunct to pricing flexibility, energy companies could offer energy "sales" wherein off-peak power (both night-time and weekends hours) is priced substantially lower than during on-peak periods. This provides the customer with choices in consumption patterns based on affordability and value of energy at any point in time, and further encourages the most efficient utilization of energy.

CONCLUSION

Electric utilities have communicated with their customers and field facilities for several decades via the establishment of electronic meter reading, demand-side management applications, direct telecommunication services, and distribution automation. The electric utility industry is also one of the most capital intensive industries in our society. The capital investment per customer in the electric utility industry is often 6-7 times greater than the investment of contemporary communication services to the home. The opportunity to defer new capital requirements through cost-effective conservation and efficiency programs is an attractive alternative for many companies. As a result, implementation of cost-effective demand-side management applications by the electric utility industry can bring significant improvement in the communications infrastructure within the United States. We are confident PowerView can play that role.

EXHIBIT 1—POWERVIEW

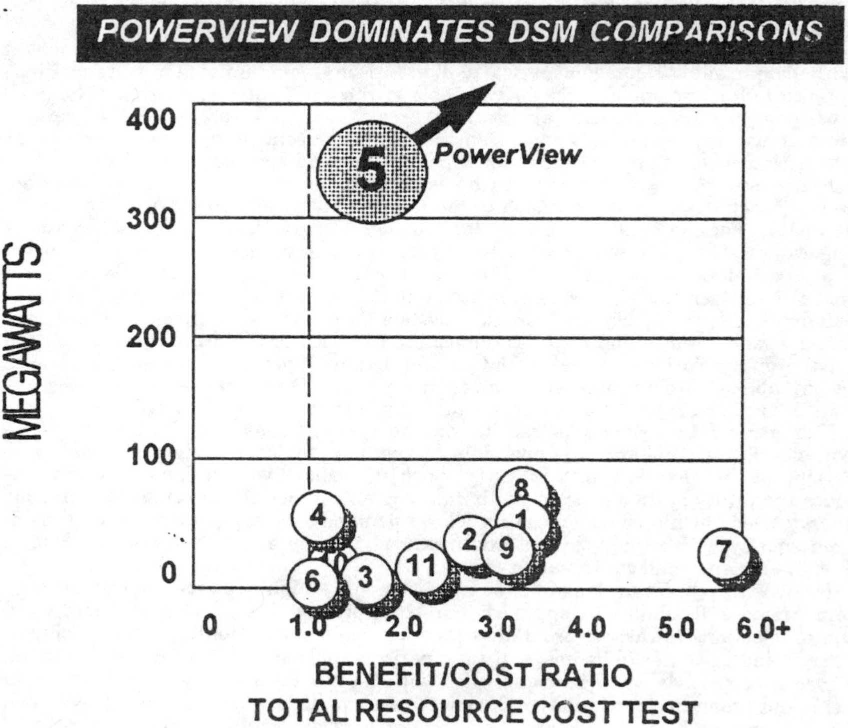

EXHIBIT 2—POWERVIEW

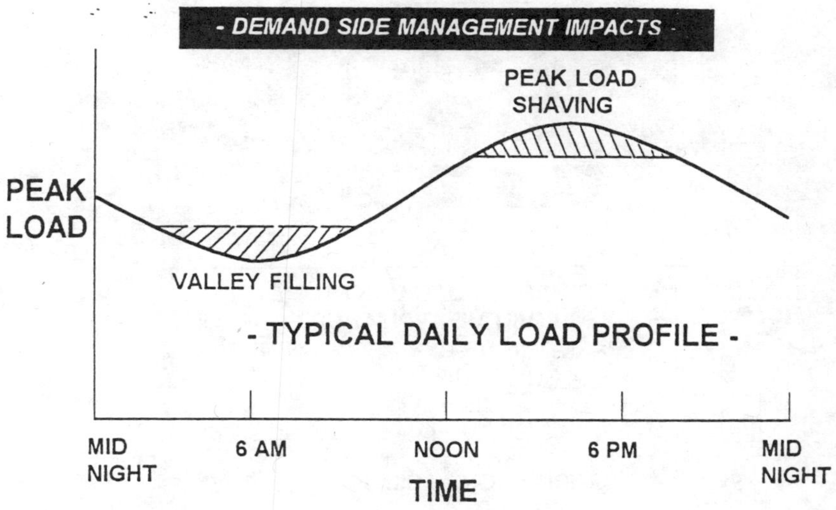

EXHIBIT 3—POWERVIEW

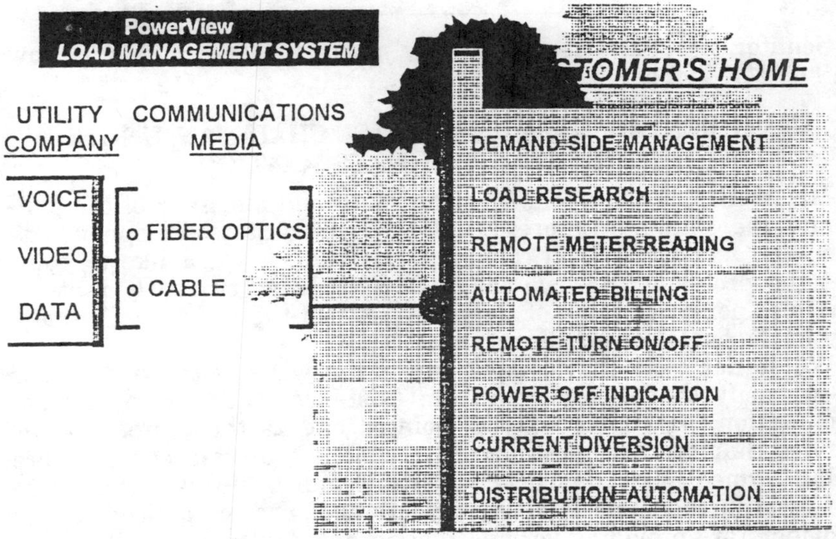

EXHIBIT 4—POWERVIEW

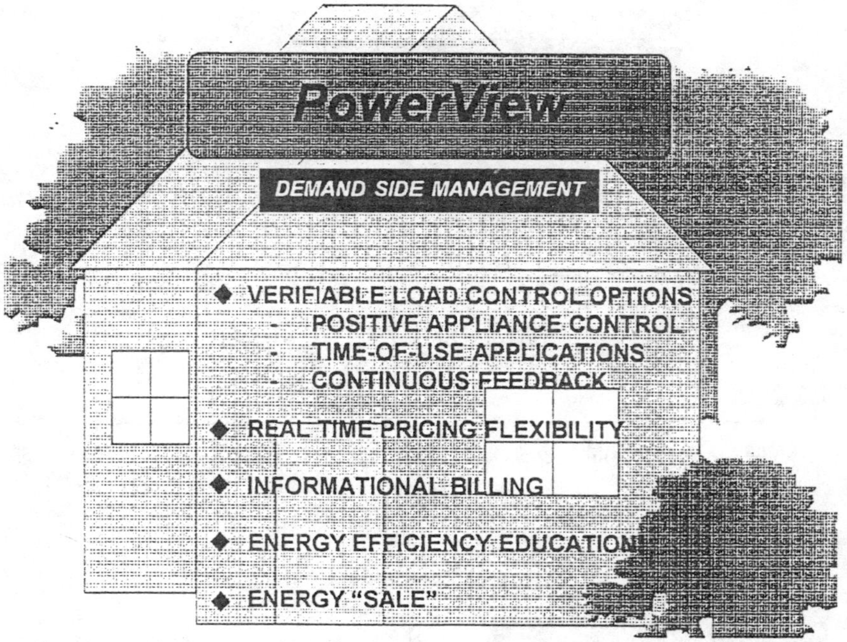

Senator INOUYE. Thank you very much, Mr. Niggli. May I now call on Mr. McBee.

STATEMENT OF GARY McBEE, VICE CHAIRMAN, UNITED STATES TELEPHONE ASSOCIATION

Mr. McBEE. Thank you, Mr. Chairman and members of the subcommittee. I am here representing the United States Telephone Association and 1,100 local company members. I would like to point out the problems that we see with S. 1086 as currently written, and also point out some of the areas we feel the legislation is moving in the exact correct direction.

We do not deny that competition is the wave of the future, nor do we deny that it is currently in existence. There is substantial competition, as Mr. Crowe has pointed out, in those areas of our service that are the most lucrative. There are coalitions that are being formed every day between traditional businesses that were not even in the same industry 5 years ago. The competitive access providers are providing very good service and I think that was well spelled out.

The problem is the way the current pricing structure is set up in the United States that creates an umbrella for them to come in and underprice services that are provided by the local companies and gain substantial benefits from that service. I think Congress would, in fact, enhance consumer opportunities all over the country, small consumers as well as large, by freeing the exchange carriers to do more services, as you have spelled out that you intend to do with cable.

No. 3, I think the changes must be made to the current pricing structure in the United States. S. 1086 overturns the current regulatory regime in terms of how the pricing would be done. But I do not think that it adequately addresses what the replacement might be, nor is there an understanding of how that might be accomplished in a cost-effective way.

Universal service across this country has been made possible by regulatory pricing strategies that, in fact, shift huge amounts of revenue from low cost areas to high cost areas, from business customers to residential customers, from long distance users to nonlong-distance users. That process has been put in place over time, with an idea in mind that all customers in the United States ought to have access to low priced telecommunications service or universal service.

I think the FCC has attempted in the past to try to deal with the whole issue of changing price structures, and as they have done that I think they found that it will not work the way they did it originally. Their intent initially was to put the cost for universal service on access charges or the off ramps, as was referred to earlier. And that is where competitive access providers see an opportunity to move into the marketplace and, in fact, do. And that is a reasonable result of high priced services such as we are offering. That price structure has to be redone in some way so that you do not artificially increase the price of some services and introduce competition where it need not be.

To try to point this out or spell this out and make the whole idea of pricing within the industry better understood by the committee, USTA has asked to have a study done—it was completed by strategic policy research—to look at what are the excess revenues that are currently included in only two services—one of them being access, the other being long distance that is provided by the local companies. That study, which will be released within the next week or two, indicates that the pricing anomaly there represents a low number of $18.3 billion nationally and a high water mark of $21.1 billion, and that is price above cost—just price above cost. Assuming that you were to fix that pricing anomaly and charge the individual end-user customer the appropriate price for their service, every customer in the United States would have to pay approximately $12 more per month just to take care of that anomaly.

That, of course, does not even take into consideration the problems that you have in rural communities where the costs are substantially higher, even, than they are in the rest of the country. The rural customer, in some cases, as pointed out very adequately before, have a very cost because of the low density. And the cost to serve in those areas is extremely high, and without a mechanism to fund the service in those area, the rural customer would be required to pay substantially more than they are today. One example, in California we have exchanges within Pacific Bell where the rate to provide service would be $105 a month in Baker, CA. Now, I do not think that we would ever anticipate Baker residents paying $105 a month, but that is the type of anomaly that is in the pricing structure today.

One of the reasons that USTA is opposed to the bill is that the FCC is required—would be required in some way—to provide for universal service, to maintain lifeline and linkup program, and the lifeline program in California alone has 1,800,000 customers that have lifeline service in California. They pay one-half of the local service rate for that State. That has to be taken into consideration, I think, as we move forward. Because of these reasons and our great concern for what might happen to the rural customers and the rural exchanges, currently USTA will not support S. 1086 as it is written. However, we do think that there are a number of things in the bill that are quite positive. And I will point out some of them.

We believe that progress is being made on the repeal of the cable cross-ownership—progress.

We think that seeking to share the cost of universal service among all the providers is an excellent step forward, but the question remains how is that done and how long will the argument be for how much that is and who is going to pay for it?

I think you are moving in the right direction of having competitors have equal rules applied to them. I think that is a definite positive, and the statement for pricing flexibility is absolutely necessary in a competitive world. I think the legislation certainly has building blocks for building a future telecommunications structure in this country, and USTA will look forward to working with the committee members in producing such legislation.

Thank you.

[The prepared statement of Mr. McBee follows:]

PREPARED STATEMENT OF GARY MCBEE

Good morning Mr. Chairman, members of the subcommittee, I am Gary McBee. I am a director of Nevada Bell and just recently retired from Pacific Bell where I served as executive vice president of external affairs. I am the vice chairman of the United States Telephone Association (USTA) on whose behalf I am testifying today.

On behalf of all the members of USTA, Bell and independent, small, medium and large, my testimony seeks to identify problematic elements of S. 1086, while highlighting the leadership the legislation demonstrates in other areas.

With more than 1100 local exchange carriers as member companies, USTA promotes the general welfare of the telephone industry, provides information and assistance to member companies, and serves as the industry's liaison with the federal government and regulatory agencies.

I. INTRODUCTION

Mr. Chairman, my testimony seeks to convey three messages:

• First, competition to local exchange carriers exists today and will increase regardless of federal legislative action. Traditional state and federal regulatory policies increasingly expose local exchange carriers to competition while advances in technology erase not only the lines between enhanced and basic services, but also traditional industry boundaries. One of the indications of these changes are the daily announcements of strategic business alliances between corporations that were previously thought to be in separate businesses.[1]

Another barometer of the increase in competition for access services, a sort of pick up and deliver service LECs provide to inter exchange carriers, is the percentage of access traffic that is exposed to collocaters. A recent USTA survey, a copy of which is attached to my testimony, confirms that collocation, the linchpin for competition to every service provided by LECs, exists or has been requested at 957 local exchange carrier' central offices. While these offices represents less than 14 percent of all LEC wire centers, they serve almost 80 percent of the industry's access traffic.

Congress will not capture the benefits of competition for all consumers unless regulatory policies which advance the interests of competitors to local exchange carriers

[1] Examples of these alliances include AT&T & McCaw, British Telecom and MCI, TCI/Cox/Teleport, US West/Time Warner and TCI/Time Warner/Micro Soft.

at the expense of LEC customers and shareowners are preempted. Local exchange carriers must have regulatory parity, that is they must be treated the same as those with whom they are competing.

• Second, Congress would enhance consumer choice in voice, switched video and high speed data services by freeing local exchange carriers to compete in the offering of enhanced services, such as S. 1086 would do for cable. Being free to provide such services would provide LECs with incentives to modernize the public switched network.

Additionally, providing LECs with the statutory authority to share network functionalities and facilities as provided for in S. 570 would ensure small rural exchanges can provide their consumers the same service choices available to urban consumers.

• Finally, while USTA believes changes must be made to current pricing regulations S. 1086 overturns the traditional utility model of regulation without ensuring sufficient support for universal telephone service. S. 1086, we fear, fails to appreciate the "seismic intensity" the collapse of current LEC pricing structures would have on universal telephone service.

Contributions to the universal service fund, alone, are not sufficient to ensure the universal availability of affordable telephone service.

II. UNIVERSAL SERVICE

Regulators have consciously built universal service support flows into LEC rate structures. S. 1086 directs the Federal Communications Commission (and states acting in conformity with the directives of the Commission) to take on the monumental task of ensuring universal service by providing the following in a fully competitive market:

• Support mechanisms such as the high cost fund (which flows to companies with high nontraffic sensitive costs) Lifeline and Linkup Programs (which benefit low income individuals);

• Support flows between customers for the same service through rate averaging (such as support from high volume to low volume users, and from low cost urban areas to high cost rural areas); and

• Support flows between services; (such as from access charges to residential service and from business services to residential services).

To assist policy makers appreciate the extent of this last support mechanism, that is "support between services" such as access and intra-LATA toll services, USTA commissioned a study by Strategic Policy Research (SPR).

The SPR study, which USTA plans to release upon completion, will establish the margin between revenues and costs of providing access and intra-LATA toll services. This margin embodies a contribution which has been traditionally used to support the provision of telephone service to small business, residential and rural customers.

Preliminary results of the SPR study indicate these contributions, on an annualized basis, to be no less than $18.3 billion, and perhaps as high as $21.1 billion. To put these numbers in perspective, they are twice the annual earnings of the entire local exchange industry.

USTA also asked Bellcore to compile data from studies previously prepared by their client companies to assess the amount of contribution received by local residential and business customers for access to telephone service. The Bellcore numbers mirror those of SPR in identifying support for local telephone service.

USTA will share both of these studies with the subcommittee upon completion.

Absent contributions from access and toll services on the order of $18 billion to $21 billion, residential consumers would be required to pay the full costs of their local telephone service. On average, this would require residential customers to pay twice as much for local telephone service.

And Mr. Chairman, this estimate does not take into consideration the additional support necessary for provision of local telephone service in rural areas.

Because of the gap between costs and revenues in access and intra-LATA toll services, it should come as no surprise that it is in this area cable companies, competitive access providers and other competitors, including interexchange carriers are concentrating their efforts. And you will remember I indicated that collocation either exists today or has been requested at the central offices which provide almost 80 percent of the access traffic handled by the industry.

While S. 1086 requires all telecommunications providers to contribute to universal service, the SPR study demonstrates such contributions alone will not be sufficient. Legislation which mandates competition must also detail the way in which it will provide safeguards for residential and rural consumers.

USTA is further troubled that S. 1086, by preempting state and local franchising authority, deprives those jurisdictions of the tools they have traditionally employed to ensure the universal availability of affordable telephone service.

III. TELECOMMUNICATIONS ENVIRONMENT: CONVERGENCE OF TECHNOLOGY AND BLURRING OF MARKETS

Dramatic changes are underway in the telecommunications marketplace. Regulatory policies are increasingly opening local exchange carriers' central offices to competitors, as advances in technology erase not only the lines between enhanced and basic services, but also traditional industry boundaries. The result of these changes are strategic business partnerships which are announced on a daily basis.

These alliances between traditional long distance, cellular, competitive access and cable providers reflect the coalescence of technologies and markets. Each of these parthe rships adds to already significant competition for local telecommunications services—services which have traditionally supported universal service.

If these market and technological developments are to serve the public, and not just serve competitors to local exchange carriers, public policy makers must ensure:

- Universe service and ubiquitous infrastructure development do not become the victims of competitors' ability to selectively serve only the most profitable portions of the local telecomuunications market; and
- Local exchange carriers are freed from a regulatory paradigmrich was crafted to govern a monopoly environment that no longer exists.

S. 1086, as currently drafted, does not fully achieve either of these goals.

IV. LECS MUST BE FREED TO FOLLOW WHERE THE MARKET PLACE AND TECHNOLOGY LEAD

As competitors enter low cost, high volume markets, LECs must be free to respond to this competition both in the markets we currently serve and those where advanced telephone technology provides us with new opportunities.

Congress would enhance consumer choice in voice, switched video and high speed data services by freeing local exchange carriers to compete in the offering of enhanced services, such as S. 1086 would do for cable. Being free to provide such services would provide LECs with incentives to modernize the public switched network.

A. Cable

S. 1086 recognizes the cable cross ownership ban of the 1984 Cable Act as one such outdated limitation.

But even in removing the cable restriction, the legislation establishes an asymmetric regulatory regime for provision of video programming. S. 1086 would:

- Require a telco to provide video programming transport as a tariffed common carrier, while no other provider of video programming, including incumbent cable operators, have a similar requirement.
- Prohibit a local exchange carrier from purchasing more than a 5 percent interest, joint venture or partner with a cable company in the same service area.

USTA believes if the price of basic telephone and cable service is regulated, in the absence of effective competition, shouldn't these two telecommunications providers be free to pursue the same economies of scale every other combination of telecommunications services are free to pursue?

Additionally, the legislation would mandate the requirement that a telco provide video programming through a separate subsidiary that would be prohibited from jointly marketing its telephone and cable services. while this prohibition appears also to apply to cable companies which seek to provide telephony services within the same area in which they provide cable, USTA believes the prohibition to be bad public policy.

Steps the Congress should take to free LECs to provide consumers with the benefits of a fully competitive market place include:

- Freeing the regional Bell operating companies to manufacture telecommunications equipment under the terms and conditions of S. 173, (Senator Hollings' legislation of the 102nd Congress);
- Directing the Federal Communications Commission to restructure its approach to regulation so as to eliminate disincentives to invest and take risks in the provision of new services. The Commission's focus should be on regulating the price of services which are not subject to competition;
- Rejecting any efforts such as those of section 11 [Information Services) and section 12 (CPNI Requirements) of S. 1086 to limit the ability of local exchange carriers to provide their consumers with the fullest array of information services at prices which need not bear the costs of onerous regulatory burdens; and

• Amending section 229(h) of the legislation to mandate regulators permit pricing flexibility.

Finally, Congress should provide local exchange carriers with the statutory authority to share network functions and facilities with each other so as to maximize the availability of the benefits of the public telephone network. This may be accomplished by enacting legislation providing for infrastructure sharing as articulated in S. 570, the "Local Exchange Infrastructure Modernization Act."

V. ADDITIONAL AREAS OF DISAGREEMENT WITH S. 1086

USTA further opposes S. 1086 as it would:

• Impose information service requirements on the Bell operating companies—requirements federal courts and the FCC have already found are not justified to protect competition;

• Impose a prior written approval requirement for access to customer information that local exchange carriers might otherwise utilize to inform customers of public switched network services; and

• Mandate requirements such as number portability and interconnection, at any point technically feasible, with insufficient consideration of the economic impact of, or demand level for, such requirements.

VI. AREAS IN WHICH S. 1086 DEMONSTRATES LEADERSHIP

S. 1086 does include provisions which would promote development of the nation's public switched network. among these provisions are:

• *Cable Relief.* The recognition LECs must be free to more fully utilize their networks by repealing the cable cross ownership ban including the ability of Bell operating companies to operate a cable headend and provide cable service across LATA boundaries.

• *Sharing of costs of universal service.* Mandating all providers of telecommunications services contribute to universal service.

• *Joint Network Planning.* Requiring all telecommunications providers to participate in joint network planning and comply with mandated standards for interconnection.

VII. CONCLUSION

Mr. Chairman, for the reasons outlined above, as currently drafted, USTA can not support S. 1086.

Still, there is much in S. 1086 which USTA applauds.

Progress on the repeal of the cable cross ownership ban, seeking to share the costs of universal service among all telecommunications providers, attempting to fashion a regulatory environment in which providers of like services are treated alike, and providing for pricing flexibility are significant building blocks for a national telecommunications policy which properly addresses future telecommunications markets.

USTA looks forward to not just the remainder of this hearing, but to the work which follows, in which such a national telecommunications policy might be crafted.

Thank you, and I look forward to your questions.

USTA DATA REQUEST—INTERCONNECTION EXPOSURE

A USTA study of 6818 telephone switching locations, representing approximately 33 percent of all such locations in the U.S., shows that competitive access providers collocating in only 14 percent of the 6818 locations can address 80 percent of all access traffic in those locations.

These statistics show that CAP's compete only in those areas where high volumes of traffic exist. This allows them to skim off the most profitable LEC business while not incurring costs to serve in locations having small quantities of traffic.

In order to measure the loss of access traffic LEC's may face in a competitive environment USTA gathered the attached data. The data were requested from all LEC's required by the FCC to provide collocation. Nine (9) companies responded.

The study shows that the responding companies represent 6818 wire centers, about 113 of the total wire centers in the US. Nine hundred fifty seven (957) or 14.0 percent of the 6818 wire centers have collocation, request for collocation or a customer has expressed interest in collocation.

To measure the risk of lost access traffic the companies reported the number of DSI equivalents in total wire centers, compared to the number of DSI equivalents in collocation wire centers. These reports.were separated between switched access and special access.

A DS1 is a transport facility representing 24 voice channels. For purposes of this study a DS1 equivalent is considered to be 9,000 MOU per circuit per month (216,000 MOU per DS1). This is the formula used by the FCC in several proceedings including CC Docket 91-213, Local Transport. In order for a CAP to have an economic incentive to aggregate traffic there must be adequate volume. LEC's are vulnerable because where competitors are collocating is where traffic volume is highest.

Therefore, as this study indicates almost 80 percent of the access traffic, is at risk even though collocation is present in only 14 percent of the offices.

ROLL UP OF 9 COMPANIES REPORTING

Number of collocation wire centers, 957; Total wire center, 6,818; Percent (1/2), 14.0; Switched DS1 equivalents in collocation wire centers, 132,666; Total switched DS1 equivalents, 200,964; Percent (4/5), 66.1; Special DS1 equivalents in collocation wire centers, 344,117; Total special DS1 equivalents, 397,061; Percent (7/8), 86.7; Combined switched and special in collocation wire centers, 476,783; Total switched and special DS1 equivalents, 597,755; and Percent (10/11), 79.8.

Senator INOUYE. Thank you very much, Mr. McBee. Mr. Frischkorn.

STATEMENT OF ALLEN R. FRISCHKORN, JR., PRESIDENT, TELECOMMUNICATIONS INDUSTRY ASSOCIATION

Mr. FRISCHKORN. Mr. Chairman and members of the subcommittee, I am pleased to appear here today on behalf of the Telecommunications Industry Association. TIA's nearly 550 members collectively provide the bulk of the physical plant and associated products and services used to build and maintain and improve the telecommunications infrastructure in the United States.

I am pleased to have the opportunity to present to you TIA's views with respect to the critical public policy issues raised by S. 1086. TIA commends you, Mr. Chairman, Senator Danforth, and both your staffs, for having the courage, foresight, and energy to tackle the single-most important and challenging issue facing our industry today. As your bill recognizes, in order to ensure that our Nation's telecommunications infrastructure remains second to none, it is absolutely essential that we develop and implement a national policy which promotes competition in the provision of all telecommunication services. Without such a policy, our Nation and the American people will be denied the full benefits available through the use of advanced telecommunications technologies in areas such as education, transportation, manufacturing, health care, financial services, entertainment, and other equally vital sectors of our economy.

Technological developments, regulatory initiatives undertaken over the past several decades, and the structural changes affected under the terms of the AT&T consent decree, have dramatically increased the level and scope of competition in virtually all sectors within the U.S. telecommunications industry. However, legal and regulatory barriers continue to limit the ability of alternative providers to fully enter and compete on a fair and equal basis in the local telephone service market.

These barriers, together with the statutory and regulatory restrictions on the provision of video services by local telephone companies, impose serious impediments, in our view, through the deployment of new technologies and systems that are capable of efficiently delivering a broad range of services which cuts across the existing legal and regulatory boundaries.

TIA believes that removal of these artificial barriers to competition in the provision of all types of local telecommunications services; that is, voice, data, and video, will yield substantial benefits to consumers, the industry, and our Nation's economy as a whole in the form of lower prices, better service, increased innovation, and a more diverse and efficient array of products and services and service providers.

The introduction of S. 1086 represents a welcome first step toward the establishment of a national policy which is designed to ensure that these potentially enormous benefits are available to American businesses and consumers alike. In particular, TIA would like to express its support for proposed section 229(a) included in section 5 of S. 1086, which provides for the elimination of State and local statutory and regulatory provisions "that prohibit or limit in a manner inconsistent with Federal regulations or with this act, the ability of any entity to provide interstate or intrastate telecommunication services" including local exchange and exchange access services.

TIA also generally supports the imposition of interconnection requirements which will ensure that incumbent monopoly providers of local telecommunications services do not withhold access to "bottleneck" facilities that are essential to the ability of alternative providers to enter and become viable competitors in the market for such services.

In this regard, TIA recognizes that removal of explicit legal barriers to entry does not guarantee that meaningful competition will automatically develop in the provision of local telecommunication services. TIA therefore urges that careful attention be given to the need to strike a balance which recognizes and addresses the risks of anticompetitive conduct by the current monopoly provider of such services as alternative providers attempt to enter and compete in the local services market, while at the same time making provision for appropriate relaxation of regulatory constraints on the incumbent provider when and where competition, in fact, emerges.

In addition to supporting competition in the local telephone service market, TIA endorses the basic provisions contained in section 8 of the bill, which would amend section 613(b) of the Communications Act to eliminate the current restriction on the provision of cable television service by local telephone companies within their telephone service areas. In this area, as well, TIA urges the sponsors to continue their efforts to develop a balanced, workable, procompetitive policy approach that provides reciprocal opportunities for telephone companies and cable companies to expand the scope of their current operations under appropriate terms and conditions.

Achieving a legislative consensus on these difficult issues will not be easy. But TIA strongly believes that the development of a sound national policy in this area promises to yield enormous dividends to our society for years to come.

Thank you, Mr. Chairman, for the opportunity to appear here today, and I would be pleased to answer any questions you may have.

[The prepared statement of Mr. Frischkorn follows:]

PREPARED STATEMENT OF ALLEN R. FRISCHKORN, JR.

My name is Allen R. Frischkorn, Jr. I am President of the Telecommunications Industry Association (TIA), a national trade association whose membership includes nearly 550 manufacturers and suppliers of all types of telecommunications equipment and related products. IA's members are located throughout the United States and collectively provide the bulk of the physical plant and associated products and services used to build, maintain, and improve the telecommunications infrastructure in the U.S. In addition, TIA's-members are involved on an ever-increasing basis in providing telecommunications equipment and related products in other developed and developing nations around the world.

I am pleased to have the opportunity to present to you TTA's views with respect to the critical public policy issues raised by S. 1086, "The Telecommunications Infrastructure Act of 1993." TIA commends you, Mr. chairman. Senator Danforth, and both your staffs for having the courage, foresight, and energy to tackle the single most important and challenging issue facing our industry today. As your bill recognizes, in order to ensure that our nation's telecommunications infrastructure remains second-to-none, it is absolutely essential that we develop and implement a national policy which promotes competition in the provision of all telecommunications services. Without such a policy, our nation and the American people will be denied the full benefits available through the use of advanced telecommunications technologies, in areas such as education, transportation, manufacturing, health care, financial services, entertainment, and other equally vital sectors of our economy.

Let me assure you that TIA shares your commitment to the promotion of "investment, growth, and competition" throughout the telecommunications industry. several years ago, in response to a Notice of Inquiry issued by the Commerce Department's National Telecommunications and Information Administration (NTIA), TIA offered a number of recommendations for public policy initiatives to promote the continued advancement of the domestic telecommunications infrastructure. In its April, 1990 comments, TIA specifically urged that "the cornerstone of any effort to configure a comprehensive public policy framework designed to spur the future growth and development of the telecommunications infrastructure in the U.S. must be a continuing commitment to the promotion and preservation of open, competitive markets in which telecommunications manufacturers, equipment suppliers, and service providers are free to compete on a fair and equal basis." NTIA's October 1991 Infrastructure Report generally concurred in this view, concluding that "[t]he soundest way for government to encourage efficient infrastructure development is through removal of unnecessary regulation and promotion of a competitive telecommunications marketplace."

Technological developments, regulatory initiatives undertaken over the past several decades, and the structural changes effected under the terms of the AT&T Consent Decree (the Modification of Final Judgment or "MFJ") have dramatically increased the level and scope of competition in virtually all sectors within the U.S. telecommunications industry. Competition in the equipment manufacturing and long distance service sectors of the industry has resulted in substantially reduced prices, more rapid innovation, and a vastly expanded range of products and services tailored to meet the specific needs of American consumers and businesses.

Until recently, it was generally assumed that the provision of local telephone service was a "natural monopoly." However, recent technological advances and marketplace developments have combined to challenge this assumption. Still, while the FCC and a number of states have taken some significant steps to facilitate the growth of competition in the provision of inter and intrastate local access services and, to a much lesser extent, in the provision of certain local telephone services, legal/regulatory barriers continue to limit the ability of alternative providers to fully enter and compete on a fair and equal basis in the local telecommunications service market. These barriers, together with the current statutory and regulatory restrictions on the provision of video services by local telephone companies, impose serious impediments to the deployment of new technologies and systems that are capable of efficiently delivering a broad range of services which cuts across the existing legal/regulatory boundaries.

TIA believes that removal of these artificial barriers to competition in the provision of all types of local telecommunications services, i.e., voice, data, and video, will benefit consumers, the industry, and our nation's economy as a whole. Consumers stand to reap substantial benefits in the form of lower prices, better service, and a broader range of products, services, and service providers. The telecommunications equipment and service sectors of the industry will also benefit from the new market opportunities created by competition in local service markets. Given the dramatic growth which the industry has experienced in other sectors where competition is already well-established, it seems reasonable to expect that market opportunities for new entrants and existing service providers will expand in a competitive local telecommunications environment. In addition, as TIA indicated in its last appearance before this subcommittee, the emergence of meaningful, market-based constraints in the Regional Bell Operating Companies' ability to abuse their monopoly power, as competition develops in the local telephone service markets which they now dominate, would provide a viable basis for the eventual removal of the current restriction on RBOC entry into the telecom manufacturing business. lore generally, creation of a competitive marketplace would reduce or eliminate the need for regulations designed to address the risks of anticompetitive behavior associated with the current monopoly environment, thereby reducing the overall cost of regulation at the federal and state level. Finally, businesses in other industries, particularly those which are highly dependent on the use of telecommunications, will benefit from the cost reductions and the broad array of new and improved services available in a fully competitive local telecommunications marketplace. Most importantly, they will have the ability to select the technologies and service providers that are most efficient and responsive to their needs.

The introduction of S. 1086 represents a welcome first step toward the establishment of a national policy which is designed to ensure that these potentially enormous benefits are available to American businesses and consumers as quickly as possible. Accordingly, TIA strongly supports the purposes articulated in Section 3(2) and Section 3(3) or the proposed legislation, i.e., to "ensure the availability of the widest possible range of competitive choices in the provision of telecommunications and cable television services" and to "eliminate the existing regulatory barriers to competition in the provision of telecommunications and cable television services."

More specifically, TIA wishes to express its support for Proposed Section 229(a), included in Section 5 of S. 1086, which provides for the elimination of state and local statutory and regulatory provisions that "prohibit or limit in a manner inconsistent with Federal regulations or with this Act the ability of any entity to provide interstate or intrastate telecommunications services." TIA also generally supports the imposition of interconnection requirements which will ensure that incumbent monopoly providers of local telecommunications services provide access to "bottleneck" facilities that are essential to the ability of alternative providers to enter and become viable competitors in the market for such services. TIA recognizes that certain aspects of Proposed Section 229(c), which as introduced would extend interconnection obligations to other providers who arguably do not control "bottleneck" facilities, as well as certain other provisions contained in Section 5 of S. 1086, have proven to be somewhat controversial. TIA urges the sponsors and their staffs to work with the affected parties to address the concerns that have been raised with respect to these provisions in a manner consistent with the pro-competitive purposes of the legislation.

In this regard, TIA recognizes that the removal of explicit legal barriers to entry does not guarantee that meaningful competition will in fact develop in the provision of local telecommunications services. Accordingly, TIA urges that careful attention be given to the need to strike a balance which recognizes and addresses the risks of anti-competitive conduct by the current monopoly provider of such services, as alternative providers attempt to enter and compete in the local services market, while at the same time making provision for appropriate relaxation of regulatory constraints on the incumbent provider when and where competition in fact emerges.

In addition to supporting competition in the local telephony market, TIA endorses the basic provisions continued in Section 8 of S. 1086, which would amend Section 613(b) of the Communications Act to eliminate the current restriction on the provision of cable television service by local telephone companies within their telephone service areas. TIA at this time defers further comment on the specific details of the terms and conditions under which telco entry will be permitted, with the exception of the anti-buyout provisions embodied in Proposed Section 613(b)(1)(A), which TIA supports. In this area as well, TIA urges the sponsors to carefully review the concerns raised by the parties most directly affected by this proposal (i.e., cable and telephone service providers), and to make every effort to develop a balanced, workable, pro-competitive policy approach that provides reciprocal opportunities for telephone companies and cable service providers to expand the scope of their current operations, under appropriate terms and conditions. Achieving a legislative consensus on these difficult issues will not be easy, but TIA strongly believes that the development of a sound national policy in this area promises to yield enormous dividends to our society for years to come.

Before concluding my remarks, I would like to briefly address one other issue of considerable concern to TTA. As an ANSI accredited standard-setting organization, TIA is actively Involved in efforts to develop voluntary standards for the telecommunications industry in a wide range of areas. There are currently more than 1,000 individuals serving on more than 50 committees and subcommittees involved in TIA-sponsored standardsetting initiatives. Accordingly, TIA has an strong interest in ensuring the continued success of the current, industry-led standard process.

TIA believes that generally speaking this process, which emphasizes broad-based participation in an open forum in which all participants have ample opportunity to articulate their views and to contribute to the development of a consensus view as to the appropriate standard, has served our industry and our country well. Forward-looking technical standards for telecommunications can best be written by the experts who are working on a daily basis at the leading edge of that tecnnology. In most cases, those experts reside in the private sector. As a matter of principle, we believe that unwarranted government intervention in this process would pose substantial risks to the continued growth and dynamism of the telecommunications manufacturing and services sectors of the U.S. economy.

Certainly, government has a useful role to play in the standard-setting process, e.g., a participant in the formulation of standards that implicate the government's interest as a major user of telecommunications equipment and services, and as an interested observer in cases involving the development of standards that may affect the FCC's performance of its statutory responsibilities under the Communications Act (e.g., with respect to the allocation and use of radio spectrum). However, before permitting the government to assume a more active role in this process, care must be taken to ensure that 1) there is in fact a "problem" that clearly cannot be addressed through the established, industry-led standards development process, and 2) that the benefits of government intervention in the standardsetting process outweigh the potential costs.

Section 6 of 5.1086 includes several provisions, contained in proposed Section 230(d)(1), which would require the FCC to take certain steps to "encourage" and "assist" the industry in developing standards to "ensure interconnection and interoperability of telecommunications networks," and would authorize the Commission to "establish" standards "when industry participants fail to reach agreement." While TIA is not opposed to government "encouragement" or "assistance" per se, we are concerned that the line between "assistance" and direct government involvement in the establishment of a standard is somewhat uncertain. Accordingly, we would urge the bill's sponsors and other members to consider at a minimum, including language that expresses a clear preference for industry-led standards development, with any form of direct or indirect government intervention to be considered only as a last resort in situations where the established process has clearly broken down, where there is a compelling need to establish a standard, and where it is clear that the benefits of government intervention outweigh the costs. TIA would be pleased to work with the sponsors of S. 1086 and other interested members to develop legislative language which addresses TIA's concerns in this area in an appropriate fashion.

In closing, I would again like to thank you, Mr. Chairman, and Senator Danforth for your efforts to craft legislation which secures for our nation the substantial benefits which can only be realized through the development of full and fair competition throughout the telecommunications industry. TIA looks forward to working with you and other members of Congress as the legislative process moves forward. appreciate the opportunity to appear before the Subcommittee and I would be pleased to answer any questions you may have.

Senator INOUYE. Thank you, Mr. Frischkorn. I have several questions for all of the panel members. However, I will limit myself to one or two and, if I may, submit the rest for your consideration and response.

So, Mr. Crowe, according to your testimony you already are competing rather successfully with the phone company. If that is the case, why do you need this bill?

Mr. CROWE. We are competing in a very, very narrow range of services, Mr. Chairman, specifically access for large businesses. My friend from the USTA mentioned that we are—and this is a term that is often used—we are cream skimming, if you would, competing for profitable services at the exclusion of less profitable services. I would tell you that the answer there is to let competition reign and see what happens.

Today, I would submit to you that most of the profits are in areas that we do not compete in. Today, we are not allowed to compete for the switched access that the gentleman mentioned. Today, we are not allowed to compete for intra-LATA long distance, They are monopolies. The areas where we compete are clearly the most competitive, and we would tell you probably priced below cost.

The difficulty in all of this, I would think for you, is to determine whether in fact there are legitimate subsidies or whether the regulatory structure is simply subsidizing 100 years of inefficiency. Are we subsidizing legitimate goals, or are we simply subsidizing costs that ought to be far less in a competitive environment? And I would tell you that the latter is our view. If you open up more markets, that is, the other 99 percent of the local loop, we will bet the same kind of benefits we have seen in the narrow 1 percent we can compete in.

Senator INOUYE. So, you want to get into the 99 percent?

Mr. CROWE. We are working as hard as we can. Yes, sir.

Senator INOUYE. Thank you, sir.

And now, if I may ask Mr. Roberts, if the Congress is successful in passing this measure, how soon can we expect to see companies providing competition for local exchange telephone service?

Mr. ROBERTS. Our own company did a demonstration last year of the technology, which is obviously the first step. In September, we had a five-way conference call using an interconnection of cable, cellular, PCS where we had an experimental license in personal communications, and alternative access facilities here and in the United Kingdom. The five-way conference call did not use any local exchange facilities here or abroad, and the connections were perfect.

Our view on all of this is that there is not going to be a new network where somebody goes out and builds the whole country, but rather, already, a lot of the infrastructure is in place. Cellular is now available in almost all of America. Cable is in 95 percent. Alternate access is building a fiber highway.

So, I think our answer would be it is coming very soon, probably within 2 years you would begin. The construction has been announced by several companies to accelerate. There is a race on to come up with applications that will help fund this, things like home shopping which our company is involved with—such as QBC with Barry Diller—and some of the electronic retailing initiatives.

So, I believe that there is great promise, and that this bill is of real importance because you will see an acceleration of such an infrastructure with your support.

Senator INOUYE. Do you believe that PCS can become a competitive alternate to the telephone service?

Mr. ROBERTS. I do not know that we know yet. We are a cellular company, as well, at Comcast. There are 50 million cordless phones today. And mine does not really work too well when you go from the living room to the kitchen. You get a lot of static as you get away from the base station. But there is a tremendous demand for an untethered telephone in your home. There are only 10 million cellular phones, just to put it into perspective.

I think our view of what PCS could be is interconnecting with other carriers to provide maybe a local cellular service in your neighborhood. Someday PCS could be more products than that, but that might be a targeted use because you have to build very many cells for PCS, and we think that that would be logical to do in local neighborhoods, interconnected perhaps by cable or alternate access facilities, and of course into the local switch network.

Senator INOUYE. I thank you very much, sir. If I may now ask Mr. Ware, Mr. Ware, we have listened to your testimony with great care, and so I wish to assure you that it is not the intent of the authors to disturb existing subsidy flows set by the FCC. In fact, we hope to give statutory backing to the FCC regulations.

Now, if we are able to clarify that in this bill, would you still oppose the measure?

Mr. WARE. Mr. Chairman, thank you, and we do appreciate your concern and the concern of the other subcommittee members for the need to maintain universal service.

To answer your question, we do have some other concerns in that the bill basically severely restricts the authority of the State commissions, but then gives them the responsibility for universal service. So, with some crafting or recrafting of the language in the bill it possibly could be shaped to meet those concerns.

We are concerned about the effect of the limits on the State regulatory bodies and the direction to directly targeted support flows rather than the broader general support flows in place today which have done such a very good job.

Senator INOUYE. We will get together with you, sir.

Mr. WARE. Thank you, sir.

Senator INOUYE. Mr. Coe, as a large user, would it matter to you as to who provides the network service?

Mr. COE. I think as a large user, indeed what we are looking for are alternatives. The largest component cost of credit card transactions is indeed the last mile, and we also have limited technical options in that area. So, we believe that providing competition there would lower the cost and give us some alternatives.

Senator INOUYE. So, you are hoping that competition will give you the lowest cost alternative?

Mr. COE. Absolutely, sir.

Senator INOUYE. Mr. Niggli, why can the phone companies not provide your company with energy management capabilities you need at this moment?

Mr. NIGGLI. Well, there are two reasons, Mr. Chairman.

Senator INOUYE. Is fiber the only way you can do it?

Mr. NIGGLI. Pardon?

Senator INOUYE. Is fiber the only way you can do it?

Mr. NIGGLI. It may be from the standpoint of having the verifiability and the response rate that we need to ensure that we have response for the control actions that are taken. If you think about the energy management situation, our system operators have a choice when load is increasing on a hot day like today. They can push a button and call up a generator.

Well, they need to also on the demand side be able to push a button and bring down load for energy management. We have to have a feedback that says that has happened and those kinds of things have occurred and will stay off the line. Otherwise, we jeopardize some of the reliability of our system.

So, it is a speed-and-response issue right now that if we poll the large number of consumers that we believe will be on this system and all those responses come back, the network cannot handle it right now.

Senator INOUYE. Thank you, very much.

Mr. McBee, does your organization support the provisions that would permit telephone companies to get into cable?

Mr. MCBEE. Yes, sir.

Senator INOUYE. And what is your position on separate subsidies for electronic publishing?

Mr. MCBEE. Separate subsidiaries, in our opinion, have been dealt with very adequately by the FCC and recently by the appellate court, and I believe even in the beginning of July of this year the FCC has strengthened the cost allocation rules. We think those are adequate. There have been virtually no complaints about cost allocation—an awful lot of discussion of it but very few complaints. So, we think they are not necessary.

Senator INOUYE. Do you have any thoughts on long distance waivers in the bill?

Mr. MCBEE. I am sorry, sir?

Senator INOUYE. We have waivers on long distance here. Do you have any thoughts on that?

Mr. MCBEE. No, sir.

Senator INOUYE. I see. Thank you, very much.

Mr. MCBEE. If I might add one thing, Mr. Chairman——

Senator INOUYE. Yes.

Mr. MCBEE [continuing]. I did not use the term "cherrypicking," but the idea of the interexchange carriers and the alternate access providers moving into our central offices is demonstrated by a study that is attached to my testimony where they are in 957. They have asked for location in 957 offices of the local exchange carriers or collocation. Those 957 offices only represent 14 percent of all the offices in the United States, or at least that were in the study, but they have access to or they provide access to 80 percent of the access traffic.

So, I do believe that there is a business demand that says you go after the most lucrative markets first, and I think that question was brought up earlier. That is what business is going to do, and I think that is what they are doing.

Thank you.

Senator INOUYE. Thank you.

Mr. Frischkorn, has the manufacturing sector become more productive since the breakup of AT&T?

Mr. FRISCHKORN. Yes, it has indeed become more productive. There are fewer employees putting out more equipment than ever before. Since the AT&T breakup, prices of products have declined, the quality of products have improved and variety of products has increased. We probably all remember 10 or 12 years ago when the rotary dial phone was standard. Now you have such a variety of options. So, all these things have been spurred in part by the competition that was engendered by the AT&T breakup.

Senator INOUYE. Now, if we are going to have the benefits of high technology for all of our users and consumers, it would take the deployment of some network infrastructure. Do you have any idea as to how much that will cost?

Mr. FRISCHKORN. Well, with respect to one segment of it, I have information. There are some estimates that having a fully operational fiber optics network by the year 2035 will cost $450 billion plus.

Senator INOUYE. $450 billion.

Mr. FRISCHKORN. $450 billion. Well, initially, that may seem like a big number. [Laughter.]

That might seem like a big number even to you here in Congress. Let me say this—telephone companies spend roughly $20 billion a year on new equipment and cable companies somewhere over $2 billion. That is about $22 billion. So, in 20 years telephone and cable companies will spend about $450 billion on the network. And that spending will increasingly be on new products and services. As those new products and services come on line their prices will decline. Of course, you probably all remember the VCR and the handheld computer. They were very expensive at first, but the more you produced and sold, the more the price declined. And that will happen also with fiber optics.

One of our member companies, Corning, Inc., estimates that an interactive broadband network including fiber to the curb will be universally available around 2035 under a purely cost-driven scenario. However, if we want to accelerate the deployment of this network so that most Americans can be served within, say, 20 years, an incremental cost will be involved. Corning estimates that incremental cost to be about $23 billion in 1992 dollars over the next 20 years. While that number seems high, the $23 billion could be easily recovered through the pricing of new services which are going to be coming on line.

So, the bottom line is $450 billion may seem daunting at first, but it is not so daunting when you look at the overall economics of the telecommunications industry.

Senator INOUYE. Thank you, very much.

Senator Danforth.

Senator DANFORTH. Mr. Ware, my understanding of your position is that in the case of rural telephones companies, competition might not necessarily be beneficial to consumers.

Mr. WARE. Mr. Chairman and Senator Danforth, yes, sir, that is our position. And I think it is very reasonable. When you approach the competitive environment you envision sharing markets between competitors. And when you come into the sparsely populated rural area and you try to share a 500 customer market, for example, the economies of scale are not there. You may not have either participant viable once that happens.

Senator DANFORTH. Now, other than Mr. Ware, do any of the other witnesses on the panel—and I thank each of you for being here—do any of you think that competition in local telephone service is an undesirable objective?

[No response.]

Senator DANFORTH. Do all of you believe it is a desirable objective?

Mr. FRISCHKORN. Yes.

Mr. McBEE. If I might, Senator, I think it is a desirable objective but the decision as to how that services like in Mr. Ware's territory can be maintained if, in fact, you have a very small business, small telephone company, one of your customers represents 30 percent of your income. The competition would more than likely come after that one customer and leave Mr. Ware with the balance of the customers and 70 percent of the revenue.

So, I think that there are some considerations but we are not saying that it will not happen. I think it is less likely to happen in the rural areas, but if it did it could be disastrous for some of those companies.

Mr. WARE. Mr. Chairman and Senator, if I might, as I indicated in my remarks that we are asking the committee to not mandate competition in rural areas at this time, but to rather defer and allow the State commissions to open up the local exchanges to competition when they feel there is evidence that it is viable and that consumers will not be injured by it.

Senator DANFORTH. All right. Other than a possible concern with respect to—well, it is not a possible concern, it is a real concern from those who expressed it. With respect to small communities and rural telephone service, is it the view of this panel that competition is desirable in local telephone service?

Mr. ROBERTS. If I might, I think if you look to the long distance market as a recent example of what happens. Our understanding is that rates have come down 40 percent, AT&T is healthier than ever, there are many competitors who are doing well, and the consumer has been the big winner. And the same is true in equipment, that we do not just have to buy the black rotary telephone. We now have fax machines and all sorts of devices that are easily purchased and the prices have come way down.

So, I think you can look to that and say that if the whole network is competitive at all stages with open architecture and interconnections, you are going to have a very robust, rapidly accelerated marketplace.

Mr. CROWE. Senator, I might make a comment, too. I think some of the most lucid and carefully thought out argument's concerning the benefits of competition in the local loop have been submitted by the Bell companies over in the United Kingdom to their equivalent of the Federal Communications Commission. They make the argument, probably because they have a lot more lawyers, far better than we could. I would certainly commend those writings to anyone who is interested. They are very well drafted.

Mr. McBEE. Senator, if I might add, the person originally from Pacific Bell who made that comment, I wish he were here today. And as far as your term about desirable, that would not be my choice of words. I think it is inevitable, and I think this committee, the one thing you cannot do is to create legislation that would stop competition.

Senator DANFORTH. But under the present situation, it is stopped.

Mr. McBEE. I understand that it is stopped, yes, sir, but it will not be over time. Just like every market gradually is being opened to competition. And our position is that that is going to happen but you clearly have to understand the support flows that are necessary to provide universal service before unleashing that.

Senator DANFORTH. Thank you all, very much. Thank you, Mr. Chairman.

Senator INOUYE. Thank you. Senator Burns. Oh, I am sorry, Senator Packwood.

Senator PACKWOOD. Mr. Niggli, in your service area, I take it you serve both urban and rural areas?

Mr. NIGGLI. That is correct.

Senator PACKWOOD. Is there cross-subsidization from urban to rural or from business to residential in the electric business?

Mr. NIGGLI. That probably depends on which jurisdiction you are looking at. We have cost of service rates that we file. The last rate case we had in our service area was about 1985. So, it has been quite a while since we have had a specific rate case.

Senator PACKWOOD. Well, I do not mean as a specific rate case, as a matter of philosophy. In the old days the argument was we use long distance to subsidize residential use, business, to subsidize rural, phones and—do you have that in the electric business?

Mr. NIGGLI. Yes. I believe you would find it in many of the State commissions. They have set it up so that certain classes are subsidizing others and we do not have a true cost of service.

Senator PACKWOOD. Do you by and large make money on residential customers or at least break even on them?

Mr. NIGGLI. We make money on residential customers, yes.

Senator PACKWOOD. OK, now, Mr. McBee, on page 7 of your statement you say absent this $18 to $21 billion subsidy, residential consumers would be required to pay the full cost of their local telephone service. Why should they not?

Mr. MCBEE. I think that the FCC are trying, in fact, to do that, Senator.

Senator PACKWOOD. Do you have any objection to them paying their own way?

Mr. MCBEE. No, I do not have an objection to them paying their own way, if that were the only problem. The problem is that it is not just $12. I mean, assuming that everyone paid $12 more you would still have the problem of the urban areas at that point would be substantially providing support to the rural areas. And therefore, you have a pricing structure in place in the urban areas that encourages competition because they are priced above cost, and the rural areas are at the same price, although if you eliminated the price subsidy, the real prices would go up substantially.

Senator PACKWOOD. Well, here, I am not thinking about rural so much as residential, because you are talking about a subsidy of $18 to $21 billion residential consumers, so I assume you do not mean rural. You mean all of us living in the center of the city.

Mr. MCBEE. No, sir, I mean residential customers or consumers in general. That may be poorly stated there, but it is customers in general, residential or home customers.

Senator PACKWOOD. Well, all right. But why should telephones be different than electricity? Unless you are saying that the telephone is more critical than electricity.

Mr. MCBEE. Well, today I certainly would not say that because it is so hot. [Laughter.]

I am suggesting that the price differential between urban and rural is substantial; and if, in fact, we could say that every rural customer could in fact pay the cost of their service, that would probably be the best, economically efficient way to price. But I think it also then really runs in the opposite direction of universal service. I think a lot of people cannot afford it.

Senator PACKWOOD. I guess I am not getting my point across. I am talking about business versus residence, not urban rural.

Mr. MCBEE. Do I think business customers should pay their fair share, their full amount? Yes, sir.

Senator PACKWOOD. Well, apparently in the electric business, Mr. Niggli manages to make money on residential customers; I do not know if he makes money on rural or not, but on residential customers in New Orleans. And yet, I sense that you are saying telephone customers are subsidized in these cities.

Mr. MCBEE. What I am suggesting is that we make money on residential service also; but we made the money on residential service through those who are heavy users, not those who just sit there and have the telephone for emergency purposes.

Senator PACKWOOD. Well, that may be true of Mr. Niggli. My hunch is he may make more money on somebody with an air conditioner in New Orleans than somebody who does not have an air conditioner in New Orleans.

Mr. MCBEE. Yes, that is correct, I would assume. But I am just suggesting that the competition for the long distance market, if, in fact, opened without appropriate mechanisms to support the rural communities, would create a problem.

Senator PACKWOOD. Do you agree with Mr. Crowe's statement, that the problem is inflated costs, and that is one of the reasons that if we had competition, the costs would go down? And regulatory system lends itself to covering costs that are unnecessary?

Mr. MCBEE. I think that is a bit of a misnomer. I have heard a number of times that the telephone companies are guaranteed a rate of return. What they are guaranteed is the opportunity not to earn more than a rate of return. Not to earn that rate of return. So, I believe that the efficiencies have been gained within the industry over the last few years; we have all changed since 1982.

I think the telephone industry is much more efficient than it was in those days; and I do not believe that it is, I do not believe it is a problem of inefficiency. I think it is a problem of past depreciation rates, and the lives on some of the equipment that is in place today.

Mr. CROWE. Senator, could I make one comment? Because I believe there may be some confusion about the subsidy flows we are talking about. I know there is confusion; I am not sure anyone knows where they go to, or where they come from.

But I think there is general agreement, and it goes directly to your point, that residential users of long distance are the source of the subsidy. It is not big business. Our customers most likely pay at or below cost. It is the residential user, making payments to AT&T, MCI, and Sprint, who in turn make payments to the telephone companies that are the sources of those subsidies. It is not big business; it is small users, it is residential users.

Senator PACKWOOD. By small users, you mean occasional users?

Mr. CROWE. I mean small business.

Senator PACKWOOD. But using a lot of long distance?

Mr. CROWE. The folks who cannot afford a telecommunications staff, the folks who cannot afford to go out and go through the difficult effort of finding an organization like my own that can afford to deal with them on a custom basis. It is medium and small business, where all the jobs are created, and it is residential users, that provide the subsidy; if, in fact, that subsidy is real and not simply covering inflated cost. I guess we do not know the answer there; although I would point out that the average of American telephone companies' return on equity is now far higher than the average competitive American business.

Senator PACKWOOD. Mr. Roberts, let me ask you a question. As you know, I am a supporter of cable television; so the question is not malevolent.

I read your statement, and I am intrigued with the marvels of fiber and everything that cable is doing. And, given its extraordinary advancements, why does my cable go out more often than my telephone?

Mr. ROBERTS. That is a very fair question. It depends on your cable company. There are many reasons. But obviously, if somebody hits a telephone pole a mile away from your house, and it breaks the connection, and you do not necessarily pick up your phone that minute, you might not know it is out. But your TV's are on 7 hours a day, the average American TV set is on 7 hours a day. I do not think we use our telephone 7 hours a day.

But I think a broader answer is that the cable industry is improving on service. There are now national customer service standards. And I believe that the national customer service standards are the same as the models for the local telephone companies. There is no question, cable industry service has got to be at the same standard of excellence, or it is not going to be competitive. And what we are talking about here is allowing people the chance to compete.

And in the United Kingdom, we are building a system as a cable company, and we are finding 20 percent of the telephone users over there want to buy our cable telephone products.

Senator PACKWOOD. And you were allowed entry in the United Kingdom, without restriction?

Mr. ROBERTS. Yes. And, in fact, as was pointed out, so were the American telephone companies, who are over there as well. And they have said that, in order to create this level ballfield, we need to give time for this to nurture; but eventually, British Telecom will get into television, if that is what they choose to do.

But the point is, we have had to learn how to give good service, or nobody is going to buy our products. And I think the marketplace will ultimately force better service, and that is good. And I would like to submit for the record some more on that, if there is an opportunity to study what the actual outage ratios are—I do not know, off the top of my head—between one industry and another. But the perception may be higher because your TV is on all the time. And just a 5-minute outage is very annoying.

So, we think we will bring our standards up very quickly.

Senator PACKWOOD. I do know people who talk on the phone 7 hours a day. Mr. Crowe, if you were given the opportunity, do you want to go into the business of serving high cost, rural areas; and consumer telephones, generally?

Mr. CROWE. Well, I think I can answer that question, as a matter of fact. Our parent companies recently announced plans to acquire a rural telephone company. They are a for-profit organization, and obviously believe that that is a business that makes sense.

My own company—I will step back and say, we are firm believers that with competition comes a necessary focus on customers. That, like any competitive business, the phone business will start to focus on customers, and organize around customers.

Our focus is business; and we do not think that a single organization can be everything to everybody at the start of a new industry and a new era. Our focus is business; that is where we plan to spend our time. But we do have an affiliate that is in that very industry.

Senator PACKWOOD [presiding]. Thank you, Mr. Chairman.

I believe we are down to Conrad Burns, if I read this list right.

Senator BURNS. Thank you, Mr. Chairman. I appreciate that. Thank you for your excellent testimony, all of you today. And we have been wiggling around with this issue now for about 3 or 4 years; it looks like, after a while, we would get it all ironed out and we would all understand it.

Mr. Ware, I am interested in your comments. And by the way, I want to congratulate the rural telephones; because in my State you have been very cooperative, especially in distance learning and this type of thing. We have set up some terrific educational pods, and we are doing some things now with Indian reservations that has had a very positive effect on our State. And it has all been done through the creative thinking of cooperative and rural telephones. And so we appreciate that.

I would, I want to ask you though, just for the matter of the record and how we would look at this: How would you identify rural exemption? Where do you draw the line? Whenever we start putting this into, and it is going to have a lot to do with how you operate, and how you deal with your local PUC's on rural exemption. Could you give me a, where would we draw the line, and give me an idea?

Mr. WARE. Senator, I do not know that I can give you the line, or the location to draw the line, as far as size or density. But there are some precedents out in the States, where they have exempted the telephone companies that they felt were rural and locally responsive, and some of them have drawn that line at 50,000 access lines; some have drawn that line at other levels.

So, I think that would need some research; but I would be very happy to provide some followup information for you.

Senator BURNS. And Mr. McBee, in your view on effective competition, that test for determining when telephone companies have provided greater freedoms from regulatory and legal situations. Would you want to comment on that, what level you think that would be? Can you identify what effective competition would be?

Mr. MCBEE. I do not think that I can. I can tell you, though, that I believe that it would definitely vary by who the requested party would be. If you wanted to look at effective competition in terms of how much of our access revenues are subject to competition, you could say we have effective competition there.

I believe that you have heard testimony from Mr. Crowe that that is not the case; he has got, I forget what the percentage was, but a very low percentage of the market.

If the intent is to wait until there is competition in each of our central offices, each of USDA's central offices, that is not likely to occur for a very, very long period. And the majority of those central offices currently will not have interconnection, nor do they have requests for them. So, I think it is a very difficult question to answer.

Senator BURNS. And give me, what kind of pricing reform do you specifically feel is necessary, if we in Congress are to make a policy decision to open local telephone exchanges to competition? And be as specific as you possibly can.

Mr. MCBEE. There are a couple of things. I think the FCC again, several years ago, was on the right track in terms of trying to shift costs from the long distance service to the end user. And they wanted to do that in a systematic way, and there was—Congress did not like that, and held a number of hearings. And so that process was stopped.

I think a process like that has to go on. I think the way that the other providers of service contribute to the maintenance of universal service also is something that has to be dealt with, and it has to be in a way that the revenues from that source cannot be avoided.

Senator BURNS. Do you see, Mr. McBee, do you see us redefining what universal service is, in the future? Do you see that changing?

Mr. MCBEE. No, I do not. I see universal service being what it is today; but I expand that into saying that that is the service that would continue to receive some sort of a subsidy or a support. But in fact, all customers in the country ought to have access to more enhanced services; or at least, access to it at a price.

Senator BURNS. Mr. Roberts, I would ask you the same question I asked Mr. McBee. If you could define effective competition or the competition test for exemption to regulatory or legalese, we might say?

Mr. ROBERTS. Thank you, Senator. First of all, I would want to state that the goal of NCTA here is to try to work constructively to find the solutions here; and so I do not have one answer that has to be the only answer.

But I would perhaps say that in the Cable Act, it used a definition that said, if 50 percent of the market is covered by a competitor, and 15 percent of the market is actually buying from the competitor, then there is effective competition. And that is the video effective competition test that was just passed last year; so perhaps that is a starting place.

Senator BURNS. I am very interested. You know, last year when we did reregulate the cable industry under the guise that most of the service was not only part of that but, in other words, there was a great outcry because rates were continuing to grow on an accelerated rate.

And the other day, if you do not think it is not costly, not only did our rates go up in cable, we also passed a little supplemental appropriation of $11 million for the FCC just to write the rules and regulations, and to regulate the industry. So, instead of just cable users going to be paying this increase for regulation, everybody, whether you take cable or not, is going to pay for some of this. So, I guess that pretty well tells you where I am coming down on this. I thank you for your testimony today, and we will be in——

Mr. ROBERTS. Can I just add to that point? I think that the goal of the Cable Act last year was that eventually we want an effectively competitive marketplace. The rules go away, that $11 million, whatever it is going to cost, is going to go away if there is effective competition. And I think the goals of this bill seem to be thinking the same way, which is, if you can encourage competition, then down the road all these regulations ought to go away. The goal ought to be to get rid of regulations, when you have effectively competitive markets. And I think that is a very intelligent model.

Senator BURNS. I would ask the power company, Mr. Niggli, are there any Federal laws now that prevent you from pursuing your plan? And I would tell you that I am sort of interested in that. And State laws, are you asking the Federal Government to preempt State law?

Mr. NIGGLI. Right now, there are no Federal laws that we are aware of that would keep us from implementing the power view concept, especially for demand side management applications within our industry.

So, we are not aware of any that would preempt us, and frankly, one of the reasons that we are here today is to remind you that we bring an economic justification for these types of facilities; and we do not want to be preempted in the future from having the ability to communicate with our customers for demand side management purposes.

Senator BURNS. Thank you, Mr. Chairman.

Senator INOUYE. Thank you. Senator Exon.

Senator EXON. Mr. Chairman, thank you much. I want to thank all our witnesses. This has been a very fascinating and informative hearing. I will have some additional questions for the record. I want to ask Mr. Frischkorn a question, then a question or two for Mr. Crowe. And then I will have additional questions for the record.

Mr. Frischkorn, I notice in your testimony on page 2, you state that your organization commends the sponsors of S. 1086 for their efforts to craft legislation which secures for our Nation the substantial benefits which can only be realized through the development of full and fair competition throughout the telecommunications industry. I agree with that.

My question has to do with something that has not been addressed here this morning, but I could not help passing up the opportunity to ask you and the organization that you represent, as to another matter that is going to be facing the Congress in the very near future; and that is the North American Free Trade Agreement.

I am surprised that some of you have not emphasized this morning how important further telecommunications is in the United States of America. We obviously have not done a very good job. There was a front page story in the New York Times the other day,

day before yesterday, I believe it was, that said that CBS and the New York Times did a poll of Americans. And 49 percent of the Americans had never heard, nor did they know anything about, the North American Free Trade Agreement.

I thought that was a startling statistic. It shows me that we have an awful lot of entertainment, and a lot of information, but it must not be very precise, because if it is true in that poll, that one-half of the people in the United States had never even heard of the North American Free Trade Agreement—which is something that we are going to be hearing from them on in the very near future, and we are hearing from them on right now.

But I want to get, specifically, and see if you could answer a question that is uppermost in many people's minds that are contacting my office these days, Mr. Frischkorn, and that is: When we have this expansion, when we have this explosion, which we are going to have, there is some concern that the North American Free Trade Agreement may force a substantial portion of the new jobs thus created to Mexico, or to other countries around the world, less developed countries, basically with reference to cheap labor rates.

Can you tell me what your opinion is of this? What percentage of the manufacture of the firms that you represent is done in the United States today? And what percentage of the manufacturing jobs that will be created by this explosion in technology are likely to be in the United States? Or are we going to see more and more of these jobs being shipped elsewhere?

Mr. FRISCHKORN. First of all, Senator, let me assure you that TIA as an association supports the North American Free Trade Agreement. We have been lobbying Congress and working with the administration on getting it implemented. It would provide a great deal of benefit to my member companies by opening up that huge new market in Mexico. I do not know if you have ever been to Mexico, but if you have you will know that their communication system is not as good as the system in the United States. And there is a lot of room for improvement, and that improvement can come through American products and technology.

With respect to your question on the creation of jobs, there currently exists under U.S. law provisions relating to customs duties, that have resulted in the creation by Mexico of the maquilladora program. Under that program, American companies can manufacture in Mexico, and ship the products back to the States, without incurring duty. A number of my companies, member companies, take advantage of that.

So, I think the point is, to the extent that American companies have wanted to move production offshore to Mexico, or to other parts of the world, they probably have already done so. There is now no restriction on such moves. So, I do not see the North American Free Trade Agreement as resulting in any loss of jobs from the United States. The Mexican market is so large, over 100 million people who need modern telecommunications equipment. Moreover, Mexico is on the same standard as the United States—the North American standard. So, I see every reason that the jobs that will result from an expansion of that market will most likely be in the United States.

That is not to say that there will not be some jobs moving down to Mexico. It does make sense, from a purely business point of view, to be close to your customer. So, American telecommunications companies may want to establish manufacturing joint ventures in Mexico; enter into distribution arrangements in Mexico; or set up sales offices in Mexico. Clearly, there will be some jobs that move down there. But that is part of the territory. You cannot attack a market that big, without having local people to deal with the market.

Senator EXON. You do not think we should be concerned, then, about the cheap labor market that is available to American industry in Mexico today, and certainly is expected to be available in the future? You do not think that is a major concern?

Mr. FRISCHKORN. Well, there is no impediment to manufacturing there today, and so the NAFTA is not going to change that. If American companies had felt the need to go to a low labor area for whatever reason, they are already there. Again, I do not see NAFTA changing that.

Senator EXON. Mr. Crowe, let me turn to your excellent testimony.

Let me ask you, because I am sure you have heard the concerns expressed by Mr. McBee and Mr. Ware and others, people sitting up here, with regard to rural America and the unfairness that might be created, or that we are fearful might be created. I am not saying it is going to be created.

You and I in some conversations we have had previously on this matter have explored this matter in some detail, and you came up with what I thought were some rather interesting proposals that I do not believe are addressed in your written statement. Would you care to expand on some of the concepts and ideas that we talked about?

You heard earlier, I made a statement with regard to, maybe, to help out the rural areas we could have an REA program, if not an REA program, maybe a low interest lending program as we have with REA, or some of the rural telephone companies, also the service to certain areas and certain individuals that maybe could not afford that. Would you care to expand on some of the conversations that we have had in this area?

Mr. CROWE. Yes, Senator. First, I would start with the statement that I do believe that the definition of universal service needs to be looked at, and looked at hard, that otherwise we are in danger of becoming a nation of information haves and have-nots, which I believe is unacceptable, and my company has supported that since its inception, as have other competitors in the market.

We have come down firmly and publicly in support of appropriate subsidies for those parts of America that do not have access to affordable communications. The only difference, I believe, between our position and that of the USTA is those subsidies ought to be out in the open. They ought not to be buried in an accounting structure that no one understands, that no one has any idea of what its ultimate impact is. It ought to be pulled out. It ought to be administered by a body like the FCC, and the necessary charges for that kind of a subsidy ought to be levied on all industry participants in a way that is competition-neutral.

With respect specifically to the rural parts of America, I think that all of what is going on needs to be viewed as a process, not an event, and it is very dangerous to preconclude where that technology revolution will take us.

We have quite a number of technologies in the wireless area which eliminate much of the problem with distance when it comes to providing modern new services, so one ought not to prejudge the incredible pace of technology and the benefits it can bring. To have done that at the beginning of the computer industry would have eliminated many of the benefits we see from that industry today. No one had the imagination to understand where we would be today 10 years ago. We think the same is true in the communications industry.

So, I think it needs to be viewed as a process, and an impartial body like the FCC has to have the authority to raise and target necessary subsidies to avoid a Nation of information haves and have-nots.

Senator EXON. Are you then saying, Mr. Crowe, that you recognize—and I see most of the people seated at the table shaking their heads on the statement you have been making. The Federal Communications Commission is another one of the bureaucracies that costs a great deal of taxpayers' money to operate. You all seem to feel that regardless of what we do we are going to have to have some kind of regulation of the industry by an agency of the Federal Government, and I suspect that the Federal Communications Commission or something like it would be the logical body.

I go back to the situation that I referenced previously in this area, and that is that we have a regulation and examinations of our financial institutions. Some past histories have indicated that we have not had as good an examination as we should have had, but in any event, the banks and the financial institutions pay a fee when they are investigated, and by and large that offsets much of the expense of the regulatory functions that the law provides as necessary for the common good.

Would you feel that the industry as a whole, and you certainly can speak for your company, but would your company, and in your view the—as a whole the industry support some type of a basic fees that would be paid by the industry for the enforcement of the regulations that the bill before us suggests, or something like it, would put in place?

Mr. CROWE. Senator, we have 100 years of regulation which all parties agree have greatly distorted the signals and cues that come from the marketplace, and the structures of the industry. That is not going to be fixed in 6 months. It is going to take a period of time.

It is going to have to be administered by a technically competent, impartial organization. I believe the FCC is in the best position to do so. It is going to cost money, and we would support competition-neutral levies to fund an organization like the FCC's efforts.

As far as the industry as a whole, I am not even going to try to guess if there would be consensus on that. It is getting too hard to define what industry as a whole is any more.

Senator EXON. Mr. Roberts.

Mr. ROBERTS. I would like to just echo that. Speaking for the NCTA, we wholeheartedly support this idea in terms of recognition that universal service is important and there is an obligation on anybody that wants to compete to pay your fair share, and I think Mr. Crowe has given some terrific suggestions, as you pointed out, and this could be shaped by the FCC, and people have to pay their fair share.

Senator EXON. Mr. Ware.

Mr. WARE. If I might just comment, Senator—Mr. Chairman, the subsidy or support flows have been characterized as some hidden mechanism, and I guess I would like to take issue with that, because I believe the FCC has a very good handle on the mechanisms for the support flows that exist today, and knowing the scope of those flows, and so I do not believe it is the hidden process that it has been characterized as.

If I might just go on, I think also there have been comments indicating that some new pressure such as competition is needed in order to have the exchange carriers focus their efforts and attentions on the customers, and I think that is a mischaracterization, because I believe we do that very well, and recently our company surveyed our customers and overwhelmingly the response indicated that our service was good to excellent.

So, I think while there may be some merit in these observations that have been offered, they should not be applied in blanket fashion to all of the participants or all of the carriers.

Senator EXON. Gentlemen, thank you, and thank you, Mr. Chairman.

Senator INOUYE. Thank you. Senator Pressler.

Senator PRESSLER. Mr. Chairman, I shall be fairly brief in view of the fact that some of my questions have been asked, but let me ask a question of anybody, any volunteer.

As you know, S. 1086 permits all telephone companies to provide cable television service within their telephone service areas and rural telephone companies are allowed to provide cable.

Now, many of the RBOC's are already investing heavily in cable systems outside their service areas in this country and overseas. In fact, I visited one of NYNEX's cable TV activities in Portsmouth, England, sometime ago.

Now, those telephone companies currently are permitted to provide common carrier transport of video programming and other nonownership relationships. However, they cannot have more than a 5-percent interest in programming. Some argue the telephone companies must have the opportunity to own programming before they will invest in upgrading the networks. Is this incentive necessary, and why would a telephone company invest abroad or outside its regions and not in its own back yard?

Mr. CROWE. Senator, I might take a quick shot at the latter part of the question, are those incentives necessary, and I would simply point out that for a number of years I believe it is accurate to say the phone business in the United States, the local phone business, invested about depreciation.

The excess cashflow spilled over into every business but the one that was understood by the participants—overseas real estate, banking, et cetera.

Recently, within the last 2 years, U.S. West, Bell South, Ameritech, Bell Atlantic, and others have announced multibillion dollar programs to upgrade their facilities—$2, $5, $13 billion in one case—without any regulatory change, without any prompting, and I would submit to you it is the result of competition and the fear of competition.

Mr. McBee. Senator, I would just like to add that I think that a restriction on competition is inconsistent with opening the market, and I think incentives, or the opportunity to provide the programming, is consistent with a competitive marketplace.

Mr. Roberts. I would just quickly say that what is—if you just say, open up all the markets, and gee, that seems like a logical solution, you have to remember, as I am sure you do, that there is an essential facility for all of us who want to compete in telephone, which is the public switched network, and what the United Kingdom did in their Portsmouth example is they gave access at competitive rates to interconnection facilities, and it is that kind of encouragement that is necessary.

I also listened earlier to testimony that talks about $450 billion to build out fiber, and again, I for one look at the whole size of the cable industry and the whole video marketplace as $22 billion a year. So, are we going after too much money here, and who is going to pick up the tab if it does not work?

I think our proposal, as the cable industry, is something more like $20 billion in the next 5 years, and we think we get a great network that can compete, and then at that point it may be time to get rid of some of the common carrier regulations.

Senator Pressler. Another question, S. 1086 would bar a telephone company from buying out existing cable companies in its own service areas unless permitted to do so by the rural exemption.

I understand the rationale for this provision is to preclude one monopoly service provider from replacing another and stimulate competition. However, some areas may not have the economic base to support two competing systems.

Some of the small cable operators in my State who are worried about their economic viability under the new cable rules have complained that they will not be able to sell their systems to the most likely buyer, the local telephone company.

I realize that a rural cable ownership exemption would address their concern in some situations. Nonetheless, there may be others where it does not, so the question is, Do you have any thoughts or concerns about this problem?

Mr. Roberts. As I said in my testimony, we definitely feel that that is a very legitimate concern, and it is a daunting proposition given the size of the regional Bell operating companies, so it is not just a rural problem.

I think that one could have a model that said the FCC has to protect consumers in certain markets where there is not competition and it does not need to be as regulatory in markets where there is competition, but I do not think that the bill should completely just prejudge this issue. We would suggest instead to defer it.

If you have strong regulations in instances where this might occur, it addresses some of the antitrust issues you kind of alluded to. I think that rather than prejudge where all this is going in terms of the technologies, it would be better to leave it with a regulatory agency to consider it as we go.

Mr. WARE. Mr. Chairman, Senator, it might be an area that should be explored as to raising the rural exemption, because as your comments indicated, there may be systems out in smaller communities, even though they are slightly above the existing rural exemption level, where the service would be improved and the consumers benefited by a purchase of that cable TV system by the local company, particularly if their facilities are antiquated and need to be rebuilt.

We have heard a lot of conversation about cable putting in fiber, but there are a lot of cable systems that do not have fiber and do not have modern facilities and are looking at replacement.

Senator PRESSLER. My final question is a practical one, and I will take any volunteers on it. My State of South Dakota recently was ranked as the best place in the country to locate a credit card business. Would this bill make it even better, and how?

Mr. CROWE. My State of Nebraska is the second best. We have got a large credit card processing facility in Omaha, and I think it is for many of the same reasons. Both States enjoy a strong telecommunications infrastructure as a result in no small part of the defense buildups of previous decades.

I think that the upgrade of the infrastructure eliminates what you might think of as the tyranny of geography in business today, allowing businesses to relocate away from small concentrated areas where employment and communications are adequate to the parts of the country like South Dakota that look attractive under that scenario.

Mr. COE. Senator, I would like to say that we feel that the support of the bill would indeed open up many areas across the country that today are probably not viable locations to add credit card processing. If we move forward with S. 1086, I think there will be ample opportunity in ample locations.

Senator PRESSLER. Mr. Chairman, I have some additional questions for the record. I thank you very much.

Senator INOUYE. Gentlemen, thank you very much. You have been very patient with us and we will submit further questions to you if we may.

Our final panel for this hearing consists of the following. The president of the Bell Atlantic Corp., Mr. James Cullen; president of Lebanon Publishing Co., representing the national Newspaper Association, Mr. Dalton C. Wright; the president and publisher of the Pottsville Republican, Mr. Uzal H. Martz, representing the Newspaper Association of America; the president of Comptel, Mr. James M. Smith; and the vice president and general counsel of Information Industries Association, Mr. Steven Metalitz.

On behalf of the committee I wish to thank you for your patience in waiting all these hours, but I will be here until everything is completed, so your testimony will be well received. May I first call upon Mr. Cullen?

70-334 O - 94 - 4

STATEMENT OF JAMES G. CULLEN, PRESIDENT, BELL ATLANTIC CORP.

Mr. CULLEN. Thank you, Mr. Chairman. I do appreciate the opportunity to be here this afternoon to testify on Senate bill 1086 on behalf of the seven regional holding companies.

First I would like to recognize your leadership, that of Senator Danforth, and the subcommittee in addressing these extraordinarily complex telecommunications issues.

I know it does seem like we have been debating these issues for at least two decades now, and probably in fact we have. Certainly at this point there is not much to be said that has not already been said before somewhere sometime. And the frustration led earlier this morning to Senator Breaux observing that perhaps we should just let everyone into everything, and perhaps that is a worthwhile objective although I think we do need to recognize that competition inevitably will drive prices to costs, and that has serious implications.

However, I do think it is fair to say that everyone here today would agree with the primary goals of S. 1086—to promote real infrastructure development, to permit fair competition, and to protect the public interest.

The problem here from an RBOC perspective is that the bill as written fails to promote all of these goals. In fact, in many ways it discourages them, and that is why we oppose the bill.

However, I do want to say and be very clear that I am not here today to speak against increased competition in telecommunications. Clearly, our industry is moving very rapidly toward full competition in all aspects of the business. We see it every day and we see it in particular in the profitable parts of our business.

I am here, though, to address the important policy issues based on the facts of our industry and not the myths. One myth that I do see reflected in this bill is that the local exchange carriers are not investing enough in the telecommunications infrastructure, and that investment somehow needs to be jump started by injecting competition into the local exchange business.

In fact, the local carriers are investing in a variety of new technologies—investing more, in fact, that anyone else the industry. Since divestiture, for example, the regional companies have deployed roughly twice as much fiber optic cable as the long distance carriers. Bell Atlantic alone has deployed almost as much fiber as AT&T even though AT&T is three times as large.

There is a second myth that I see reflected in the bill and that is that AT&T, MCI, Sprint, McCaw, and the cable companies all need to be protected from competition by the local exchange companies. As this subcommittee knows and as your bill recognizes, the cable companies have no competition and customers have been clamoring for it for years. To think that any of these billion dollar companies, cable or long distance, need any added protection from local exchange companies is ludicrous.

When it comes to entry into the long distance business one contention we sometimes hear, and in fact heard earlier today, is that the regional companies can somehow discriminate against unaffiliated long distance carriers. Among other things, this contention totally ignores the FCC-mandated carrier selection process put in place at great expense in the 1980's which absolutely guards against any form of discrimination. Every RBOC customer, every customer, is free to select his or her long distance carrier, and no RBOC has any way to interfere with that selection.

And despite the barrage of expensive TV advertisements, there really in fact is precious little price competition in the long distance market. In fact, over the last 3 years average prices of the three major long distance carriers have gone up while their costs of access supplied by the local exchange carriers, a very big part of their total costs, have gone down.

As you can see in the Merrill Lynch report, which is attached to my written testimony, just 2 weeks ago the local exchange carriers reduced their access charges to AT&T by $250 million and AT&T passed along none of this reduction to its customers. In fact, as the Merrill Lynch report happily points out, this is another sign of oligapolistic tendencies in the long distance market.

It is ironic that less than a week ago our President flew 6,000 miles to Japan, urging policymakers there to open up markets to U.S. companies, while here in Washington, DC, my company, Bell Atlantic, is still trying to open up important American markets in long distance, in manufacturing, and in video services.

By the way, these markets are open to telephone companies so long as the telephone companies are foreign owned. British Telcom and Cable & Wireless, for example, are completely free to provide all these services in Washington, DC today with no restrictions.

The RBOC's should be permitted to provide not only long distance services but also video services and, of course, the manufacture of telecommunications equipment as approved overwhelmingly by the Senate in its passage of Chairman Hollings' bill last year.

What the RBOC's are talking about here are real investments in new technology, in new services, American jobs, and at no cost to the taxpayers. And we simply want to do it right here at home.

Thank you very much for the opportunity to appear here, Mr. Chairman.

[The prepared statement of Mr. Cullen follows:]

Prepared Statement of James G. Cullen

Good morning. Mr. Chairman and members of the subcommittee, my name is James G. Cullen, and I am the President of Bell Atlantic Corporation. I am pleased to have the opportunity to testify on Senate Bill 1086 on behalf of the seven Regional Holding Companies("RHCs"). I also want to recognize the leadership and efforts of Senators Inouye and Danforth in addressing these complex issues.

The principal purpose of S. 1086 seems to be to stimulate infrastructure development—a laudable goal because it holds the promise of improving our lives in the information age; improving global competitiveness; and creating jobs. Infrastructure development involves upgrading our distribution networks with fiber optic cables and other new technologies to carry more information; deploying digital switches to route traffic quickly and enhance transmission quality; and developing software systems to provide new services and service capabilities.

As drafted the RHCs oppose S. 1086, not because we oppose local competition but because it fails to create the equitable competitive opportunities and regulatory parity needed to encourage further infrastructure investment from the RHCs. While we recognize that our industry is moving toward full competition in all aspects of the business, the acceleration of that competition should be done in a more even handed way.

The purpose of my testimony is two-fold. First, I want to make sure that this subcommittee separates the widely held myths about the local telephone companies from the facts. An understanding of the real facts will enable the subcommittee to better assess what steps are really needed to encourage infrastructure investment. Second, I want to describe how S. 1086 can be improved to stimulate even more infrastructure development while not destroying the incentives that already exist for the RHCs to invest in the network.

I. THE RHCS ARE INVESTING FAR MORE IN INFRASTRUCTURE THAN THE LONG DISTANCE CARRIERS ARE

The first myth which S. 1086 reflects is the assumption that infrastructure investment by the RHCs lags behind investment in the infrastructure made by long distance companies, and that this lag is due primarily to competition in the long distance market. The plain facts, however, are that the RHCs already have deployed more fiber and made more infrastructure investment than all the long distance carriers combined.

For example, according to a recent FCC report, "Fiber Deployment Update-End of Year 1992," the regional RHCs had deployed 4.9 million fiber miles by the end of 1992—roughly twice as much as the interexchange carriers, even though the annual revenues of the RHCs and the long distance carriers are approximately the same. Bell Atlantic's fiber deployment, for example, is roughly the same as AT&T's, even though AT&T is three times as large as Bell Atlantic.

Moreover, while the interexchange carriers' fiber optic investment has leveled off to a mere 3 percent growth in fiber miles per year, the RHCs' fiber investment continues at a healthy 25 percent growth per year. Also, the RHCs have recently announced in a White House visit their willingness to invest up to $125 billion in the infrastructure by the year 2000.

As I mentioned earlier, infrastructure development means more than just fiber optic deployment. The RHCs are also moving rapidly to replace old switches with digital ones that can process more information more quickly. In the past three years, the RHCs, have invested nearly $10 billion in digital switching capabilities. In addition, significant investments have been made to further upgrade our switches with a new network arrangement called 557 that speeds call delivery and provides easy access to information services.

In sum, the myth that the RHCs are lagging behind the long distance carriers is wrong. The plain facts are that the RHCs far outspend the long distance carriers on infrastructure development.

II. S. 1086 WILL UNDERCUT INCENTIVES TO INVEST IN INFRASTRUCTURE BECAUSE IT INCREASES THE RHCS' RISK WITHOUT ALSO INCREASING THEIR OPPORTUNITIES

The second myth I want to address is the notion that even-handed regulation will not work and that, therefore, the RHCs must be subject to market exclusions and other onerous restrictions in order to stimulate fair competition for new entrants and accelerate infrastructure development. These new competitors are not "mom and pop" carriers. Instead, the primary beneficiaries of the bill's entry provisions are large, well-entrenched carriers, such as AT&T, and large cable companies, such as TCI, Time-Warner, and Cox Cable, as well as foreign carriers such as British Telecom and Cable & Wireless. Given the financial resources of these types of competitors, there is no reason to impose new burdens that apply only to the RHCs. for example, the bill gives the RHCs very limited long distance relief, despite the fact that the long distance market already is supposed to be competitive. As a result, the RHCs will only be able to provide local service, but our competitors will be able to provide both local and long distance service.

Second, the bill places substantial restrictions on telephone company entry into cable and electronic publishing, but places no restrictions on entry by cable or other electronic publishers into the local telephone business.

Third, the bill imposes substantial requirements on the RHCs to open up their networks so that other carriers can connect to them, but neglects to impose comparable burdens on cable companies or on the dominant long distance carrier, AT&T.

There is no justification for these kinds of one-sided burdens and restrictions. If cable companies, which Congress has found to be unregulated monopolies, and dominant long distance carriers like AT&T are free to enter the local telephone business, then the RHCs should be free to enter the long distance and cable businesses on the same terms.

It is very important to recognize that today there is a system in place for customers to choose a long distance carrier. That system has worked well in the past and will work just as well if the RHCs are allowed to be another provider that customers can select as their long distance carrier. Since the RHCs will enter the long distance market with zero market share, there is no possibility at all that they could hurt competition in this market.

The RHCs have been doing their part to modernize the nation's telecommunications infrastructure. But they cannot be expected to continue much less increase those efforts in the face of legislation like S. 1086 that increases their risk by singling them out for special burdens and obligations. In creating competitive opportunities for the long distance carriers to compete in the local markets, the bill must address both the new opportunities for RHCs in the long distance market, as well as the regulatory obligations that should apply to long distance carriers in the local market.

III. S. 1086 IS ALSO BASED ON THE ERRONEOUS ASSUMPTION THAT LONG DISTANCE COMPETITION HAS REDUCED ORDINARY LONG DISTANCE RATES

The third myth in S. 1086 is that it clearly assumes that long distance competition and investment in fiber has lowered ordinary long distance rates. I know that you sometimes hear that competition has reduced long distance rates by 40 percent or more. The fact is that, while competition probably has reduced rates for the largest business customers who buy customized network solutions, the reduction in ordinary long distance rates is more than accounted for by reductions in the rates that RHCs charge to long distance carriers for the local links used to complete long distance calls. The FCC has noted, for example, that:

> [T]he single force most responsible for driving down long distance rates over the last several years has been the reduction of access charges long distance companies pay to local exchange carriers.—Policy and Rules Concerning Rates for Dominant Carriers, 4 F.C.C. Rcd 2873, 3054 (1989)

Not only have our access rate reductions accounted for the reductions long distance rates, but long distance carriers have failed to pass all of these reductions through to consumers, as one would expect in a truly competitive market. For example, since 1984 AT&T's annual access charges have been reduced by approximately $10 billion. AT&T's annual rates to its own customers, however, have been reduced by only $8 billion. The remaining $2 billion has remained with AT&T. This means that despite our success in cutting costs and reducing prices, the consumer does not get the benefit of those reductions. Instead, AT&T's shareholders get the benefit of those dollars to the tune of $2 billion annually.

This lack of price competition in the long distance market has not gone unnoticed by industry analysts. Attachment A to my testimony contains a Merrill Lynch report noting that, in the most recent round of rate filings at the FCC, although AT&T received access charge reductions of $250 million, it raised some of its long distance rates and made no rate reductions whatsoever. Also, Attachment B shows the lack of genuine price competition in the face of access charge reductions.

IV. WHAT CONGRESS CAN DO

It is clear that if we look at the facts and not the myths, in many states there are already incentives in place to promote infrastructure development. The challenge then becomes how do we encourage additional infrastructure development that will mean new jobs and will lead to the development of new technology, while at the same time ensuring fair competition. To meet this challenge, we recommend:

- Remove the long distance restriction in its entirety.

Both the Department of Justice and AT&T agreed at divestiture that the long distance restriction should be removed when long distance carriers were permitted into the local exchange market. If this bill permits long distance carriers to compete in the local market, fair competition demands that the RHCs be able to compete in the long distance market.

In addition to the obvious fairness of such an approach, there are other reasons to let the RHCs into the long distance market. The long distance restriction both limits our ability to provide modern networks that meet customer needs and imposes unnecessary costs on consumers. There were supposed to be several processes for reconsidering these restrictions, but these processes have broken down entirely. There has been only one of the promised three-year reviews of competition in the industry despite the fact that divestiture occurred nearly ten years ago. Even minor waivers of the restriction can entail years of delay.

If the RHCs are permitted into the long distance market, we will make it easier for consumers, especially residential customers, to deal with the increasingly complex long distance marketplace. For example, long distance relief would give us the ability to provide software services to help consumers select the long distance services to meet their individual needs. Today, long distance companies offer a wide variety of prices and service options—different rates for calls to different places, different rates on different days of the week or at different hours of the day. Maybe you've seen the Sprint commercial featuring Candice Bergen comparing long distance rate plans to the blueprints for a nuclear submarine. As a practical matter, it is difficult for an ordinary consumer to figure out which carrier and service, or combination of carriers and services, is the best and the cheapest for his or her individual needs. Long distance relief would permit customers to choose from the various long distance companies to find the best combination of services. Instead of being "locked in" to a single long distance carrier, consumers would automatically be able to take advantage of the best prices the competitors have to offer.

In sum, long distance relief would permit us to offer a wider range of services at lower prices to meet the needs of the customer. Long distance relief would give us the opportunity to offer additional services to our customers in our regions—the same opportunities now available to foreign carriers, such as British Telecom and Cable & Wireless, who are now operating in our territories.

• Remove the video programming restriction immediately.

The 1984 Cable Act bans the RHCs from providing video programming to their hometown customers. In 1990, this committee urged the FCC to consider whether this restriction should be retained. The FCC, after developing an exhaustive administrative record, recommended that the ban be repealed. The FCC stands by this recommendation, even after Congress passed the 1992 Cable Act.

The cable relief provided for in S. 1086 is several years away because it is contingent on several time-consuming, bureaucratic requirements. This restriction should be removed in its entirety immediately.

• Remove the manufacturing restriction.

Two years ago the Senate voted by a more than two-thirds majority to remove the manufacturing restriction. Surprisingly, this bill does not include that relief. The manufacturing restriction ties up the financial resources of seven of the nation's largest 35 corporations and severely limits the activities at Bellcore, one of the world's premiere telecommunications research and development organizations.

The manufacturing restriction is an absolute bar to job creation. While companies with 60 percent of the telecommunications industry's assets watch from the sidelines, over 60,000 American manufacturing jobs have been eliminated and replaced with 20,000 foreign manufacturing jobs. As a result, the restriction necessarily slows the development of new telecommunications products and services and also puts this country at a severe competitive disadvantage with our foreign competitors, none of whom handicap their telephone companies as this country does. This committee should again vote to eliminate this restriction.

• Ensure consistent regulatory and universal service obligations on all providers of local exchange service.

Although the bill is very clear about the need to remove barriers to entry into the local telephone market, it is silent on the need to ensure that all providers of local service are subject to consistent regulatory obligations, whether that means or less regulation. For example, the RHCs are obliged to serve all customers in their service areas, no matter how expensive it is. A similar universal service obligation should extend to all local service providers. Otherwise, the new entrants will take the most lucrative business and the leave the high cost, low revenue customers to be served by the provider of last resort-usually the RHCs. Similarly, all local service providers should be required to provide operator services, 911 services, and equal access to all telecommunications providers. These are essential services that local customers enjoy today and have a right to expect tomorrow from all providers.

Finally, all providers of local exchange services should be required to support universal service at affordable rates. This obligation should not be left entirely to the existing local exchange carriers.

CONCLUSION

Contrary to many of the myths that exist in our industry, the facts clearly show that:

1. The RHCs are already making substantial investments in the nation's telecommunications infrastructure, outspending the long distance carriers by a two-to-one margin;

2. That imposing one-sided burdens on the RHCs will lead to less infrastructure development and not more;

3. Fair competition requires that all providers have the same access and obligations in the same markets;

4. The MFJ long distance and manufacturing restrictions that hamper further the RHCs' ability to compete, reduce consumers' costs, and provide new U.S. manufacturing jobs should be eliminated; and

5. The video programming restriction imposed by the 1984 Act should be eliminated.

Thank you again for this opportunity to appear before this subcommittee and I will be available to answer any questions you may have.

[Attachment A—Merrill Lynch Telecom Services may be found in the committee's files.]

ATTACHMENT B

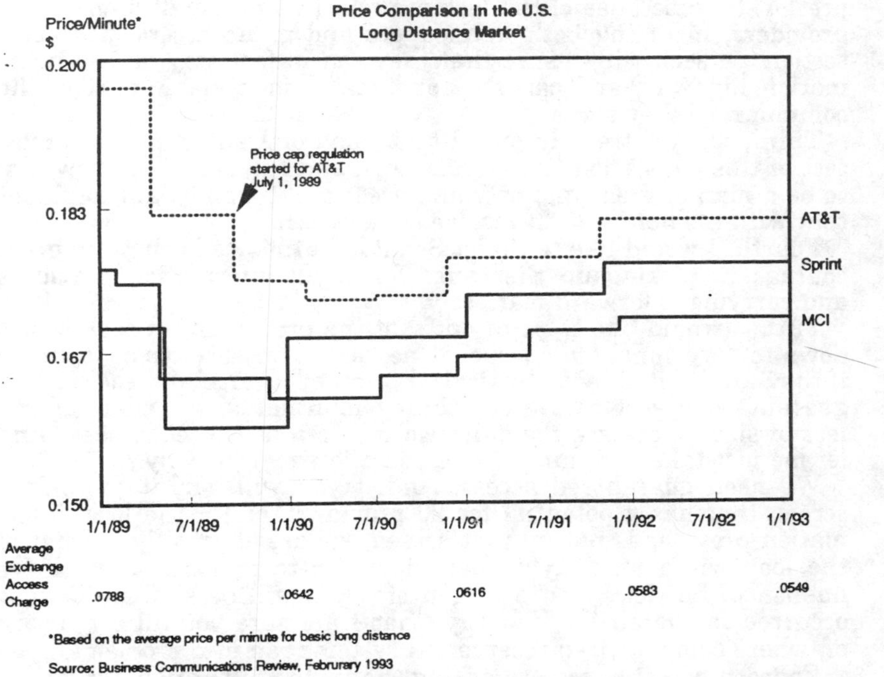

Senator INOUYE. Thank you very much, Mr. Cullen. Mr. Wright.

STATEMENT OF DALTON C. WRIGHT, PRESIDENT, LEBANON PUBLISHING CO., REPRESENTING THE NATIONAL NEWSPAPER ASSOCIATION

Mr. WRIGHT. Mr. Chairman, my name is Dalton Wright. I am from Missouri. I am the president of Lebanon Publishing Co., and publisher of a small daily newspaper, the Lebanon Daily Record.

I serve on the board of directors of the National Newspaper Association. I am also, Mr. Chairman, a member of the Missouri Press Association, which has endorsed my comments today, so I will speaking on their behalf as well.

Two distinguished members of the National Newspaper Association in the room today are Chairman Frank Garred, publisher of the Jefferson County Leader of Port Townsend in the State of Washington, as well as Jack Fishman, chairman of the Government Relations Committee, President of Lakeway Publishers Inc., Morristown, TN, and Publisher of the Citizen Tribune in Morristown.

I am going to depart, Mr. Chairman, from my prepared text. I have taken some notes this morning and I am going to try to be more brief and possibly more to the point.

First, I think that Mr. Chairman and Senator Danforth have produced a very fine bill, one that would move our Nation closer to the dream of a national telecommunications infrastructure. It will bring great benefit to us all.

Second, it does pain me personally that the National Newspaper Association and its 4,600 members cannot support the bill at the present time because of the lack of protection for small information providers, including both newspapers and broadcasters. This protection we seek will ensure that they can in fact stay on the information highway, and can stay and maintain a viable part of the communities they serve.

Third, I am extremely proud that Senator Danforth is an architect of this legislation. We in Missouri, and I personally, know him to be a man of great integrity and great ability, and I am confident that with his help we can resolve our concerns.

Fourth, I would like to thank Senator Burns. He is showing great courage in picking up the cause for small information providers and carrying it forward fearlessly.

Fifth, I would like to point out that the protection we seek comes down to very simply two issues. One, access for services or facilities if they are limited by technological or other means, and second, the guaranteed lowest rate given to any information provider regardless of size. These are the only two things that we really seek. And let me just take a minute to give you a few reasons why.

We need guaranteed access. And if you will, sir, if there is a screen that has a potential for 20 providers, and if the local information provider is not on that screen, we are effectively left out of the loop without a way to participate or to compete. Also, if we happen to be No. 21 on a screen of 20, it will be similar to what occurred in the airline industry, that I am sure you all are aware of, where computerized reservation systems ran into problems.

Concerning rates, we are asking that if the big national information service providers get volume discounts, that the small local provider get the same rate so his prices are not undercut by the big guy which, believe me, would have a very negative impact on our ability to serve those communities that we live in.

I know there is a lot of language, and very good language, about nondiscrimination, but that really is not the issue that brings me here today. The nondiscrimination issues will not take care of our major concerns of access and rates.

Sixth, this is not a grab for favoritism. I know that a lot of people feel that may be the case, but we are not asking for anything that one or more of our competitors will not already have.

Seventh, Congress has traditionally had a concern for protecting small, local providers against sudden technological change, particularly in communications policy. We are asking that Congress continue to protect localism as a central goal in its telecommunications policy.

Eighth, the reason Congress has shown this concern in the past is that small, local information providers are not only vulnerable, they are also very valuable. They perform an essential consumer service. Through them neighbors learn about neighbors. They learn to care for their neighbors.

Without a local newspaper a community is much less a community. I could also say without local broadcasters a community is much less a community. The National Newspaper Association members serve more than 30 million households in thousands and thousands of large and small communities around the Nation.

Ninth, what we are asking for would impose no significant burden, in that there will not be a lot of local providers clamoring for access. We see it more as an evolution on the local level rather than a revolution. Even with this protection they will be slow in entering the information services market. As small providers neither the rate nor the access provision would impose any hardship on the telephone company, and certainly in no way cause any problems for the large national information service providers.

Tenth and last, I think it was Tip O'Neill that said that all politics is local. The same could be said for news, information, and advertising. What we publish is local and that is what makes it important to us, to our friends and neighbors, and important hopefully for the Congress of the United States to protect. And I ask you to do that by including the Burns amendment or similar language in this significant and much needed telecommunications legislation.

Thank you, Mr. Chairman.

[The prepared statement of Mr. Wright follows:]

PREPARED STATEMENT OF DALTON C. WRIGHT

Mr. Chairman, members of the Subcommittee, my name is Dalton C. Wright, and I am pleased to be here this morning representing the National Newspaper Association (NNA). I am president of Lebanon Publishing Co., of Lebanon, Missouri, and publisher of the Lebanon Daily Record. I serve on the Board of Directors of the National Newspaper Association and I am also a member of the Missouri Press Association, which joins me in this statement today.

With me here today is Frank W. Garred, Chairman of the Board of Directors of the National Newspaper Association and publisher of the Port Townsend Jefferson County Leader of Port Townsend, Washington. Also joining me is R. Jack Fishman, president of Lakeway Publishers Inc. and publisher of the Citizen Tribune of Morristown, Tennessee, who is a Director of the National Newspaper Association and Chairman of its Government Relations Committee.

I would like to thank the Subcommittee for inviting us to testify this morning and for taking the time to consider our concerns. We applaud the Subcommittee and its staff for being aggressive on the issue of crafting fair and competitive public policy that will shape this country's telecommunications infrastructure. We are extremely pleased to see a commitment to bringing effective competition to local telephone service and maintaining safeguards on the telephone monopolies until the day competition really takes root. We join others on this panel in urging this Subcommittee to move vigorously toward strong safeguards.

We are here today to ask the Subcommittee to do two things:

- First, that Congress continue to include "localism" as one of the central tenets of national telecommunications policy; and
- Second, to give assurance that local newspapers and broadcasters, as small, local information providers, will receive statutorily guaranteed access to facilities or services and rates equal to the lowest rates that are provided to large, national information providers.

The National Newspaper Association is a national trade association representing more than 4,600 small community newspapers throughout the United States, including almost 700 dailies and nearly 4,000 weeklies. Most members are community newspaper publishers such as myself. Our members concentrate on the news of their home town—what is happening in county and town governments, school boards, local universities and local businesses. Our social pages cover school events and church festivals. Our sports pages cover high school and local college games.

Advertising in our papers is done by local retailers, whose economic survival depends upon their friends and neighbors in their own communities. Our editorial pages focus on matters of local interest. Few, if any, local bond issues, public works projects or local elections pass without extensive coverage by our members. From engagements and births to deaths and obituaries, our news is local news. Like the local clergy, we hatch 'em, match 'em and dispatch 'em. In our papers, regular people are as important as Hollywood stars.

Small newspapers also serve a very important social role in American society, for our members are generally the focal points around which our country's communities build and maintain their identities and their sense of communal self. We play a particularly important role in rural areas for we are the institutions that maintain a sense of identity and social cohesion in rural towns and villages. When a rural town loses its local newspaper, the town frequently loses its identity, and often shrivels up and dies.

When considering public policy affecting media, it is often common to conjure up an image of very large newspapers, television personalities and magazine publishers, but it is tremendously important to consider the impact of public policy upon thousands of members like ours. In shaping our nation's telecommunications infrastructure, one cannot ignore, either in the marketplace of ideas or in the marketplace of advertising, the thousands of local publications that exist throughout the country—publications that reach more than 30 million people every week.

The Inouye-Danforth bill (S. 1086) to promote a national telecommunications infrastructure represents, in many ways, a positive legislative response to the rapidly changing business and technological environment for the dissemination of information in our society. Yet, the proposed legislation is deficient in one crucial respect: it does not assure that local information services—that is community newspapers and local radio and television broadcasters—will be able to serve their communities by means of the new infrastructure.

The need to preserve local information sources should be self-evident. Many of today's societal problems and their consequences—crime substance abuse, mental illness, child neglect, welfare dependence, to name but a few—have sadly become hallmarks of our high-tech, depersonalized society, particularly in urban centers where truly local information services are not widely disseminated. It is difficult for most people to care about their communities—or their neighbors—if they have no idea what is going on in their own neighborhoods.

Tip O'Neill's famous maxim that "all politics is local politics" also applies. Where there is little media coverage of local government issues, there is little public awareness, low voter turnout, and unresponsive government. The success of our democracy depends on citizen involvement. In the first instance, that involvement is in the debates played out in our neighborhood communities, our town councils, our political parties' local caucuses, and in local petitions, referendums and initiatives. Local information services meet the need for information to spur citizen involvement in community events and issues. Local publishers and broadcasters should be allowed to continue to play this important role.

The United States today stands at the threshold of the much-discussed "information age," which will most likely be marked by the convergence of communications and computer technology and the replacement of the term "telephone service" with "telecommunications service." However, we don't know what the future of electronic information dissemination is going to be. We can see fascinating glimpses of it on the horizon as we read of new technologies and new concepts of communicating information to the public. Nevertheless, we don't really know what lies ahead and this is why it is so important for Congress to provide a regulatory framework for telecommunications that will insure competition and flexibility as we move to the future.

The telecommunications infrastructure, on which a modern information services market depends, is transforming itself into a national market. Within that national market, the firms that have nearly unlimited access to homes and businesses, i.e., telephone and television cable companies, will be the dominant players. A national information services market, offering near-instantaneous, interactive access to virtually all sources of information, up to and including full-motion video programming, will be built upon this infrastructure.

Informative coverage of local news and community events is simply not the focus of national or even regional or metropolitan media outlets, nor should it be. Community newspapers are unique because our job is collecting information that no one else collects. Throughout the country community newspapers are the basic assemblers and publishers of local information. Community newspapers are subsequently in a unique but very vulnerable position with respect to new telecommunications technologies. We are vulnerable because we're local in scope and small in size. If appropriate regulatory mechanisms and policies are not in place, we can be squeezed out by larger, nationally-based providers.

We are equally vulnerable because the assumption is often made that all potential providers will be able to gain access to the network of the future. S. 1086 includes a general nondiscrimination provision that has a premise that access will be available to all. For the foreseeable future, however, access will be limited functionally in a variety of ways, but primarily technological and financial. The large Bell operating companies and other telephone companies with a monopoly over the local telephone loop will be developing and managing these electronic information networks. Because the Bells control the infrastructure, local providers face threats not only from Bell-owned information services but from deep-pocketed private providers who can strike lucrative deals for low rates and prime access.

Any legislation concerning the telecommunications infrastructure must contain protection for equal rates and guaranteed access for local information providers, if our local communities are to preserve the quality of life that community newspapers provide.

The National Newspaper Association has been working to have the concept of "localism" included in telecommunications policy and we seek assurance that local newspapers and broadcasters, as small, local information providers, will receive guaranteed access to facilities or services and rates equal to the lowest rates that are provided to large, national information providers. The National Newspaper Association's localism initiative is sponsored by Senator Burns, whose proposed amendment to S. 1086 would provide for such guarantees while placing minimal burdens on telecommunications carriers.

First, there is the issue of guaranteed access.

The proposed Burns amendment calls for "guaranteed nondiscriminatory access to local information providers in any situation where the facilities or services are limited by physical or technological barriers." This simply means that whenever information service providers are required to "queue up," local providers will be guaranteed a place in the line. To take an example that is familiar from the computerized airline reservation battles, a screen menu on a telephone company "gateway" system might have space for twenty listings on its first screen. Local information service providers would be guaranteed some portion of those listings. This requirement is no different in principle from the 1992 Cable Act's requirements for a minimum number of leased access channels that can be used for local programming. Without this requirement, telephone companies—themselves monopolies—would likely reward their best customers—and themselves—with the most favorable listings, just as some airlines were thought to have given preference to their own flight listings in the computerized reservation systems that they control.

Second, there is the issue of guaranteed equal rates.

The Burns amendment requires "that a local information service provider may be required to pay a nondiscriminatory rate no higher than the rate offered to any other information provider in the market." Without this specific provision, a telephone company could offer volume discounts to its largest customers while charging higher rates to smaller, local information service providers. Such discrimination would likely be considered "reasonable" under applicable legal precedent and would thus be permitted under a general nondiscrimination provision. The result could be devastating for a local provider.

As one example, a regional telephone company might offer discounts based on volume commitments to a national information service provider, enabling that provider to pass through its lower costs to attract subscribers. This would undercut competition from local information service providers, who would face higher communications costs. Larger, non-local information service providers could thus "crowd out" local competition in targeted markets, such as advertising and buying services or home financial services.

It's important to remember that the National Newspaper Association's localism initiative as embodied in the Burns amendment would place minimal burdens on telecommunications carriers. The demand for nondiscriminatory access and service rates would be limited by the small size of the geographic areas affected and the relatively small number of local providers in each area, which would in turn be limited by the same market forces that limit the number of local providers today.

Finally, adoption of the Burns amendment would foster the goals set forth in the Inouye-Danforth bill's findings that "the national welfare will be enhanced if community newspapers are provided ease of entry into the operation of information services disseminated through electronic means primarily to customers in the localities served by such newspapers at reasonable, nondiscriminatory rates to such newspapers."

Without protection for equal rates and access, community publishers will not be able to support the legislation. There is much in S. 1086 that is important and valuable—the safeguards contained in the bill significantly reduce the ability of the telephone companies to abuse their monopoly power—but we believe that local information providers don't have the protection they need to compete fairly in the expanding marketplace that this bill would encourage.

We urge the Subcommittee to work toward ensuring that this nation's small rural communities have an advanced, accessible and affordable information infrastructure, and that community newspapers—and other local information providers—receive this statutory assurance and protection so they will not be prevented from playing an important role in our nation's new information highways.

Senator INOUYE. Thank you very much, Mr. Wright. Mr. Martz.

STATEMENT OF UZAL H. MARTZ, JR., PRESIDENT AND PUBLISHER, POTTSVILLE REPUBLICAN, ON BEHALF OF THE NEWSPAPER ASSOCIATION OF AMERICA

Mr. MARTZ. Mr. Chairman and members of the subcommittee, my name is Uzal Martz. I am president and publisher of the Republican in Pottsville, PA. I appreciate the opportunity to appear before you today to present the views of the Newspaper Association of America, NAA, on S. 1086. I am secretary and a member of the NAA board of governors.

NAA represents approximately 1,250 newspapers in the United States and Canada. The majority are daily newspapers that account for more than 80 percent of the daily circulation in the United States.

In the interest of time I would just like to summarize my written statement by starting out first and foremost to commend the chairman and Senator Danforth on the clear focus in S. 1086 on preserving and promoting the diversity and competition that has developed in the information services marketplace, and in the means of delivering these services.

Both of these are essential if the United States is to continue to lead the world into the Information Age, and consistent with this goal NAA has reviewed S. 1086 and finds much in the bill to commend.

There are three major points that I would like to make which are elaborated on in my statement. First and foremost, American newspapers have a special stake in the development of a vibrant telecommunications infrastructure as envisioned in this bill. For 300 years a free press has been central to the political, economic, and social life of this country. Telecommunications technology from computers to satellites have enabled newspapers to gather and disseminate news more effectively, with greater responsiveness to our readers, and will help ensure that this tradition will continue into the next century.

For more than two decades, newspapers have been very active participants in the development of electronic information services. Today hundreds of newspapers, from large and small dailies, metropolitan areas, small communities provide a growing menu of services. And we have listed in this directory all of the services that are presently being provided by U.S. newspapers.

More than 500 newspapers offer more than one information service on stock quotes, time, temperature, weather, news, homework hot lines, electronic classifieds, and minimovie reviews; 400 newspapers offer 900 information based services; and 140 provide full, on-line access to full-text data bases of their newspapers.

My own newspaper in Pottsville, PA, which has a circulation of 30,000 offers our community 295 different information categories in our audiotext system. This service is advertiser supported, and free to the caller.

There are 24,000 other entities who provide information services today in one form or another on everything imaginable, which makes the United State today a world leader in the information services marketplace with more than half of the information data bases available, and provides a positive international trade balance.

The second point I would like to make is that NAA strongly supports S. 1086's commitment to fostering competition in the local loop. A competitive local exchange marketplace is the best safeguard against anticompetitive abuses by local telephone companies. The ability of both information providers and information seekers to choose from alternate transmission facilities will enable both to enjoy the fruits of competition of many suppliers competing on price and quality.

There is no doubt that the RBOC's continue to exercise monopoly control over the local telephone facilities that competing electronic publishers must use to reach their customers. Absent an entry test, which is based upon evidence of competition, strong safeguards are essential to ensure that safe transition that we have been hearing about today from the monopoly telecommunications facilities to truly competitive ones.

The third point is that NAA is pleased that S. 1086 includes a strong starting point in this transition from where we are today in the provision of electronic publishing services by the RBOC's, in requiring separate subsidiaries, mandating arm's-length dealings between the RBOC's and their subsidiaries, requiring cost allocation and other provisions.

We believe that these safeguards should be strengthened. And I dare say your bill, S. 2112, is a legend in its own time and set standards that we feel should be considered for incorporation into S. 1086.

As you may know, representatives of NAA and the RBOC's have been meeting for several months now in an effort to reach agreement on a package of safeguards to help prevent competitive abuses by the RBOC's when they enter information services. These talks are ongoing and have not produced a result yet, so I cannot comment on the substance of the talks. Nevertheless, I am hopeful that at some point they will make a positive contribution to your efforts to protect consumers and enhance competition in the information services market.

Mr. Chairman, every newspaper large and small, and we fully support the concerns that the National Newspaper Association has just voiced, recognizes and appreciates your untiring efforts on behalf of free competition and information services. We look forward to working with members of the committee and the staff in helping to ensure that this exciting time of diversity of information and new infrastructures comes to fruition.

Thank you.

[The prepared statement of Mr. Martz follows:]

PREPARED STATEMENT OF UZAL H. MARTZ, JR.

Mr. Chairman and members of the subcommittee, my name is Uzal Martz, and I am President and Publisher of the Republican in Pottsville, Pennsylvania. I appreciate the opportunity to appear before you today to present the views of the Newspaper Association of America ("NAA") on S. 1086, the Telecommunications Infrastructure Act of 1993. I am Secretary and a member of the NAA Board of Governors.

The Newspaper Association of America represents approximately 1,250 newspapers in the U.S. and Canada. The majority are daily newspapers that account for more than 80 percent of the daily circulation in the U.S.

America's newspapers have a special stake in the development of a vibrant telecommunications infrastructure. For three hundred years, a free press has been central to the political, economic, and social life of this country. That tradition must, and will, continue into the next century. Telecommunications technology from computers to satellites has enabled the newspaper industry to gather and disseminate news more efficiently and with a greater responsiveness to our readers.

The rise of electronic information services also provides us with a wide range of new business opportunities. For more than two decades, we have been active participants in the development of these services. Today, hundreds of newspapers, from large metropolitan dailies to those serving small communities, provide a growing menu of information services. For example, more than 500 newspapers nob offer at least one voice information service, providing information ranging from stock quotes and weather reports to homework hotlines and brief recordings by musicians performing in local concerts. Many newspapers offer electronic classified ads, minimovie reviews, and a variety of other services. At least 400 newspapers are offering 900-number information services. More than 140 newspapers provide on-line access to full-text databases of their newspapers.

Our newspaper in Pottsville, for example, which has a circulation of 30,000, offers our community 295 different information categories in our audiotext service. The service is advertiser-supported and free to the caller.

In addition to newspapers, no less than 24,000 other entities provide information services today. These services offer everything from complex legal libraries to reports on local surfing conditions, and are run by cable companies, broadcasters, database services, and start-up entrepreneurs. The information services industry in the U.S. has flourished, making us a world leader in the information services marketplace. This industry produces more than half of all databases available worldwide and enjoys a positive international trade balance.

If the U.S. is to continue to lead the world into the "information age," we must preserve and promote the diversity and competition that has developed in the information services marketplace and in the means available for delivering those services. Consistent with this goal, NAA has reviewed S. 1086 and finds much in the bill to commend.

First, NAA is pleased that the bill includes safeguards to govern the provision of information services by the Regional Bell Operating Companies ("RBOCs"). Unfettered participation by the RBOCs in electronic publishing, which has been authorized by recent court decisions repealing the MFJ's information services restriction and by recent FCC decisions, will harm competition, information diversity, and consumers.

There is no doubt that the RBOCs continue to exercise monopoly control over the local telephone facilities that competing electronic publishers must use to reach their customers. The RBOCs' control over the sole means of distribution for competitors gives them the ability and the incentive to undermine competition in the information services marketplace, if they choose to do so. They could favor their own services by denying or impeding access to the network by others, or they could use revenues from telephone rate payers to subsidize their entry into information services.

Because the RBOCs retain monopoly control over essential local telephone lines and the ability and incentive to abuse this power, legislation imposing tough and effective safeguards on RBOC participation in electronic publishing is urgently needed to preserve diversity and promote competition in the information services marketplace. S. 1086 provides a strong starting point by permitting the RBOCs to provide information services only through separate subsidiaries; mandating arm's-length dealings between the RBOC and its subsidiary; requiring the establishment of a cost allocation system to prevent cross-subsidization; and imposing certain other regulatory safeguards. NAA believes that these safeguards should be strengthened, and we look forward to working with the Subcommittee to improve them.

As you may know, representatives of the NAA and the RBOCs have been meeting for several months now in an effect to reach agreement on a package of safeguards to help prevent anticompetitive abuses by the RBOCs when they enter information services. These talks are ongoing and have not produced a result as yet, so I cannot comment on the substance of the talks. Nevertheless, I am hopeful that at some point they will make a positive contribution to your efforts to protect consumers and enhance competition in the information services marketplace.

Finally, let me make it clear that NAA supports fully the efforts of NNA and community newspapers to achieve the goals they have described today.

In addition to supporting the bill's inclusion of safeguards, NAA also strongly supports S. 1086's commitment to fostering competition in the local loop. In many ways, a competitive local exchange marketplace is the best safeguard against anticompetitive abuses by local telephone companies. Best of all, with the ability to choose from alternate local transmission facilities, electronic publishers will be able to enjoy the fruits of competition—that is, many suppliers competing on price and quality.

As you know, Mr. Chairman, we have long supported policies to encourage local loop competition. S. 2112, the bill you introduced in the last Congress and which we strongly endorsed, recognized the critical nexus between growth of local competition and information diversity. An infrastructure comprising a "network of networks," rather than one that relies on a single provider or a particular technology, will encourage innovation and offers the best opportunity for thousands of large and small businesses to provide the widest possible range of electronic information services. We are encouraged by the growing evidence in the marketplace that such an infrastructure is not only desirable, but technically and economically feasible. We welcome these developments.

We must caution, however, that the potential for local competition is not the same thing as competition itself. The removal of legal and regulatory barriers will open the way for the development of competition, and we are hopeful that competition will develop swiftly. As this Subcommittee well knows, however, predictions of the rapid growth of competition in any given sector of the telecommunications industry often prove optimistic. Until there is effective competition in the local exchange—and the bottleneck controlled by the RBOCs is broken—the need for meaningful, effective safeguards remains.

Mr. Chairman, every newspaper, large and small, recognizes and appreciates your untiring efforts on behalf of fair competition in information services. We look forward to working with you and your staff in this exciting time to fashion a teleconunications infrastructure policy that ensures a future of information diversity and competitive choices for the American public.

Thank you.

Senator INOUYE. Thank you very much, Mr. Martz. Mr. Smith.

STATEMENT OF JAMES M. SMITH, PRESIDENT, COMPTEL

Mr. SMITH. Thank you, Senator Inouye. I have a prepared statement for the record and I would be happy to answer questions on it, and in fact I would ask leave to supplement that statement before the hearing record closes, but allow me to settle for four or five key points.

My name is James M. Smith. I am president of CompTel, the Competitive Telecommunications Association. We are the national industry association for over 120 small- and medium-sized long distance carriers and their suppliers, and I am happy to have the opportunity today to testify on behalf of the Nation's long distance industry. Mr. Cullen devoted most of his statement to long distance entry, and I will do the same.

Point 1 is that S. 1086 is close to being an excellent bill, indeed, a visionary bill, because it will do several important things. It would help induce meaningful local competition where today there is none.

This is a chart generated by AT&T which shows that 99.86 percent of its 1992 local access payments went to the monopoly local exchange carriers. That leaves fourteen one-hundredths of 1 percent of its access payments to all competitive access providers combined.

Competition in the local loop would bring greater local infrastructure development, which has not happened yet to the extent that it should in the local exchange as it has happened in the long distance industry under the spur of competition.

Since the year after divestiture, 1985, AT&T, MCI, Sprint, the NTN Consortium of CompTel companies and many others have laid over 2.4 million miles of fiber in this Nation.

Mr. Cullen says that the local exchange carriers have built more fiber. Of course that is correct because of the intensity of the local networks, as opposed to the long distance networks, and because there is a difference between a fiber that is operational and a fiber that is not, or is unlit.

There is no greater spur to modernizing a network than to risk that a competitor will beat you to it. That happened—exactly that happened in the long distance industry in the late 1980's. It should be required to happen in the local loop as well. Local competition and local infrastructure development go hand-in-hand.

Point 2—but the bill gets badly mixed up and in fact I think corrupted by including MFJ long distance restriction relief for RBOC's in several sections of the bill. That would be extremely counterproductive to the purposes of the bill and harmful to the competitive long distance industry for two reasons.

One is that it would incent and encourage the local exchange carriers, and particularly the RBOC's, to concentrate their investment not on local infrastructure development but on the development of RBOC long distance networks, which would be, as I have said, basically useless since the long distance infrastructure, the long distance digital information superhighway, is already in place. What is needed is the on-and-off ramps, the local first mile and last mile, to complement that long distance infrastructure.

Second, long distance entry that is contemplated in the bill seems to be rather narrow. But Mr. Cullen obviously is espousing a complete long distance entry, and the Bell companies are espousing that in their lobbying efforts as well. That would cause grievous harm in an industry that has been the paragon of competition in the last 10 years.

It is not true that long distance rates on the whole have gone up in the last couple of years. They have continued to go down, and studies by USA Today and TRAC have testified to that. Since divestiture long distance prices have gone down over 40 percent in absolute terms, and in CPI terms effectively 53 percent over the last 10 years.

There are now 400 long distance carriers in the Nation. And, as I said, there are 2.4 million fiber miles in the long distance networks, all the result of competition and because of the successful public policy that this Congress and the courts exercised in early and mid 1980's.

The Bell System was broken up because of discrimination and cross-subsidization concerns. It was proven that the Bell companies, when they were the unified Bell System of long distance and local exchange, discriminated against would-be long distance competitors and did not allow competition to take root in that market. It would happen again. A GAO report released just early this year concluded that the FCC's efforts to control cross-subsidization have not succeeded and cannot succeed.

Point 3: Do the Bells need interexchange entry? I think the answer is absolutely not. As I have shown, they control 99.9 percent of access dollars, 99.75 percent of total local exchange revenues. They have record revenues and profits. Indeed, Bell Atlantic recently reported record first quarter revenues of $4.6 billion and record profit of $414 million. That is in the first quarter of 1993, 3 months in a sluggish economy.

Next, they are already in the long distance business to an astonishing but really unappreciated extent—45 cents of every long distance dollar goes to local exchange carriers in the form of access payments. When you include short-haul long distance, which the Bells provide—intra-LATA toll—the percentage of long distance dollars that are enjoyed by the local exchange carriers is 54 percent of all long distance dollars.

Their record on jobs has been dismal. Tens of thousands of jobs have been lost in the United States at the hands of the RBOC's at the same time that they have spent $11.3 billion overseas for overseas investments in 1990 through 1992 alone. Meanwhile, local rates have gone up as long distance rates have gone down.

Point 4: So, what you have here, Mr. Chairman, is the Bell companies' failure to modernize their local infrastructure adequately, and asking for new benefits in return for doing it now, and those new benefits would divert their investment and their resources from the job of developing local infrastructure. It is not only like the proverbial child who kills his parents and then begs for the mercy of the court as an orphan, it goes further and applies to the court for orphan's benefits.

This may sound harsh, but let me be clear, the Bell companies provide good local telephone service, but they have a duty and a responsibility while they are still monopolies, and while they do enjoy 99 percent, 99.9 percent of the marketplace, to modernize their networks. They have not sufficiently, so clearly you need local competition to spur them to do it.

The Bells should come back when they have done their duty to modernize their infrastructure and there is true local competition in the local loop, and then should apply for abolition of the interexchange restriction.

Until then, to paraphrase Admiral Stockdale during the Vice Presidential debate, why are we here? The long distance restriction really should not be on this table. The USTA representative a few minutes ago declined your invitation to even speak to that restriction, and I do not think it should be on the table just in a backdoor way. MFJ relief has no place in a network infrastructure, local infrastructure and local competition bill.

Thank you very much, Mr. Chairman.

[The prepared statement of Mr. Smith follows:]

PREPARED STATEMENT OF JAMES M. SMITH

Thank you, Mr. Chairman. My name is James M. Smith. I am President of the Competitive Telecommunications Association. CompTel is the national industry association of over 120 small and medium-sized long-distance telecommunications companies and their suppliers. I appreciate the opportunity to testify before the Subcommittee on behalf of the nation's long-distance industry.

We believe that, with needed changes, S. 1086 can represent a visionary effort to introduce genuine competition and private investment into local telecommunications markets, where little or no competition presently exists. In particular, the introduction of real competition in the local loop should result in lower local telephone rates and lower long-distance rates, for at last there would be price competition for the local access charges that comprise 40-50 percent of a long-distance company's costs. The long-distance industry—which has become the paragon of such competition as a result of immensely successful pro-competitive public policy by Congress, the Judiciary and six Presidential administrations—strongly supports affirmative legislative action to foster competition in present monopoly telecommunications markets, without causing harm to the beneficial and still-developing competition that has characterized the interexchange marketplace in the past decade. But there is the rub, because as now drafted S. 1086 would significantly harm competition in long-distance even as it seeks to create competition in the local exchange.

For its part, the competitive long-distance telephone industry has already contributed enormously to the modernization of the nation's telecommunications infrastructure since the Bell System breakup just nine years ago. The spur of this competition has resulted in the expenditure of $8-9 billion since the mid-1980s by consortia of CompTel member companies as well as AT&T, MCI, Sprint, and many others, to lay over 2.4 million miles of fiber transmission capacity throughout the nation. As Vice President Gore and many others have recognized, the digital information superhighway already is in place with vast capacity in the long-distance portion of the system. And it has been financed entirely by the long-distance companies themselves, without any assistance, give-backs or promises by the government. Nor has it been borne on the backs of captive local ratepayers.

The problem is that this great long-distance superhighway can't be utilized to the full advantage of the nation unless and until the on-and-off ramps—that is, the local telephone networks now dominated by Bell Company monopolies—are modernized. Local "digital driveways" are the final critical links to finish the superhighway.

That's why local competition legislation like S. 1086 is so important—so that the heat of competition will compel the Bell monopolies to dramatically improve their local telecommunications infrastructures, as it has so successfully in the long-distance marketplace. There is no greater spur to modernizing one's network than to risk that a competitor may beat you to it. The question of tremendous importance to the long distance industry, then, is: What does allowing the monopoly Bell companies into the competitive long-distance industry have to do with local infrastructure development?

In our view, these provisions of S. 1086 reward the BOC monopolies for their failure to modernize their infrastructure by starting down the path toward lifting the core prohibition on Bell company provision of long-distance services that enabled the successful development of long-distance competition in the first place. More to the point, why give the Bell companies this huge plum while they are still monopolies, enjoying a virtual stranglehold over the local exchange bottleneck through which all telephone calls must pass? In other words, lifting the restriction in any fashion rewards an entrenched monopoly for its record profits, record layoffs, record investment in foreign countries, and failure to build local facilities to complement the long distance infrastructure that we have built four times over.

The 1984 divestiture of the Bell System and the consequent opening of long-distance to competition has been tremendously successful and beneficial to the American people. Long-distance providers will tell you that competition is imperfect and still developing, but make no mistake that there is real competition in the long-distance marketplace. One need only look at the number of long-distance providers—over 400 by most counts—or at their prices, which have plummeted by more than 50 percent in real terms over the past nine years, or to the laying of those 2.4 million miles of fiber caused solely by competition. S. 1086 seeks to introduce that kind of competition into the local telephone marketplace, and that's good. But it makes absolutely no sense to strengthen the hand of the monopolist with one hand—and, critically, to divert its attention from the task at hand, namely local infrastructure development—even as you subject it to competition with the other. Passage of this bill as currently drafted will result in a huge diversion of Bell Company funds to long-distance development, to the detriment of infrastructure development.

So then: Why implicate the MFJ interexchange prohibition in this bill? The Bell companies will variously tell you that their bottleneck has been shattered, that local competition is here today, that they cannot survive unless they are allowed to provide interexchange services, that American jobs are at stake. But let's look at some facts, which the Bell companies may dodge but cannot deny:

Fact 1: By any estimate the Bells control over 99 percent of the local exchange marketplace. In terms of revenues, in 1992 the Bells had 400 times the revenues of all local access competitors combined. Bell Atlantic recently reported record 1st quarter revenues of $4.16 billion and record income of $414 million. That's in 3 months in a sluggish economy.

Fact 2: In the most bottom-line sense—money—the Bells are already in the long-distance business to a massive degree, because between 40-50 cents of every long-distance dollar goes to local exchange companies in the form of access payments. Local access is the largest operating expense of every long-distance company and one of the largest sources of Bell Company revenues.

Fact 3: The BOCs' record for creating jobs for U.S. workers in recent years has been dismal—in fact, they have decreased their U.S. payrolls by hundreds of thousands of workers in the past 8 years even as they invested $11.3 billion overseas in the past three years alone. A recent study by New Networks Institute concluded that the Bells now serve their investors first and their ratepayers second.

Fact 4: Local rates have gone way up even as long-distance rates have been cut in half. The same study found that average toll charges across the U.S. cost consumers and businesses $4.9 billion dollars more than if long-distance companies offered the service. It costs 90 percent more to make a one-minute call from Manhattan to Montauk, Long Island 75 miles away than to call 2,900 miles to California. Compare the long-distance experience: in 1984 AT&T charged $2.70 for a 5-minute daytime call from Washington to L.A. Now the charge is $1.25—and the smaller carriers generally charge even less.

So when the Bells tell you that there is no bottleneck, or they can provide lower long-distance rates in any form, please look at history and demand that they show you how based on their historical experience. Stripped down, the Bell companies' argument for MFJ relief is: If you're going to take away our monopolies you have to first free us to provide interexchange services. Or: give us MFJ relief now and then we'll show you local competition. But they have it backwards. Make them show you that there is real, measurable, broad-based competition, not that there might be someday. Then—and only then—should the Bells be allowed to provide interexchange services.

Bell entry into the interexchange arena could only harm long-distance competition. The core reason for the government's breakup of the Bell System was that the integration of monopoly local exchange telephone service with long-distance service bred cross-subsidization and discrimination that was making it impossible for fledgling competition to take root in the long-distance market. Well, this February a GAO Report (OAO/RCED-93-34) found that FCC regulation cannot prevent RBOC cross-subsidization. And as for discrimination, consider that every long-distance carrier must depend on the monopoly local carrier for its access to customers. In order to get that access, it must give the Bell company its most proprietary customer lists and information. Now imagine if the RBOC that utterly controls that access and that customer information is directly competing with the long-distance carrier. That is the very antithesis of a level playing field.

Some may say that S. 1086 doesn't let the Bells too far into long-distance, rather that it opens the door only a little, in the areas of cable and cellular. The answer is simple: you yourselves have heard from the Bells that they want much, much more, if not this year then next. MFJ waivers have routinely been granted for cellular and the like. They don't need legislation for that. Their explicit agenda is significant long-distance entry. If the Senate wishes to carve out these two narrow exceptions, it should codify the rest of the MFJ long-distance restriction until there is real, substantial local competition.

I'd also like to address one other important provision of S. 1086: in the wake of a federal court decision just last week, the FCC may now have no flexibility to relieve non-dominant long-distance carriers of the burden of filing tariffs to cover every single rate they charge. CompTel strongly supports a strengthened Section 5(h)(1) of the bill to preserve the Commission's ability to maintain flexibility in its tariff regulation of smaller, non-dominant carriers.

In summary, critically needed local infrastructure development will come with local exchange competition, just as it did with long-distance competition. That's why S. 1086 has great merit as a local competition bill. But there is no earthly reason why long-distance should be implicated in this bill. In fact, it would only give the Bells a perverse incentive to turn their investment to businesses other than a state-of-the art local network. If the Bells can't show you that even 1 percent, much less 10 or 20 percent of their traffic or revenue has gone to the competitors; if they can't show you that their monopoly revenues and profits have even suffered a dent; if they can't show you that they have created even one net U.S. job in the past 5 years—because they certainly have created thousands overseas; then I ask you why they should be rewarded now with even limited long-distance entry.

Make them show you something more than illusion. If they don't, then the MFJ provisions of S. 1086 are just a giveaway that will make S&L deregulation pale by comparison. No lawmaker wants to harm irreparably a market that has become a model of competition thanks to wise public policy—namely, the U.S. long-distance market. With regard to long-distance, our plea is: if it ain't broke, don't break it.

[Appendices may be found in the committee files.]

Senator INOUYE. Thank you very much, Mr. Smith. Mr. Metalitz.

STATEMENT OF STEVEN J. METALITZ, VICE PRESIDENT AND GENERAL COUNSEL, INFORMATION INDUSTRY ASSOCIATION

Mr. METALITZ. Thank you, Mr. Chairman, for this opportunity to present the perspectives of the Information Industry Association on S. 1086.

IIA represents some 500 companies that are involved in the creation, distribution, and use of information products, services, and technologies. Our member companies are dedicated to meeting the information needs of American business, professional and consumer markets.

Our association is also uniquely positioned, I think, among all those you have heard from today, to offer a balanced and comprehensive perspective on this legislation, since our members include not only many of the leading information providers and electronic publishers but also some of the leading telephone companies, both local and long distance.

Mr. Chairman, when you and Senator Danforth introduced this legislation last month, you called it "a landmark day in the history of the Senate's consideration of telecommunications infrastructure policy." We think that is an accurate assessment. This legislation does break new ground. It anticipates the new realities of a marketplace that is changing quite dramatically. It helps to shift the terms of debate to a more constructive plane.

We have heard a lot of discussion of frustration in this room this morning, but I think there are also some broad areas of agreement and some grounds for hope that diverse interests can work together to achieve greater benefits for the American public.

Mr. Chairman, this legislation achieves this breakthrough because it reflects three fundamental realities about the way ahead in the telecommunications marketplace. First, it recognizes that more competition is the key that will unlock the potential for developing and deploying advanced infrastructure. In particular, bringing competition to the local telephone loop is an essential precondition for the American people to enjoy the full benefits of the Information Age.

Second, the bill calls for a strong Federal leadership role in opening up competition throughout the network. That leadership role includes preemption of State rules and regulations that stand as obstacles to local competition.

Finally, the bill recognizes that strong safeguards are needed to ensure a level playing field.

I would agree with Mr. Martz that competition is the best safeguard, but it is not the only safeguard that is needed in this situation. Monopoly advantages do not disappear overnight simply because an administrative agency says they are gone. Structural separation and interconnections guarantees, as you provide in the bill, are among the safeguards that are needed.

IIA strongly supports all three of these elements of the framework of S. 1086. They are fully consistent with the positions that we have taken in our paper on telecommunications infrastructure objectives which we have attached to our testimony. That paper reflects a consensus viewpoint among our diverse membership, and we hope it is useful in a very practical way to you and your staff as you work on refining this bill.

The bill does contain the broad outlines of a constructive legislative solution, and we commend you and Senator Danforth for your leadership in this regard. We do have some comments and suggestions about some important aspects of this legislation. We also agree with some of the statements that have come before us this morning. Let me just summarize a few of our concerns.

First, this is sweeping legislation. It is intended to be so, and in that situation there needs to be some close attention to the definitions that this bill writes into the Communications Act for the first time. In particular, the bill provides a rather complex definition involving telecommunications carriers, information service providers, and enhanced service providers.

The FCC has made a longstanding distinction between basic and enhanced services. It is a binary distinction. It is easy to understand, it is relatively easy to apply, and it has been very valuable in demarcating the limits of regulation in this area.

We would urge the subcommittee to reexamine whether that simple definition might be more effective than the more complex definition in the bill. We do have some concerns that without some change in that regard, there is a risk that some companies other than true telecommunications carriers will be classified as carriers under this bill. That is not what we think the bill intends to achieve.

Second, we would suggest that the legislation should reaffirm the viability of antitrust law in this area. Strong antitrust protections remain essential for deterring and punishing anticompetitive behavior by any party in the telecommunications marketplace.

Third, we think Congress should legislate more specifically with regard to interconnection and access requirements, including a requirement for complete unbundling of services and for cost-based pricing. Putting some more detail in this legislation will help ensure that the network does become truly available, truly open to all competitors on equal terms and conditions.

Finally, we would suggest in some areas a clearer statement of Federal preemption. We think this legislation does send a very strong and constructive signal to the States, but that that could be spelled out more clearly.

In conclusion, Mr. Chairman, we applaud the steps that you and Senator Danforth have taken in introducing this legislation. We know that considerable work remains in filling in the broad outlines that it sets forth, and we are ready and willing and eager to assist you and the subcommittee members in that task.

Thank you very much for your invitation to testify.

[The prepared statement of Mr. Metalitz follows:]

PREPARED STATEMENT OF STEVEN J. METALITZ

My name is Steven J. Metalitz and I am Vice President and General Counsel of the Information Industry Association. IIA welcomes the opportunity to provide its perspective on S. 1086, the Telecommunications Infrastructure Act of 1993. IIA is a trade association of more than 500 companies, which celebrates its 25th anniversary this year. Our member companies are involved in the creation, distribution and use of information. IIA's membership includes entities of all sizes providing enhanced services, telecommunications and associated equipment, print media and electronic publishing services, as well as some of the largest providers of common carrier services. IIA's members are in the forefront of the development and implementation of new information products and technologies.

IIA supports S. 1086 for its vision of an advanced, affordable, ubiquitous telecommunications infrastructure driven by market forces and competition. Such an infrastructure will speed the benefits of the Information Age to Americans. IIA has endorsed this vision in its Telecommunications infrastructure Objectives and Implementation Principles.[1] Now, whether through increased local competition while preserving universal service, improvement of services in rural areas or removing the regulatory uncertainty between local and federal authorities, the Telecommunications Infrastructure Act of 1993 can go a long way in bringing the best of the Information Age to the doorstep of America.

IIA has always been a supporter of a regulatory and marketplace environment that will ensure the maximum opportunities for the development and enhancement of information-related services. We believe that the availability of efficient, economical and ubiquitous telecommunications services is critical to the free flow of information throughout our society. We commend the Subcommittee for holding this hearing to examine such important issues.

It is important to note that IIA's members include not only independent enhanced service providers, but also regional Bell Operating Companies (RBOCs), interexchange carriers (IXCs), independent telephone companies (ITCs) and local exchange carriers (LECs). This unique membership mix gives IIA a balanced perspective on the important issues addressed by S. 1086. IIA consistently has urged the establishment of fair ground rules for the participation and protection of all users, competitors and carriers alike in the telecommunications marketplace.

IIA appreciates the complexity of the issues contained in S. 1086. Within our own organization, we also have had to grapple with a fundamental principle of fairness in a marketplace of diverse technological capabilities and rapid business and technological growth. It has not been an easy task and as stated in our letter to the principal sponsors of S. 1086 on June 22, 1993, we applaud the effort this bill represents "in tackling such difficult issues as telecommunications infrastructure, privacy, and network development." This is a bill that offers many positive solutions to current issues facing telecommunications providers, enhanced service companies and users while leaving open the opportunity for growth and advancement in the future.

Despite our broad interest in S. 1086, we will limit our comments to a few critical areas of greatest importance to our members. First, we note that the testimony you hear today shares a number of our concerns with regard to the creation, maintenance and monitoring of appropriate safeguards to assure market fairness and prevent discrimination, cross-subsidization or other anticompetitive behavior by local exchange carriers.

In addition, we agree with others about the Importance of customer proprietary network information (CPNI). Not only from the perspective of necessary privacy protections, IIA believes that opportunities for the use of and access to CPNI unfortunately have been skewed to the advantage of the basic telephone service providers who both create much of this information and hold the keys to its first and easiest use.

Further, as this bill sets the stage for improved and widespread availability of digital transmission facilities for all American users, we applaud what S. 1086 can do to make more and valuable services accessible. For example, open and voluntary standards should be encouraged to benefit all parties, particularly end users. In many respects, S. 1086 can encourage the use of voluntary standards as providers, users and regulators embark upon improvements in information highways with high service quality and network reliability—features so essential to the competitive success of our country's information services.

With this background, we are pleased to offer comments on four issues that should be resolved if this legislation is to achieve its objectives:

- The scope and implication of the definitions contained in S. 1086;

[1] A copy of these principles and objectives may be found in the committee's files.

- The appropriateness of strong antitrust protections;
- The need for strong direction to ensure cost-based pricing of basic services offered by local exchange carriers and avoidance of enhanced service provider access charges; and
- The importance of a clear and unequivocal statement on federal preemption of any state public utility and entry/exit regulation of enhanced services.

S. 1086 BECOMES A MORE POWERFUL INSTRUMENT FOR CHANGE BY ADOPTING WIDELY-ACCEPTED SERVICE DEFINITIONS

In its Computer II decision, the Federal Communications Commission 1(FCC) decided that "the provision of enhanced services * * * is not subject to Title II regulation, nor is a regulatory scheme for enhanced services * * * required to protect or promote the overall objective of the Communications Act." IIA members, and indeed carriers and ESPs worldwide have relied on this regulatory framework in the development and offering of their services. This definition clearly separates basic transmission from any enhancements which may be added to the transmission that make it a new, different, and generally improved product.

The FCC decision was based on the following definitions:

> * * * [B]asic service is limited to the common carrier offering or transmission capacity for the movement of Information, whereas, enhanced service combines basic services with computer processing applications that act on the format, content, code, protocol or similar aspects of the subscribers' transmitted infornatlon, or provide the subscriber with additional, different or restructured information, or involve subscriber interaction with stored Information.

Although IIA has not always agreed with FCC decisions, in this instance the Commission crafted a definition that works. The Commission language draws a bright line separating two distinct services: getting from Point A to Point B using a telecommunications medium, as distinct from any number of enhancements that could be employed along the way. This is a definition that allows enhanced services to be offered by telephone companies, whether LECs, RBOCs, IXCs, or ITCs, and by non-telephone companies, including computer companies, database companies, or Information providers. It Is the addition of "computer processing" applications as described in Computer II, which changes basic services into enhanced services—many of which also are information services.

Under the FCC's definition, information services are a subset of enhanced services. Companies which offer enhanced services may purchase basic transmission services from common carriers at published tariff rates set by federal and state regulators. Enhanced service providers use the capabilities of the nation's telephone networks to offer data, information and information management services. Some of these companies provide telecommunications and enhanced services—most often in separate business units—under current distinctions between basic and enhanced services. Most importantly, however, the regulatory status of the service is determined by the nature of the service, basic or enhanced, and not by the nature of the company or the other services it offers. The fact is that enhanced service providers are not regulated as telecommunications companies.

Instead of the widely-accepted distinction between basic and enhanced services, S. 1086 provides specific definitions for telecommunications services1 Information services and electronic publishing. Consequently, IIA members find themselves confused as to whether their enhanced or gateway offerings could be characterized as telecommunications services and thus raise their service requirements to those of common carriers.

For example, in addition to applying to transmission, the "telecommunications" definition includes the passage "with or without benefit of any closed transmission medium, including all instrumentalities, facilities, apparatus, and services (including the collection, storage, forwarding, switching, and delivery of such information) essential to such transmission." This portion of the definition could apply to private line networks used by online enhanced service providers, value added networks, electronic mail systems or other enhanced services. Some companies may have excess private network capacity which occasionally they offer for sale on a non-public basis. Others may have private lines for intra-company purposes associated with enhanced products they offer. Regardless, such uses do not always involve what is traditionally characterized as "information." For these companies, only by rising to the level of an "information service" would they be exempt from the requirements of telecommunications service providers under Sec. 4 of the bill.

Further, as crafted in S. 1086, the definition of information services excludes many of the added value enhancements that have made ESPs successful. Under the bill, information service is defined as "[T]he offering of a capability for generating, acquiring, storing, transforming, processing, retrieving, utilizing, or making available information which may be conveyed via telecommunications, except that such service does not include any use of any such capability for the management, control, or operation of a telecommunications system or the management of a telecommunications service." This list does not reflect the range of activities today's and tomorrow's businesses and consumers are involved in with enhanced service products.

Simply put, S. 1086 should apply to basic transmission mechanisms irrespective of any particular content. A distinction between basic and enhanced services, as enunciated by the FCC, Is all that is needed to ensure afaIr and complete regulatory framework. Other definitions unfairly corral competitive enhanced service providers within telecommunications service definitions.

STRONG ANTITRUST PROTECTIONS SHOULD REMAIN CORE PRINCIPLES OF S. 1086

In discussions of telecommunications infrastructure development, the need for strong antitrust protections must not be overlooked. It may be appropriate for S. 1086 to have an "antitrust savings clause" preserving all federal and state antitrust laws with regard to LEC activities. Such language would acknowledge the importance of universal service and also would preclude unfair advantages in the local loop. By narrowly crafting exceptions appropriate in universal service or other FCC-mandated activities, users in any area of the country could be assured of the most competitive alternative.

As noted above, IIA released its Telecommunications Infrastructure Objectives and Implementation Principles earlier this year. In that document, IIA noted:

> Market forces are the preferred means of determining telecommunications applications and services, and thus the development of telecommunications infrastructure.
>
> As telecommunications markets continue to evolve from a regulated monopoly to a competitive regime, the role of governments should be to promote competition while protecting the public interest where competitive market forces are not yet fully operative.

S. 1086 recognizes correctly that competitive market forces should drive telecommunications infrastructure development. However, particularly in regard to most LEC offerings, a monopoly situation currently exists. As time progresses, competition at the local level—undoubtedly spurred by legislation like S. 1086—will remove most, if not all, of this bottleneck. Until that time, full antitrust protections and vigorous enforcement are essential to the provision of market services.

Strong antitrust protections can ensure that local exchange carriers would not be not insulated from antitrust law and regulation by the provisions of the Communications Act this legislation would affect. Ultimately, IIA is concerned that competition not be harmed: not now as we move from a monopoly regime to a competitive one; and not in the future when full competition exists. Whenever there is anticompetitive behavior in the form of conspiracy, collusion, price-fixing or other unfair market manipulation that rises to the level of an antitrust violation, full enforcement, penalties and consequences should follow. Strong antitrust protections in this bill will facilitate the creation of simple rules governing economic, technical and financial aspects of competitive behavior In a changing technological environment.

S. 1086'S ENDORSEMENT OF FULLY UNBUNDLED SERVICE OFFERINGS CAN ENSURE COST-BASED PRICING AND MARKET FAIRNESS

For many years, IIA has argued vigorously for cost-based pricing in basic telecommunications services, offered in their simplest, unbundled form. Hand-in-hand with this argument has been the opposition to the imposition of access charges applied to ESPs on the basis of supposedly different network use by such providers. ESP access charges included in current network configurations frequently benefit carriers which own and design the network, and work to a disadvantage to many non-telephone company affiliated providers. Unfortunately, as currently administered, open network architecture (ONA) and the accompanying cost accounting rules do not constrain effectively the ability of carriers to discriminate against competing enhanced service providers.

In defining the obligations of telecommunications carriers, S. 1086 approaches the issue squarely in calling for:

> (1) interconnection * * * (2) nondiscriminatory access to any of the carrier's telecommunications facilities; (3) non-discriminatory access, where technically feasible, to the poles, ducts, conduits and rights of way owned or controlled by the carrier; (4) non-discriminatory access to network functions * * * (and) (5) telecommunications services and network functions without any restrictions on the resale or sharing of those services and functions.

However, IIA maintains that these requirements will not be enough to ensure the full and complete access to basic network infrastructure that the authors of this measure intend.

This legislation should require unbundling of network services and facilities to allow customers the option of obtaining basic functional elements. Such services could Include end-user access lines, switch-based features, inter-office transport, signaling functions, ordering, testing, maintenance and other support services, billing and collection services or gateway functions. Such elements should be provided on an actual or virtual collocation basis in a non-discriminatory manner to any provider.

The unbundling of network features and functions must be accompanied by pricing which bears a fixed, reasonable and consistent relationship to Its underlying cost. Such cost-based pricing will ensure that all underlying costs are consistently allocated by a verifiable and independently auditable methodology covering unbundled services.

In sum, S. 1086 should direct the FCC to:

- Require local exchange services be sufficiently unbundled to meet market demand;
- Prohibit LEC configuration of ONA service offerings in ways which best meet the needs of their subsidiaries while ignoring the needs of competitors;
- Grant independent enhanced service providers opportunities to participate in the benefits of equal access for service and Information of equal type and price to carriers;
- Increase collocation opportunities, whether actual or virtual;
- Require cost-based pricing for all service offerings to independent providers; and
- Mandate that transactions between the RBOCs and their affiliates be on an arms-length basis that parallels the treatment afforded independent enhanced service providers.

These are the steps to a level playing field that Congress should tell the FCC to take. Once the level playing field exists, the Commission should have the flexibility to consider expansion of the opportunities for regulatory parity among all providers. IIA looks forward to the emergence of full and effective competition and the ability of all providers to offer any service.

S. 1086 CAN STRENGTHEN ITS IMPORTANT STAND ON PREEMPTION

IIA always has argued for uniform rules to promote the enhanced services industry, especially as the industry develops on a regional and national basis. Uniform national policies will bring stability and growth to the enhanced services industry. The potential of unilateral action by a single state to deny its citizens access to an important Information marketplace will only harm the public interest. This point is admirably recognized in S. 1086.

The availability of a competitive enhanced services marketplace is repeatedly threatened both by state public utility regulation of enhanced service providers' intrastate offerings, and by state efforts to claim jurisdiction over interstate offerings. The result is an ever-encroaching national patchwork of regulations for the enhanced services industry that reduces consumer access to enhanced services by making It difficult for providers to offer services reliably on a national basis.

IIA was among 16 trade associations and companies that petitioned the FCC for a ruling preempting state entry/exit and tariff regulation of enhanced services. These petitioners are national providers of a wide range of enhanced services that would be impacted negatively by diverse and incongruent state regulation of their services. For example, a number of jurisdiction, including Florida, Georgia, Delaware and the District of Columbia, had proceedings over the last couple of years that proposed to treat ESPs as carriers subject to public utility regulation. As noted above, IIA vehemently disagrees that ESPs are basic telecommunications providers.

As these matters have evolved, we were heartened by the resolution adopted by the Federal/State Joint Board on Open Network Architecture that spoke against such regulation. However, we are aware that the Joint Board has no regulatory authority and cannot bind the states. In anticipation of continued state attempts to regulate ESPs, we believe S. 1086 sends a strong and constructive signal about the distinction between national and local enhanced services; and the supremacy of federal authority over interstate services.

We are concerned, however, that the information services definition contained in Sec. 4 of S. 1086 and referenced in Sec. 13 may limit federal regulatory authority affecting other enhanced services. As discussed above, IIA supports the existing distinction between basic and enhanced service providers.

IIA has consistently advocated that there is no valid policy basis upon which a state regulatory body can or should seek to apply public utility regulation to the provision of enhanced services. To the extent that states literally apply the guidelines of this legislation, additional confusion may result from the lack of a clear definitional standard.

For example, Sec. 229 meets IIA's broad policy goals concerning preemption of state public utility regulation of enhanced services by forbidding any "State or local statute or regulation, or other State or local legal requirement, (to) prohibit or limit in a manner inconsistent with Federal regulations or with this Act, the ability of any entity to provide interstate or intrastate telecommunications services." However, IIA believes that it will be appropriate for the record to stress that the S. 1086's allowance of state rate, term or condition regulation of Sec. 13 is limited to Intrastate offerings which do not impugn the viability of interstate services and further, any state public utility regulation of enhanced services would be wholly inappropriate.

Finally, S. 1086 provides that "The States may prescribe regulations implementing paragraphs (1) through (5) (obligations of telecommunications carriers) for intrastate services so long as such regulations are not inconsistent with those prescribed by the Commission." Although states are often the first line in consumer protection from competitive abuses by providers, IIA also has had experience with zealous state regulatory authorities disinterested in national service goals. We believe that it is important for the legislative record to reflect fully the intended scope of such state regulations. Local actions nationwide—in tax, public utility, regulatory and other areas—have shown that without guidance, many states can take steps that drive up costs and otherwise needlessly constrain the telecommunications marketplace.

CONCLUSION

As we work together, as noted in Sec. 3 of S. 1086, this bill can "encourage private investment in, and advancement of the Nation's telecommunications infrastrucuture; ensure the availability of the widest possible range of competitive choices in the provision of telecommunications and cable television services; eliminate the existing regulatory barriers to competition in the provision of telecommunications and cable television services; encourage interconnection and interoperability among telecommunications carriers; ensure the universal availability of telephone services; encourage the continued development and deployment of advanced and reliable capabilities and services in telecommunications networks; and protect the privacy interests of users of telecommunications services."

Mr. Chairman, and members of the Subcommittee, thank you once again for inviting the Information Industry Association to participate in this hearing. We would be glad to provide more detailed comments on any aspect of this matter and to respond to any questions that you or the Committee members may have.

Thank you.

[Telecommunications Infrastructure Objectives and Implementation Principles may be found in the committee's files.]

Senator INOUYE. Thank you very much, Mr. Metalitz.

Before I proceed with the questioning, I would like to assure one and all that your full statement will be made part of the record, and if you do have addendums or supplements, please feel free to submit them.

Mr. Cullen, I am certain you are aware that several proposals have been submitted by Ameritech, Bell South, and Rochester Telephone to open the networks to competition. Their proposals vary, but it seems to be a response to the demand for competition. Do you have any proposal of your own?

Mr. CULLEN. Mr. Chairman, I am familiar with the three proposals that have been made. I think they are a response both to the frustration we have heard here this morning and a response to customer requirements as well.

On behalf of the seven RBOC's, I would say that we endorse any proposal that will move us toward full competition. The Ameritech proposal does that in an environment with which we are familiar. They are proposing that that be done as a demonstration, that it be done as a trial, and that in a period of 3 to 5 years, and in fact all along the way, that it be examined to determine if it has produced the results and the benefits that are anticipated.

We advocate full competition. We are concerned, however, about the issue of universal service and of essential services. In Ameritech, they have proposed to experiment with that, and we feel that that is a worthwhile approach.

Senator INOUYE. Do you think this bill will accelerate competition and expedite it so that you can lift the ban on long distance?

Mr. CULLEN. Two points I guess, Mr. Chairman. One is competition in the local exchange is a very unbalanced competition in this bill, whether, as Senator Packwood suggested earlier, you are standing on the top of the incline or the bottom of the incline. It puts all of the obligations and none of the opportunities with the regional Bell operating companies.

Second, the relief in long distance service, where in fact as Mr. Smith suggested, it might create harm to the industry if we were allowed in, but great benefit to the consumers if we were allowed in; competition, as outlined in the bill for long distance, is nominal only, and does not even allow us to get in and resell services, which is a right that the long distance carries have here for local service.

So, it is very little and creates very, very few opportunities for the telephone companies, who again I might underscore, because I do not think Mr. Smith heard it, that the local exchange companies have in fact invested in 4 million miles of fiber optics, which has nothing to do with whether it is turned on or not turned on, this is a substantial investment and twice that of the rest of the long distance business.

Senator INOUYE. New Jersey Bell and the publishers have reached some accord. Can that agreement be implemented nationwide?

Mr. CULLEN. Mr. Chairman, I happen to have been personally involved in that accord as well as some of the talks that are underway nationally.

The New Jersey Bell accord was excellent for New Jersey. I am not sure that it would be excellent in Missouri, or Alaska, or South Carolina, but I do think it is the basis for an agreement which in New Jersey we did for the sole reason of reducing the concerns expressed by the newspaper industry so that we could go forward and begin to develop business opportunities.

I would not recommend it throughout Bell Atlantic, I would not recommend it throughout the country, but I do recommend that we open these lines of communication and assure one another that we will have the protections that we have heard about here today for newspapers.

I think as a demonstration, New Jersey will prove to be very effective. I would just point out that it does have a sunset provision of 7 years. It does not apply to information services or even all electronic publishing. It applies to the type of electronic publishing that is traditionally found in newspapers, and so I think it is an excellent start.

Senator INOUYE. Thank you very much, Mr. Cullen. We will be submitting more questions, if we may.

Mr. Wright, as you have indicated, the courts have just lifted the information services restrictions from Bell companies, and this bill would impose separate subsidiaries for electronic publishing. Are these safeguards sufficient?

Mr. WRIGHT. Mr. Chairman, our concern is, and I hate to use the term, "the playing field," which particularly concerns small publishers. Regardless of the incline, if you are from a small community like Conway, MO or Boardman, OR, your total resources on the asset side of your balance sheet might, for example, be $125,000.

You are very key to the culture of that community. And when you look across the playing field, even though it may be somewhat level, and you are looking at a multibillion dollar corporation or company that can bring one person into your community and effectively get into the information business, not only would people be put out of business in that community of Conway, MO or Boardman, OR, but you are also really disrupting the culture of that community and their information services.

So that is the reason that the National Newspaper Association has maintained that we need assurance that everyone will have equal access at the most competitive rates, sir.

Senator INOUYE. I have many other questions, but I will have your Senator do the questioning, if I may.

Mr. Martz, can you update us on the negotiations between publishers and Bell?

Mr. MARTZ. Sir?

Senator INOUYE. You are now in negotiations between Bell companies and publishers. What is the latest on that?

Mr. MARTZ. Well, as I indicated in my statement, we are not at a point where we can talk about substance. But I think this is an indication that reasonable people are trying to sit down to see if we can now iron out our differences.

I might point out that the New Jersey Bell agreement was something that came up after the fact of the legislative process; and in point of fact, the legislation in New Jersey went through a lame duck session of the legislature, and did not have the benefits of the full studied focus that this committee is putting, in terms of this bill.

But it is our hope, certainly, that as these talks progress, that we may be able to come forward with some recommendations for language that will satisfy the concerns that we have. And I think, if you go back to the dilemma that this committee is facing, you obviously are trying to balance the proper role of Government versus the role of technology and market forces. You are also trying to balance the lessons of the past with the promise of the future.

And the reality of it is that AT&T, having been broken up into the seven RBOC's and the slimmed-down AT&T, was done for a reason. And since that time, by inadvertency or otherwise, there have been continuing evidences of abuse. So, it is our feeling that, as you make this balance to get from the present of a monopoly infrastructure, with one funnel where everything has to go through that local monopoly switch, to what we all would like to see is a free competitive infrastructure, that we have to have appropriate safeguards; if, in fact, the telephone companies are going to compete with those using them.

Senator INOUYE. Why do you have an interest in the long distance industry?

Mr. MARTZ. We do not have any position on the entrance of the RBOC's into long distance. Our areas of concern are first of all, that there is a freely competitive information services marketplace; and second, that there is a competitive infrastructure.

Now, I think sometimes these two things get confused. You can have a competitive infrastructure, without the providers of the infrastructure also providing telecommunications services. And this is the content conduit dilemma: That, as long as you have a funnel through which everything must go, whoever owns that funnel, if they are going to compete with what has to go through that funnel, then we feel that safeguards, separate subsidiaries, separation, all of the kinds of safeguards that you had set forth in S. 2112, and are partially included in 1086, will help get us there.

Senator INOUYE. Thank you very much, sir.

Mr. Smith, if this bill is successful in opening up the local exchange, or the local loop, as you say, do you believe that the competition would be sufficient enough for a lifting of the long distance restriction that is now set forth in MFJ?

Mr. SMITH. That really remains to be seen. What remains to be seen is if the bill, and other forces, are successful in bringing real competition to the local exchange. If the day arrives when long distance companies and consumers have access to several alternatives—for example, long distance companies have access to different service providers which will cut down, no doubt, on the cost of access. If the local exchange competition experiment is, indeed, successful, and there is a significant market share on behalf of the alternative providers, then I would say that the day would come, if that bottleneck is broken, then yes, there could be long distance entry by the Bell operating companies.

But you have to wait for one to happen, in order to allow the other to happen. Otherwise, if the door is opened for long distance entry now or in the next few years while there is a 99-percent-plus market share and access payment share by the monopoly local exchange carriers, that is still monopoly.

And you will not find out until long distance is gone, and probably less competitive because of Bell entry and Bell integration, you will not find out until later whether local competition is a reality or a myth.

Senator INOUYE. Mr. Cullen, did you want to say something?

Mr. CULLEN. I would to add just one or two comments, if I may. No. 1, the 99-percent figure, which is frequently discussed by AT&T and other long distance providers, relates only to public network and publicly reportable data. If we look in the State of South Carolina for example, there are 75 separate companies providing private systems by satellite, by microwave. There are 44 long distance companies. So, the vast majority of the competitive inroads are not reported, and would not show up in these data.

Second, when we think about the local exchange business, it is very important, I believe, that we think about the true nature of this business. It is comprised of three essential elements: It is the pipe to the home, which can be copper, which can be fiber, which can be coax, or it can be wireless; second, it is comprised of the access charges that we are talking about here; and third, it is comprised of intra-LATA, a very short distance toll.

There is substantial competition in the short haul, short distance toll, substantial. There is substantial competition in access and in delivery. No one at Prudential or VISA or General Motors or Exxon has the slightly problem in finding alternative providers for any service they choose to provide. People in the pinelands of southern Jersey do; people in western Virginia do.

And so the issue is, we have a tremendous amount of competition. But it has gone only to those areas where we have profitable business: Access charges; short haul toll charges. Not a lot of it has gone into the local loop, because most of the local loop is underwater; unprofitable, again, by as much as 50 percent.

I'd just close by citing South Carolina, where we have some data. The studies in South Carolina show that the average cost to a residential customer for local service covers about 50 percent of the actual cost of providing the service. The rest comes from access charges and short distance toll.

So, I think it is important that we try to get the numbers as close as possible to accurate; and that, second, we understand the true nature of the local exchange business. It does have those three very important components.

Senator INOUYE. Mr. Smith.

Mr. SMITH. May I respond, too, since it is my figures that are being taken at issue.

I definitely agree that we should look at accurate comparison, and figures that are not subject to redefinition or jerry-rigging, or anything like that. The AT&T figure is the percentage of the whole, the percentage of all of their access payments that are made to monopoly local exchange carriers, and the number was 99.86 percent.

To take one other figure—and these are total overall revenues, figures that are not subject to redefinition or reinterpretation or jerry-rigging—at the recent conference of the competitive access providers, there were two Wall Street analysts there; one of whom estimated 1992 revenues of the CAP industry as a whole at about $160 million. The other said it might be closer to $220 million. So, it is within that range.

The income—the revenues, rather, of the Bell operating companies was on the order of $82 billion. That figure then, is that the CAP industry as a whole is one-quarter of 1 percent of the Bell operating companies. Those figures are difficult to quibble with; they are absolute revenue figures. And to say that yes, but our core business is being subject to a great amount of competition, it mystifies me that the core business can be that one-quarter of 1 percent.

Senator INOUYE. Mr. Metalitz, the restriction on information services was just removed by the courts. As a result of that, have members lost business?

Mr. METALITZ. I think there has been a lot of exciting development, since the information services restriction has been lifted. But we do not think that that is the ideal situation for the development of more and competitive information services.

We believe that these services would develop more robustly in an environment such as that outlined by your bill, an environment in which there are some safeguards, in which there are some stronger guarantees for connection to monopoly services, especially to local telephone services. And so, we think that that would improve the situation beyond where it stands now.

Senator INOUYE. Do you believe the safeguards are sufficient, as set forth in this bill?

Mr. METALITZ. We have some suggestions that are set forth in our written testimony, about how they could be improved. We think there should be some more specific requirements for unbundling of services, for example; and some other more specific requirements for truly making this an open system, that any competitor can interconnect with.

Senator INOUYE. Thank you. Senator Danforth.

Senator DANFORTH. Well, I want to thank the panel and all witnesses today for a long morning. And Mr. Chairman, one of the things that is encouraging to me is the fact that so many witnesses, representing so many different interests, have expressed the desire to work with us as we proceed with this legislation. And my hope is that we will proceed with this legislation.

I do not think it is possible to please everybody 100 percent; but I think that there is a wide recognition of the fact that the existing state of affairs is really not acceptable in something that is as fast-moving as telecommunications.

I am very delighted that my friend and constituent, Dalton Wright, is here, a man I have known for a long time. Mr. Cullen used the figure of speech, "underwater," to describe some part of the industry, but Laclede County is not underwater, at least to my knowledge, it is not underwater.

I would like to ask Mr. Wright just to elaborate a bit on your concern. You now believe that the local newspaper business is really an essential enterprise, as far as the community is concerned. And you have a particular fear about what is going to happen. What is that fear?

Mr. WRIGHT. In small town America, where I am from, the communication within a community concerning sports, Government, plans that the leaders of a community have, the vision and the accessibility of it within the local community is really paramount. It is one of the pillars. Schools, hospitals, all the other services are extremely important. But unless you have the flow of information, unless you have the structure of people who are familiar with the community, that can get that information and move it around, I think that without that, the glue that holds a town or community together is, in many cases, lost.

I cannot speculate on the metropolitan areas, as to whether or not that has caused any decline in the values of the community or not; but I can say that it is important to the values of the communities that I serve in Missouri, and the communities that the National Newspaper Association's members serve around the Nation.

And it is my strong belief that, if the precarious economic condition that a lot of small newspapers and broadcasters are in were upset by having other information providers come in from afar and put in, let us say, a bureau with one or two employees; and if that bureau with those one or two employees were subsidized by lower rates because of their company's size, as compared to the Lebanon Daily Record, or the Conway Chronicle, or the Richland Mirror, this could be enough to destroy the vitality of that smaller economic entity. And the result would be the loss of jobs, the loss of the culture within the community, and the loss of the most important information source that small town America finds, I feel, very vital.

Senator DANFORTH. So, you are concerned that, let us say, the Kansas City Star or, perish the thought, the St. Louis Post Dispatch, would establish a bureau in Lebanon, and that because of the fact that they are bigger, they would be given preferential treatment with respect to, to what?

Mr. WRIGHT. To rates, sir. And also access. Let us hypothetically say that, in one of our metropolitan areas that is suffering from a lot of water damage right now, of which the community of LeMay is in, if they did establish a bureau, and they had a relationship with a large telephone company, it would be very easy to accomplish that, and secure a very low rate. There would be certain synergy, certain economy of scale that we would have a hard time doing.

But again, I think that the economics of it, if you took all the employees that we presently have, for example, in Lebanon, MO, they would be replaced by a much smaller number of people; and I do not believe that they would be as efficient or as effective as our local reporters and advertising people.

Mr. CULLEN. Senator, may I just comment for a second? From the RBOC perspective, we view this dilemma as an excellent business opportunity for both of us, for all of us. Because with the kind of two-way fiber wireless coax, whatever network that we envision, we can offer interactive, broad band, what we call point cast capability, that is part of a common carrier obligation that every RBOC has; to offer every small paper, or information service provider, local delivery capability.

In fact, ultimately it can be customized information delivery to your home, to everyone else's home, based on your needs. Unlike the broadcast capability that we see today, in cable for example, where we are looking at proprietary networks, and if they decide in a cable industry, if they decide to deliver the St. Louis Post Dispatch, they can do that to the exclusion of everyone else; we cannot.

So, we view this as an opportunity, not only in the newspaper industry but, for example, in the movie and entertainment industry, just to pick another example; where we can be an alternative distribution capability, so that people do not need to leave their homes, get in a 4,000-pound metal box, drive to a store, pick up three movies, watch one, and pay a late fee 2 days later.

We can offer the movie and entertainment industry the same sort of point cast delivery efficiently, economically, in fact we are beginning to trial it right across the river in northern Virginia.

Senator DANFORTH. Are you concerned about this, Mr. Wright?

Mr. WRIGHT. No, sir. Our concern is not the highway. Our concern is access to it, and the rate at which we get access.

Senator DANFORTH. Access to it. And do you think that your concerns can be addressed in this legislation, without undoing the basic point of the legislation?

Mr. WRIGHT. Yes, sir. I think you can turn right there to the gentleman to your right; he has a good solution in the language of the Burns amendment. And I believe that the language can be worked, and I think it is workable. I certainly do, Senator.

Senator DANFORTH. Thank you very much. Thank you, Mr. Chairman.

Senator INOUYE. Senator Burns.

Senator BURNS. I want to ask the Senator from Missouri if he wants to negotiate or renegotiate the Missouri River Compact now. [Laughter.]

I do not know if the folks who come to hearings, understand it or not, but for years and years we have a running gun battle with the folks in Missouri. Of course, I was raised down there, and Joe sends his regards, by the way, up at Gallatin. But we have had a running gun battle about how much water we should take out of the reservoirs in Montana in order to float them silly boats down there in the lower Missouri. So, we have got that going all the time.

Thank you very much for this panel. I am sorry I had to miss part of it, but, nonetheless, we have been through your testimony, and I had my able assistant back here writing down some questions that we would like to clarify just for the record.

Mr. Wright, I would start with you on the protection for the small information providers access. In other words, I think it is pretty clear if we get that language in there, then you would go ahead and support this bill?

Mr. WRIGHT. Yes, sir, I personally think that the National Newspaper Association would support that legislation.

Senator BURNS. Mr. Martz, you are aware that the RBOC's and the publishers, those negotiations are ongoing now?

Mr. MARTZ. Yes, sir.

Senator BURNS. Do you foresee the day when the publishing industry takes a position in favor of long distance relief?

Mr. MARTZ. Long distance relief, as I indicated in response to an earlier question, is something that is beyond the purview of our concerns. The entry of Telco's, the RBOC's into long distance is something that is better left to those who are concerned about that particular aspect.

I think it is ironic in a certain sense that as you look at the evolution of the RBOC's desires, they will end up back in doing everything that AT&T had done before it was broken up. And I think this underscores that as we get from where we are now to that point where there is total free competition not only in information services and infrastructure, but in the other areas as well, that in that transition period there are appropriate safeguards to get us from where we are today to then.

Senator BURNS. And, Mr. Smith, the resale of the local telephone services permitted as proposed in this bill, S. 1086, should we, in turn, permit resale of long distance services by the RBOC's? If not, why not?

Mr. SMITH. I would say, Senator, absolutely not. Because reselling of long distance service—two things. One, reselling of long distance service would make the RBOC's once again the gatekeeper of long distance service. They are proposing, for example, to mate up resale with something they call least-cost routing, which would basically put all long distance companies in the position of just being a wholesaler to the RBOC retailer.

Second, that would be only one step away from the RBOC's are actually asking for, which is full-scale long distance entry. You can depend that when resale—the day that resale is granted, they will say this is absurd allowing us only to resell when we have our networks all ready to go.

You will still have, in a case of resale, the opportunity to discriminate against the long distance carriers who do not give the RBOC the best deal on bulk rates, and to discriminate in favor of the long distance company who will offer the RBOC the best rates. And they will still have the only access to the local customer, the first mile and last mile of every call.

Senator BURNS. Mr. Cullen, would you like to comment on that from your perspective? Would inter-LATA relief divert your attention from the real job at hand; namely, local infrastructure modernization?

Mr. CULLEN. Senator, I think the two are totally consistent. We are not talking about separate networks. We are talking about investment in our network, in our territory in new technology. The network is transparent to the kind of messages it carries, whether it is voice or it is data or it is video. Whether it is intra-LATA or inter-LATA, it is all transparent.

The least-cost routing issue and resale that you mentioned, really, is boiled down to one simple issue. If you look at the AT&T pricing, every 6 months changing, you will find a phenomenon for the last 3 years for all the other carriers, where they mirror that, and AT&T serves as the umbrella. It is the phenomenon that I included as an attachment, described in the Merrill Lynch report.

So, there is a pricing structure that follows AT&T. Least-cost routing is just a software package that the regional companies have proposed we be allowed to do—a very small part of the inter-LATA business. And it allows each customer to figure out which routes, which prices, which price schedule, and which carrier will meet their need day to day, hour to hour, week to week.

It is tantamount to having an automatic travel agent explain the airline travel guide. Not very many people can thumb through that and find the right rate and time and destination and package.

Least-cost routing, and even resale, will do the same thing for consumers—enable them to get absolutely the best deal.

Senator BURNS. Well, then, is this issue is simply one of money, or is it the rich versus the wealthy, or the wealthy versus the rich?

Mr. SMITH. Let me respond to that and say something about long distance prices. You undoubtedly have heard of Friends & Family. You have probably heard of 1-800-COLLECT. Those are both programs that MCI has implemented recently. And, by all accounts, those types of programs represent a significant departure and, in fact, the beginning of a price war in long distance prices.

Long distance prices have gone down over 50 percent in real terms since divestiture, and that is a fact. Of course, the smaller carriers do price below AT&T. They must in order to get the business. But, to give you just a couple of statistics, local rates have gone way up even as long distance rates have been cut in half. It costs 90 percent more to make a 1-minute call from Manhattan to Montauk, Long Island, 75 miles away than to California, 2,900 miles away. And to take the long distance side, in 1984, and closer to home, in 1984, AT&T charged $2.70 for a 5-minute daytime call from Washington to Los Angeles. Now, that charge is not $2.70, it is $1.25. And the smaller carriers charge, as everybody agrees, even less.

That is real competition.

And when you compare the history of the what the long distance carriers have done, prices going down, down, down, and the local exchange carriers' prices going up, up, up, it is pretty clear that it is not the rich versus the wealthy. It is just the rich.

Mr. CULLEN. Senator, may I just comment briefly? I do not want to get into quarrels here, but I guess we are already into quarrels, so I will continue it. [Laughter.]

Senator BURNS. We will call it a discussion for today.

Mr. CULLEN. A discussion, yes, sir, dialog.

Every dollar of long distance reduction since 1984 is more than accounted for not by anything the long distance carriers have done, but by the absolute reduction in prices charged by the local exchange companies to the long distance carrier to carry their messages.

This is not a relic of regulation. This is high-value service, and a very effective, very low price we charge the carriers. We have reduced those charges every year by several hundred million dollars. That is what is being passed along in long distance rates.

In fact, as I said earlier, it is not even all being passed along. We reduced our rates to AT&T by $250 million on July 1. They did not pass along a penny of that reduction to their customers in lower rates. So, the issue here is an important issue, and it has to do with why long distance rates came down, who is accountable for it and what role the local exchange companies played in supporting that.

Mr. SMITH. Can I just add that I would welcome an analysis of the ratio of access rate reductions to long distance price decreases. I am confident of the way it would turn out, because last summer, Chairman Markey on the House side held a hearing and the witness, who was a professor from Berkeley who was under contract to the regional Bell companies, agreed that access rate reductions accounted for a significant part, but far from all, of long distance rate reductions. So, it is simply not the case.

I cannot speak for AT&T, I do not represent AT&T. But long distance prices have gone down by more than simple access rate reductions, and I would welcome an analysis to that effect.

Senator BURNS. Well, as we start making policy on this, and I think with the information that you have given us today we can sort of sift through it and get where I think the chairman wants to go with this legislation, and I think where we want to go as policymakers, and make some things happen in telecommunications. And all of this has to be considered as we go along the way before we can make any kind of a decision.

Mr. Cullen, I notice that you did not predominantly mention the cable Telco relief in your oral testimony. Do you feel pretty confident about your lawsuit in Federal court, or was it just an oversight? [Laughter.]

Mr. CULLEN. No, we do feel confident, based on the oral argument, but we also support the direction here. And it is not so much that this would be an excellent business opportunity for us, it is that we are listening to consumers. And consumers understand why their cable TV service is out much more frequently than their telephone service.

So, if we polled consumers and asked them whether they want a choice, this would be a slam dunk answer from them, and we are reflecting that. Our only concern in this particular area of the bill is that there seem to be a number of, I will say, bureaucratic requirements before we can get into the cable industry.

And so, except for that, we certainly do support our entry into cable entertainment, video services, programming, and content business. And I may just comment that we heard earlier discussion which said we are investing so much today that we really do not need that in order to provide an investment. That is simply not true. We are investing a considerable amount today. But, to make that investment more attractive to meet the needs of our customers for full-service providers, both long distance, cable and then, of course, manufacturing, will be excellent incentives for the regional companies.

Senator BURNS. Well, we look forward from—we will probably have other questions. And I would ask the chairman if it would be all right if we can correspond with these people and have some questions for the record in response as we move this along? Because, philosophically and basically, we have been talking about this a long time. We were talking about Telco entry and the revitalization of our information infrastructure—sort of like the old song, you know, we were country when country wasn't cool. And so we started down this road 3 years ago.

I think what we have done is we have elevated the level of dialog, Mr. Chairman, to the point where I think we can start making some decisions. I appreciate the efforts that the chairman has made on this on the way we can work on it, to make it good for everybody we would hope. And I do not want the Government to get in this business of picking winners and losers, I want this Nation to win. That is who I am looking for, and the people of my State of Montana.

So, I thank the chairman very much.

These folks probably want to go to lunch. I am just guessing, of course.

Senator INOUYE. We have passed the lunch hour. [Laughter.]

Senator BURNS. Well, they can make it to dinner now.

Senator INOUYE. I would like to thank all of you very much. After conferring with the author of this measure, Mr. Danforth, the record of the proceedings will be kept open for 2 weeks, until July

28. This should indicate that the subcommittee is very serious about this bill.

In 2 weeks, we will begin assembling ourselves and preparing ourselves for the markup. We hope to have a markup of this measure before the August recess. So, may I, in the strongest terms possible, invite you to submit supplements or addenda if you do have any to your testimony, and to keep close tabs with the staff. The staff has been geared and prepared to expedite this matter.

So, with that, I thank you all, once again, for your patience and for your assistance today.

Thank you.

[Whereupon, at 2 p.m., the hearing was adjourned.]

S. 1086, THE TELECOMMUNICATIONS INFRASTRUCTURE ACT OF 1993

WEDNESDAY, SEPTEMBER 8, 1993

U.S. SENATE,
SUBCOMMITTEE ON COMMUNICATIONS OF THE
COMMITTEE ON COMMERCE, SCIENCE, AND TRANSPORTATION,
Washington, DC.

The subcommittee met, pursuant to notice, at 9:30 a.m., in room SR–253, Russell Senate Office Building, Hon. Daniel K. Inouye (chairman of the subcommittee) presiding.

Staff members assigned to this hearing: John D. Windhausen, Jr., staff counsel and Kevin M. Joseph, professional staff member; and Regina M. Keeney, minority senior staff counsel, and Mary P. McManus, minority staff counsel.

OPENING STATEMENT OF SENATOR INOUYE

Senator INOUYE. First, my apologies for being late. This morning we consider the bill, the Telecommunications Infrastructure Act of 1993, S. 1086.

I am pleased that we have so many distinguished witnesses this morning prepared to offer their comments on this bill. And I am encouraged by the strong interest and response shown by the parties who have testified and submitted comments so far.

We received a number of suggestions for improvement to this bill after the first hearing, and I want to encourage everyone to continue to work with Senator Danforth and the staff on this matter as we move along this process.

I would like to take a minute to summarize what I learned from the testimony delivered at the first hearing on July 14. First, I think all of us will agree that there is virtually unanimous agreement that the Congress has a special responsibility to enact legislation to upgrade the Nation's telecommunications infrastructure in this Congress.

Consumers, doctors, teachers, bankers, manufacturing companies, telephone companies, and many, many others all support legislation to promote advanced telecommunication services and capabilities. And Congress should not and will not shirk its responsibilities in this area.

Second, there is wide agreement that there should be greater competition in the market for communication services. Competition is the best means to promote infrastructure development.

Competition has already been introduced in the markets for telephone equipment and information services, and in long-distance services, with outstanding results. Prices have fallen, innovation has risen, and consumers and the economy have profited.

In part because of this experience, Congress last year enacted a landmark cable bill whose primary objective was to promote greater competition to the cable industry. Even the telephone companies who once decried the growth of competition for local telephone service have acknowledged that such competition is inevitable.

Ameritech, one of the seven regional Bell operating companies from whom we will hear this morning has already filed a proposal to open up this network to permit greater interconnection by competitors.

The Rochester Telephone Co., Nynex, Bell South, and Pacific Telesis have also announced plans to open their markets to competitive forces. In short, experience has shown that competition benefits consumers and industry alike. Indeed, if not for the divestiture of AT&T in 1984, many of the companies represented at this hearing would not exist.

A third lesson I believe all of us have learned, however, is that competition is not to be confused with deregulation. As the FCC knows all too well, the removal of barriers to competition in any market does not guarantee that a truly competitive market will immediately exist.

The effort to allow competition to develop at times requires greater regulatory scrutiny than ever to ensure that dominant providers of certain services do not take excessive advantage of their market power to thwart the growth of competition. Mr. Chairman, your comments, please.

OPENING STATEMENT OF SENATOR HOLLINGS

The CHAIRMAN. Good morning. I appreciate, Mr. Chairman, your chairing this second hearing on S. 1086, the Telecommunications Infrastructure Act of 1993. I also welcome our witnesses today and thank them for assisting the committee in this important debate.

As I said at the last hearing, I support competition. However, I do not support wholesale deregulation. Telephone service providers must operate with a view toward the public interest, convenience, and necessity. Competition in this industry must not be uncontrolled, unfair, or detrimental to the consumer. I agree with Chairman Inouye that these issues must be looked at, especially when the structure of the industry continues to change at a tremendous pace. The changes that have occurred since the last hearing are astounding—AT&T has announced a merger with McCaw Cellular, the largest cellular company in the country, and a Federal court has overturned, as a violation of the first amendment of the Constitution, an act of Congress prohibiting telephone companies from providing cable service.

We as policymakers have a delicate balance to achieve between imposing necessary regulation and allowing the benefits of competition. With regard to telecommunications regulation, these questions impact every one of our constituents everyday. I will study the record of these hearings carefully and continue to monitor developments in the industry. I look forward to the testimony of the witnesses today.

Thank you, Mr. Chairman.

Senator INOUYE. Thank you, Mr. Chairman. We heard from a number of witnesses at the first hearing, and we will hear from several more today. We believe that regulation can at times be essential to allow a truly competitive market place to develop.

The purpose of this hearing is to build on the lessons of the first hearing and to work more closely to develop a consensus on this bill.

This hearing will also explore a number of recent developments and new issues that the committee has not yet addressed. For instance, I am certain that AT&T's acquisition of McCaw, the court ruling overturning the cable telephone ban, and the Bell company's desire to enter the long-distance business will receive a good bit of discussion this morning.

Several other witnesses have concerns about S. 1086 the committee has not yet had an opportunity to consider. And I wish to thank in advance the witnesses for their efforts to educate the committee.

Before we call upon the witnesses, I would like to announce that we have a long day planned. And I wish to assure those who have been invited to testify that all of your issues are important to the committee and we will give you our full attention.

I would like to now call upon the author of this measure, Senator Danforth.

OPENING STATEMENT OF SENATOR DANFORTH

Senator DANFORTH. Mr. Chairman, just two brief comments. And thank you very much. Thank you for your leadership and your diligence in holding these hearings. Two thoughts come to mind, and I think they both are increasingly clear, to me at least, and have become even more clear over the last month.

The first is that times are changing. And, in communications, times are changing very, very rapidly. The policies that existed years ago and the policies that exist today were devised in an entirely different age as far as communications are concerned and, therefore, those policies must be reviewed. Inertia is not sufficient in an age of rapid change.

And the second brief point that I would make is that just as nature abhors a vacuum, so do Federal courts abhor a vacuum. That is the lesson of the last month. And it is not particularly a new lesson in the field of communications.

And we have had this industry basically regulated by a Federal judge. And we have sworn for years that that is not right, that this is not something that should be left to the judiciary, that if policy is to be made, Congress should really take the lead in making public policy.

Again it seems that with changing times the vacuum is being created, and that vacuum provides the temptation for the Federal courts. So, for these reasons, I think that it is particularly urgent that we in this committee move forward and that we move forward very rapidly.

And I continue to hope that by the end of this session in this Congress we will be in very good shape to see legislation actually reach the statute books by the next session.

Senator INOUYE. Well, I would like to assure the author of this measure that I will do everything possible to bring that wish to fruition. Senator Pressler, do you have any statement?

OPENING STATEMENT OF SENATOR PRESSLER

Senator PRESSLER. Thank you very much, Mr. Chairman. I would like to commend Senators Danforth and Inouye for their leadership on this matter.

Let me say that I think those two gentlemen have a great deal of standing in the Senate and they can make this bill pass. It is going to need some changes, but I salute them for their leadership. They are going into a very difficult area.

Let me tell briefly of an experience I had recently. I met with the mayor and city council of Aberdeen, SD, and the first subject they brought up was that they needed an upgrade of the telephone company's digital switch in their community.

We are living in a era when the first subject brought up to a Senator is the need to improve the telecommunications capability of the town in order to attract new industry, utilize their hospital facilities in connecting with bigger cities, have their universities and educational institutions in touch.

So, people are very aware that there is an information superhighway. They want to be on it. We cannot have a two-tiered system. We live in an era when mayors of towns, in meeting with their Senators, ask first for help in getting an upgrade of the digital switch or the long-distance company point of presence. That shows the need for this legislation and for us addressing these complicated issues.

Universal service has always been the touchstone for this Nation's telecommunications policies. Both Federal and State policymakers have tried to ensure that the benefits of communications technology are available to all citizens.

As Congress develops policies to encourage further modernization of our telecommunications networks we should ensure that improvements will be shared by all. We should upgrade the entire network. We should not create a two- tiered system of haves and have nots.

S. 1086 is designed to stimulate private investment in our Nation's telecommunication networks by opening the local loop to competition. Recent history shows that competition can spur infrastructure investment.

Faced with stiff competition in the long-distance market, AT&T accelerated optical fiber deployment and wrote off billions of dollars in older plant. Similarly, the opportunity to provide new services has encouraged investment in new technologies.

Let me say that this is happening all over the world. I recently met with Bill Warrick of AT&T in Hong Kong and discussed their efforts to invest in China. But we must also remember the parts of our own country that are not on the superhighway. This is one purpose of this hearing today.

Mr. Chairman, I have a longer statement to submit for the record. I care about the quality and the diversity of services available in small States such as South Dakota. If access to the information highway becomes an essential means of fully participating in society, we may need to redefine universal service.

I mentioned that I have been working with leading citizens in my State to bring them state-of-the-art telecommunications facilities.

Our small towns that are developing new businesses and new industries need an up-to-date digital switch and point of presence to get on that superhighway.

Our smaller universities, colleges. and hospitals need the same. So, Mr. Chairman, I want to join in this effort. I commend you very much.

[The prepared statement of Senator Pressler follows:]

PREPARED STATEMENT OF SENATOR PRESSLER

Mr. chairman, thank you for holding this second hearing on S. 1086, the Telecommunications Infrastructure Act of 1993. This bill represents a significant initiative designed to accelerate private investment in our nation's telecommunications networks. I commend you and the distinguished Raning Member of the Commerce Committee, Senator Danforth for your outstanding leadership. S. 1086 is a tribute to your bipartisan cooperation and tireless efforts to craft legislation in a complicated area. It's a good start.

Infrastructure surely is the "buzzword" of the 1990s. Traditionally, the term primarily was used to describe our transportation networks—moving people and products. Today the word "infrastructure" more often refers to cable, phone and satellite networks transporting information and ideas at the speed of light. Advanced telecommunications networks promise to change the way we work, educate our children, provide health care, and spend our leisure time.

Who will deliver these new services? Every day seems to reveal another strategic alliance between multi-billion dollar companies seeking to spearhead the delivery of services over the information superhighway. Earlier this year US West announced a $2.5 billion partnership with Time Warner, Inc.'s entertainment and cable businesses. More recently, AT&T announced plans to purchase McCaw Cellular Communications for $12.6 billion in stock. With all this economic activity, some may question whether Congressional action is needed. I believe it is.

Technological change is outpacing our existing regulatory framework. Policies designed for disparate specialized industries do not seem to make sense if multiple providers can offer competing services. Yesterday's rules may impede delivery of tomorrow's services. Congress has an opportunity to frame a policy for accelerating investment in the information superhighway. Our challenge is to ensure all Americans have access to an on-ramp.

Universal service always has been the touchstone for this nation's telecommunications policy. Both federal and state policymakers have tried to ensure that the benefits of communications technology are available to all citizens. As Congress develops policies to encourage further modernization of our telecommunications networks, we should ensure that improvements will be shared by all. We should upgrade the entire network. We should not create a two-tiered system of haves and have-nots.

S. 1086 is designed to stimulate private investment in our nation's telecommunications networks by opening the "local loop" to competition. Recent history shows that competition can spur infrastructure investment. Faced with stiff competition in the long distance market, AT&T accelerated optical fiber deployment and wrote off billions of dollars in older plants. Similarly, the opportunity to provide new services has encouraged new technologies. Bell telephone company joint ventures in the United Kingdom pioneered delivery of telephone services over cable systems.

Although S. 1086 relies primarily on competition to bring new technologies and services to the consumer marketplace, it anticipates the limitations of market forces in particular circumstances. First, the proposal addresses the need to ensure that competition does not endanger universal service. All carriers would be required to contribute to the universal availability of affordable phone service. Second, the bill acknowledges that competition alone may not bring the benefits of new technologies to rural markets. Third, the proposal anticipates that telephone company dominance in the local telephone market may require certain safeguards.

We need to examine carefully what effect local competition will have on universal service and the availability of advanced telecommunications services in small cities, towns and rural areas. I am concerned with access to affordable phone service. I

care about the quality and diversity of services available in small states like South Dakota. If access to the information superhighway becomes an essential means of fully participating in society, we may need to redefine universal service.

I have been working with leading citizens in my state to bring state of the art telecommunications facilities to Aberdeen and northeastern South Dakota. This area of the state has been losing population in recent years. What can be done to reverse this trend? It's a two word solution: economic development. Community leaders have been working hard to attract new businesses. Their efforts make dear that an advanced telecommunications infrastructure is a necessary ingredient for new economic opportunities. I look forward to working with Members of this Committee to develop a sound policy framework for bringing telecommunications advances more rapidly to all Americans.

Senator INOUYE. Thank you, sir. Senator Burns.

OPENING STATEMENT OF SENATOR BURNS

Senator BURNS. Thank you, Mr. Chairman. I too want to congratulate you for bringing this issue forward. As you know, we have been in this issue ever since the first day I walked into the U.S. Senate.

And if you go home and talk—like Senator Pressler said about South Dakota—all you have to do is go home and talk to your smaller towns and cities, because all of Montana is rural—that this infrastructure, this part of the infrastructure is vital to the survival of our rural communities.

We cannot attract new and adventurous ideas to our rural areas unless we have this infrastructure, the ability to move information with the speed of light, the way we educate our children, the way we deliver health care rurally, and also what it creates in our inner cities.

Those areas that have the tiny tax bases are utmost in this Congress. And Congress since the last 4 years has sort of sat on the sidelines and allowed this to happen.

The developments of the last 30 days, let alone the developments of the last year are very indicative of what is to happen in telecommunications in this country, and what I would like to think started with me 4 years ago.

But there are a couple of questions about this piece of legislation, and I will offer some amendments to it. I think there are two issues that should be addressed in this piece of legislation and that is, the provision for access of information networked to all America, all Americans, no skimming, including rural and inner city urban areas and more importantly, existing information providers.

And also, safeguards, or fireballs if you will, to preclude anticompetitive activity by dominant carriers. I think those are the two areas that we have to dwell upon as we hear from our witnesses today.

Mr. Chairman, I have a markup in the Appropriations Committee starting right now. And I have some questions, especially for the first panel. So, I will keep my remarks to that.

And thank you for your leadership in this area, and Senator Danforth. I look forward to working with you, because I think there is no other piece of legislation that is any more important to this country, and especially this committee, than this piece of legislation right here.

I intend to cooperate fully with you to make sure it gets enacted. And I thank you. And I apologize if I have to get up and leave and come back. Thank you.

Senator INOUYE. Senator Stevens.

OPENING STATEMENT OF SENATOR STEVENS

Senator STEVENS. Mr. Chairman, I support the concept of this bill, because of its strong emphasis on universal services. I would remind the committee that when I came to the Senate the telephone service in Alaska was—it was presented by the U.S. Army.

Mr. Chairman and I have worked together now for a series of years to assure that we extended not only modern communications but computerative communications into Alaska.

And I have supported this bill primarily because of revisions on page 11. I would urge all of you to take a look at it, beginning on line 11.

> The States may prescribe regulations implementing paragraphs one through five for intrastate services so long as such regulations are not inconsistent with those prescribed by the commission. Universal service, the role of telecommunications carriers. All telecommunications shall contribute to the preservation and advancement of uniform services. The States, in coordination with the Commission shall ensure the preservation and advancement of universal services. Assistance to certain persons in administering this subsection the States and Commission shall have with regard to actually assisting in individuals or entities that cannot afford the cost of the telecommunications service or equipment.

My footnote would be to ask that you put in the record here the article from The Wall Street Journal yesterday, "The Lessons of AT&T's Cellular Move."

And I raise the question for you: How can we assure universal services if the rights of the States to regulate, to assure that universal service is achieved is obliterated by this legislation.

I will support the legislation, but I urge you to keep in mind that rural America requires that the States maintain a role in assuring telecommunications delivery on a uniform basis. Thank you.

Senator INOUYE. Thank you, sir. Senator Ford.

OPENING STATEMENT OF SENATOR FORD

Senator FORD. Thank you, Mr. Chairman. I, too, want to compliment you and Senator Danforth for bringing this legislation forward. And I have listened to my colleagues talking about small town, rural America.

And we have a witness this morning, Billy J. Ray, who is a friend of mine. I know him very well. And I hope you will listen to the Glasgow story, because it is, I think, significant for small communities and what they have been able to do. I think he brings something that we all want and something that we should listen to.

And second, Mr. Chairman, after Senator Danforth moved all of those who stood in line for this hearing away from his office and downstairs, I have them all. [Laughter.]

And I think you could not have delivered a message this morning anywhere in Washington, DC, because they were all from the second floor to the basement, until it started raining outside.

They were even standing in line carrying their bicycle wheels. I don't mind that. But Senator Stevens and I have been innovative.

And so we think maybe next time we have a hearing like this, and all the money is going to be so that we are going to auction off space. [Laughter.]

And then we will try to reduce the deficit. [Laughter.]

Mr. Chairman, I look forward to this next auction. And we will see how many messengers come and bid. [Laughter.]

Senator INOUYE. I think that is a good idea. Senator Lott.

OPENING STATEMENT OF SENATOR LOTT

Senator LOTT. I am anxious to hear the panels, and so I would like to defer at this time.

Senator INOUYE. Senator Mathews.

OPENING STATEMENT OF SENATOR MATHEWS

Senator MATHEWS. Mr. Chairman, I have no opening statement this morning, except to compliment the committee chair and the ranking member in bringing this matter to the discussion stage, where it seems to me that we are deploying this technology where we better be careful that we do not throw the baby out with the bath water.

Developments are coming about so fast that is it difficult to keep up with where we are and why we are where we are. And I just look forward to participating in the discussion and helping to shape this legislation. Thank you.

Senator INOUYE. Thank you, sir. Senator Breaux.

OPENING STATEMENT OF SENATOR BREAUX

Senator BREAUX. I, too, commend you, Mr. Chairman, and Senator Danforth, for bringing the legislation to the attention of all of the Members, and obviously all of the affected industries that are out there.

I have looked over the testimony. It sort of reminds me of—the theme perhaps could be, for the hearing, something along the lines of let me in yours, but stay out of mine.

I think what we are saying is, all of the witness feel that they are being restricted, and they want to get into other areas that they are not allowed by congressional regulations to get into, but they do not want anybody to come into what they are doing.

That is the complaint we have been having and will continue to have, I guess, since 1934. I suggested a long time ago a telecommunications commission, which would really look at all of this and make a nonpolitical recommendation based on size and technology, and then let Congress make the political cut after we had the recommendations from a panel of experts as to what is the best for technology and science and industry of communications.

And then take that recommendation and then make the societal and social and political decisions that we have to make. I am just really fearful that what we are doing is giving a little, but taking a little.

And we lose the overall theme of what is best for communications policy. And that is the real challenge we have. I commend you and look forward to the witnesses' presentation.

Senator INOUYE. Thank you. Senator Packwood.

OPENING STATEMENT OF SENATOR PACKWOOD

Senator PACKWOOD. Mr. Chairman, thank you. I have a certain sense of deja vu as I look at this problem. I think this is going to be an ashes-to-ashes situation.

I think in 10 years we will be where we perhaps were in 1927 or 1928 when we passed the Federal Ratings Act to allocate frequencies because they were overlaying each other. And that is all we did.

I can picture universal service, extraordinary competition within areas, and all that we will have to do is to make sure that frequencies do not overlay each other. And maybe some modicum to guarantee universal service.

I can even picture science advancing far enough on the fracturing of the frequencies that we might not have to allocate them other than to protect them as you would a copyright or a patent or a trademark, that they would be so numerous.

Some of the issues we are going to consider are old issues. We have been around and around the track on the telephone entry into cable. We have been around and around the track on the modified final judgment in manufacturing and what should we do.

We have got some new issues—opening up the local telephone company into competition. We have not faced that before. We have got some minor long-distance issues in this bill.

But I hope we all are agreed on where we would like to come out. I mean, we would like to have universal service and universal competition and that we do not pass a bill that may deter rather than encourage that.

This is an opportunity for a once in a decade bill. And I hope that we do not rush through it too quickly, and make a mistake, and slow down what I think is inevitably coming.

I might say, the Oregon Legislature this year—and I have to give them credit—passed a bill that opens up the local telephone service to competition. And a competitor does not have to provide it in the entire service area.

He can pick out a competitive zone. But then, in that zone, the local telephone company can go head to head on price, does not have to file its rates or services with the public utility commission, and we are going to have very quickly some head-to-head competition with the local telephone company.

And that agreement was reached between the Oregon Public Utilities Commission, AT&T, US West, MCI, GTE, and a company called Electric Lightwave, which is a competitive access provider. They all agreed.

And so we are going to be into the competitive local telephone communication systems service soon. Would, that we could reach the same agreement among the same parties here.

Senator INOUYE. Thank you. Before I call upon the first panel, I am going to advise my colleagues that if I look a bit flustered, I hope you will understand. I am presently testing out a high-technology communications system, a hearing aid. [Laughter.]

And I am hearing sounds that I have never heard. [Laughter.]

Senator BREAUX. Is that long distance or just local service? [Laughter.]

Senator INOUYE. But it does help. So, with that I would like to call upon the first panel. If I am speaking too loud or——[Laughter.]

The chairman and chief executive officer of AT&T, Mr. Robert E. Allen; the chairman and chief executive officer of Ameritech, Mr. William L. Weiss; the chairman of the Electronic Frontier Foundation of Cambridge, Mr. Mitchell Kapor; and the vice chairman of Corning, Inc. of New York, Mr. David A. Duke.

Needless to say, we have a most impressive panel, all heavy hitters. I would like to advise the panel that this is the best attendance we have had of this subcommittee, I believe, in the last 10 years. So, I hope you will realized that we all consider your testimony to be extremely important. May I first call upon Mr. Allen?

STATEMENT OF ROBERT E. ALLEN, CHAIRMAN OF THE BOARD AND CHIEF EXECUTIVE OFFICER, AT&T

Mr. ALLEN. Thank you, Mr. Chairman. First of all, Mr. Chairman, I would like to commend you and Senator Danforth for introducing S. 1086.

It recognizes the real importance of the telecommunications infrastructure to the Nation, second, the importance of competition in enhancing the capabilities and the value of the infrastructure, and perhaps most significant, it provides a framework for testing the competition in the one market in the telecommunications industry, that has remained an entrenched monopoly, and that is the provision of local phone service.

Although my file testimony outlines some additions and suggests some changes that we believe are important, I applaud the thrust of the bill and the lion's share of its provisions.

As we contemplate the future of this important industry, we envision an information age, products and services, that revolutionize the way the we live, dramatically changing the way we deliver education and health care, the way we run our businesses and the way we entertain ourselves.

But we cannot be starstruck by the galaxy of possibilities that technology promises. We have to deliver the full benefits of information technology to the Nation. We have to deliver the vision of the Information Age.

Turning this vision into reality means even more changes in the telecommunications industry. The experience of the last decade in the long-distance business gives an important guide for the future.

Competition has transformed the long-distance business. While competitive policies have been smoldering for years, the breakup of the Bell System ignited the marketplace. And when the Bell System's local monopoly was separated from long distance and manufacturing, the competition really began to flourish.

Since then, competition in the long-distance business has created wider choices, more innovation, higher quality, and lower prices. I do not think the American public would want to turn back the clock.

Long-distance communications superhighways are world class and getting better. And the reason is simple. It is competition. The time has come to see if we can get the same benefits by introducing competition in the local telephone business.

There is no local phone competition now. Virtually every person in this country must answer "No" to the following two questions: Do you have a choice of companies for local telephone service, and can you switch to another company to get better service or better prices?

You compare that to long distance where 16 million customers switched long-distance companies last year, in 1992, where 9 or more companies offer service in 45 of our 50 States, where 81 companies offer service in 4 States, and customers in virtually every household in every State, rural and urban, can chose among several companies.

Much is being written about what telephone companies, cable companies, and others may be capable of doing and what they may be permitted to do. But no one should confuse what might be and what could be with what is.

What exists now is essentially what existed in 1984, a local telephone company monopoly. Today there are too many barriers to entry in the local telephone market and too many obstacles to testing what the real competition can develop.

The FCC and a few States have removed some of these barriers. But it will take the kind of provisions contained in S. 1086 to provide the freedom and the incentives for new carriers to try to make it in the local telephone market.

This might be a good time to set the record straight on some misperceptions about cellular calls and the proposed AT&T-McCaw merger. And I think this chart will help make it clear.

Ninety-nine percent of all cellular calls go through the local telephone company facilities. Only the tiniest fraction, mostly calls between cars within a cell—as you see there on the left or on the right at the bottom, only a small fraction avoid that bottleneck called the local telephone network.

In short, without the local telephone company, there is virtually no cellular service, either provided by the wireline or the nonwireline carriers. For people who claim that the AT&T-McCaw merger would cut out the phone companies do not know what they are talking about.

It is not true today and it will not be true for the foreseeable future, if ever. AT&T has no designs—I repeat, no designs—on getting back into the local telephone exchange business.

But the local telephone companies want to get into the long-distance business, even though they are still monopolies, and control access to the customer. When real competition breaks the RBOC's bottleneck, concern about monopoly abuse should end. And the RBOC's should then be free to enter the long-distance market.

Competition has made AT&T a much better company, more efficient, more innovative, more responsive to customers. Not perfect by a long shot, but much better. And I dare say that competition would do the same for local telephone companies.

The companies would benefit. The public would benefit. And I think the Nation would benefit. The economic vitality of America will depend in part on the capabilities and the efficiency of this communications and information infrastructure.

We have learned that competition and monopoly do not mix. And we have learned that competition is the best safeguard for customer interest and the best guarantor of innovation and value.

The task ahead is to apply these lessons throughout the communications industry for the good of the industry and for the good of the Nation. And that is why, subject to the important modifications described in my written testimony, AT&T is prepared to support S. 1086.

Thank you, Mr. Chairman.

[The prepared statement of Mr. Allen follows:]

PREPARED STATEMENT OF ROBERT E. ALLEN

Mr. Chairman and Subcommittee Members: My name isRobert E. Allen. I am Chairman of the Board and Chief Executive Officer of AT&T. I appreciate the opportunity to comment on Senate Bill 1086, the "Telecommunications Infrastructure Act of 1993," and on the broader issues that underlie it. We are encouraged that S. 1086 has embraced the pro-competitive message that has been the theme of evolving telecommunications policies in this country for the past 25 years. In particular, we believe that competitive forces can spur the delivery of the most advanced voice, data and multimedia services to America's citizens in the shortest time frame, without burdening taxpayers or consumers with unnecessary costs. A competitive market structure throughout the telecommunications industry will attract private investment and entrepreneurial activity, at a pace and degree that the marketplace sets and demands. We therefore applaud the Subcommittee's efforts. Subject to the important revisions outlined later in my testimony, AT&T is prepared to support the legislation.

For telecommunications to evolve into a fully competitive marketplace, government action is needed at three critical stages. First is the role of government in formulating a vision for a competitive national telecommunications infrastructure. Second is the plan by which the private sector can be challenged (and encouraged) to fulfill that policy. Last is the need for government to step out of the way once effective competition has developed in the market, so that competition will continue to flourish unhindered by artificial government restraints.

The nation's past experience with competition in the communications equipment and long distance markets is a testament to the benefits consumers can expect from a competitive local service marketplace. The customer premises equipment market was the first and easiest to become competitive. Once the FCC determined two decades ago that other manufacturers' terminal equipment should be allowed to be used with the Bell System network, the options available to customers proliferated. Instead of having a single, expensive black rotary phone as in the past, homes today routinely have several phones (both wired and cordless), an answering machine, a computer with a modem, and maybe even a fax or cellular phone—all made by a variety of manufacturers, U.S. and foreign. Government no longer regulates the price of customer equipment or the terms of its sale; its only involvement is participating with industry to develop technical standards.

The U.S. telecommunications network equipment business has also become fully competitive. Before divestiture, the BOCs purchased their network equipment from their Bell System manufacturing affiliate. But once the Decree severed that captive relationship, new suppliers with fresh ideas entered the marketplace. The RBOCs now have the ability and the incentive to establish multiple sources of supply, which they have done, and to buy their equipment from the manufacturers that offer the best combination of features and price.

In the 11 years since the Decree was announced and new conditions such as equal access spurred competition in the long distance markets, there has been extraordinary growth there as well. Hundreds of new carriers have begun providing competing long distance services at lower prices. Thousands of miles of fiber optic cable have been put into service across the country by a dozen different carriers. New, advanced technologies have been deployed in the long distance networks with greater speed and urgency.

At the same time, foreign product and service markets are not generally open or competitive and stand in sharp contrast to the U.S., which leads the global community in allowing foreign-based manufacturers and carriers the opportunity to compete. Other countries have been painstakingly slow in allowing U.S.-based carriers and suppliers comparable access to their markets. This raises important issues of global symmetry. While such international issues are beyond the scope of S. 1086, they will affect the ability of Congress and the Administration to unleash competitive forces for the benefit of U.S. consumers and the economy overall.

Within the U.S., AT&T believes that federal and state policymakers must now turn to the one segment of the telecommunications industry that has resisted competition: the local telephone exchange. Virtually every customer of local telephone service in the United States today must answer "No" to these questions:

Do you have a choice of carriers for your local telephone service?

If you want to switch carriers to obtain better local service or batter prices, can you?

Compare this experience to the choices available to long distance customers. In 1992, almost 16 million residential customers switched long distance carriers, at an average rate of 44,000 a day. Nine or more long distance carriers offer service in 45 states, 81 long distance carriers of far service in four states, and in every state every customer—rural as well as metropolitan—has a choice of multiple carriers.

If one looks back ten or twenty years, it is clear that the spur of competition has made AT&T a stronger company: more efficient, more innovative, quicker to market and more responsive to customers. A competitive marketplace is responsible for the fact that all long distance carriers and equipment manufacturers have invested heavily and successfully in a wealth of new technologies and services. Indeed, a robust system of interstate "information superhighways" has already been put in place by competing long distance firms. Competitive market pressures assure that these positive trends will continue.

Not surprisingly, in contrast, it is in the monopoly local exchanges where the infrastructure needs are greatest. There is no doubt that the LECs have already invested a considerable amount in upgrading their networks. And they will continue to have a key role in developing the local networking infrastructure. But until local telephone exchanges experience real competition, telephone companies will not have the marketplace incentives they need to invest more wisely, to become quicker and more innovative, and to be more responsive to the needs of their customers—including their long distance access customers like AT&T.

If over time, the collective efforts of policy makers and the industry are successful, and if true local exchange competition can be established, the benefits are potentially enormous. Foremost, it would assure—as it has in other telecommunications markets—the most effective and efficient deployment of new products, services and technologies that offer users wider choice at lower cost. It would lead to accelerated infrastructure development the growth of existing firms and the entry of new firms, producing significant new employment. All Americans would benefit as the capability of the nation's communications networks expands and the cost of using it drops.

Until recently, only the early stirrings of local competition could be felt in the industry's far reaches—where entrepreneurial firms such as MFS and Teleport have labored. But now there seems to be a serious collective interest on the part of policymakers at the federal and state levels and on the part of the industry to find out if, given the right set of regulatory safeguards and incentives, local competition has the potential to develop. We support these efforts.

The FCC has taken some small, but important first steps—by adopting rules ordering expanded interconnection opportunities in one segment of monopoly access: "transport"—the link connecting long distance carriers' networks to local telephone switches. These decisions, however, did not authorize competition in basic local service, the local loop or local switching. Some states, like Illinois, California and New York, have also begun to move in this direction on their own. These are important threshold steps, but much more is required to authorize full local service competition and tie these threads together into a coherent national policy.

That is where S. 1086 fits in. Your bill addresses some of the essential elements we believe are necessary before new carriers will be able and inclined to enter the local exchange market—such as eliminating exclusive franchises and exclusive local telephone company control of rights-of-way. S. 1086 is also on the right path in requiring interconnection and an unbundled set of service elements (such as local transport, local loop, switching and other service elements). The bill appropriately opens up resale and sharing of the local carrier's service offerings.

But the bill falls short by not fully addressing two important additional elements, which should be added to S. 1086's list of regulatory objectives:

Measuring Effective Competition. It is essential that, in creating regulatory and economic conditions that permit competitive entry, policymakers not assume that effective competition inevitably will occur. Many observers believe that basic local telephone service may be a natural monopoly—that the economics of the local exchange will preclude multiple firms from entering and investing to provide customers of local service with a genuine choice of providers.

We therefore must have agreed upon standards or metrics at the federal level that define effective competition—objective measurements that once met, clearly signal that the local market is competitive. If the objective test has not been satisfied within a reasonable time, it may be an indication that regulators need to try another path. It could also mean that further technological developments must occur before effective competition is possible. Most importantly, it would mean that the local exchange still is not benefiting from the operation of the competitive forces that are best able to stimulate new technologies and services. Only if we have a way to know whether local competition is developing—and, if so, how fast and how extensively—will we know if we are on the right course.

A method of measuring effective competition would also lay to rest the misperception being advanced by some in the industry: that the local exchanges already face substantial competition. In this regard, the "metrics" or statistics on local competition are stark: the LECs today exercise near total control over the local exchange and local exchange access services. The current facts are basically unchanged since the time of the MFJ: virtually all calls—local, short have long distance, cellular and interLATA long distance—traverse the local exchange carrier's network. The 1987 Triennial Review found that the RBOCs and other local exchange carriers served 99.9999 percent of customers and carried 99.9 percent of interexchange access traffic in their regions. These facts are almost unchanged today.

In 1992, for example, 99.86 percent of AT&T's access payments went to the local exchange carriers, with only 0.14 percent paid to the competitive access providers or "CAPs". Another measure of the local telephone company's impact on the long distance market is the level of these access payments: $14.2 billion for AT&T alone in 1992, or about 40 cents of every $1 of AT&T's long distance revenues. See Attachments A and B. If rules are changed at the FCC and in the states, the CAPs may well grow. But, today the local telephone monopoly remains extraordinarily powerful.

Cost-based pricing of basic network functions. There are still significant subsidies built into the local pricing structure, much of which is paid by the long distance carriers. The overall subsidy figure may not be nearly as high as is claimed, but it is significant enough that regulators need to begin working, now, to rethink how to promote the goal of universal service and protect rural and low income customers. A competitive marketplace should be no bar to the continuation of appropriate subsidies designed to promote universal service, provided they are imposed in an even-handed way, are based on financial need, and are administered effectively. In this regard, we believe that any subsidies should generally be provided directly to the end user, who will then be in a position to decide which competitive carrier should have his or her business.

Regulatory efforts to address the subsidy issue can and should proceed simultaneously with other efforts to open the local exchange to competition. All these efforts will take some time, and they should move ahead. Moreover, the LEC does not have to choose between maintaining the current subsidy system and losing profits. Rather, as regulators address this issue, the LECs have every incentive, in responding competitors, to drive the excess, unnecessary costs from their business. AT&T learned this lesson in spades from its competitors in the long distance business. Indeed, in the long distance market, government did not attempt to remove the historic subsidies built into long distance rates until a dozen years after competition was introduced. Even now, the goal of cost based pricing is elusive, and long distance carriers and their customers continue to bear the burden of substantial structural subsidies.

Let me put to rest directly any suggestion that AT&T's planned merger with McCaw will impact or otherwise diminish this local monopoly. The AT&T/McCaw merger involves a significant amount of stock, and it is certainly an important part of AT&T's effort to advance the day that wireless networks will connect customers, anytime and anywhere. But the more critical fact for purposes of this hearing—applicable to all cellular services today—is that some 99 percent of all cellular calls rely on the LECs' facilities. This makes the local exchange networks "bottleneck" monopolies for McCaw and other cellular carriers, just as they are for long distance carriers. AT&T and McCaw do not own any monopoly local exchange facilities now and will not as a result of the merger.

Add in the enormous disparity in price (with cellular service costing 400-500 percent more than local service); the relatively low calling volumes (under 1 percent of local calls are initiated over cellular phones); the pending questions about available spectrum and service quality—and it is doubtful whether cellular will be considered a true alternative to local landline telephone service any time soon. Surely it is not in the foreseeable future—as the RBOCs themselves have advised the Justice Department, stating:

"It has been suggested, however, that mobile services are converging with landline services * * * *Given the vast discrepancy in both price and present levels of penetrations, direct competition [with landline services] is nowhere near imminent.*"[1]

There has been media speculation, nevertheless, that the AT&T/McCaw merger might put AT&T in a position to bypass the LEC facilities in the future and link directly to its long distance network. The economics simply do not justify this today. As illustrated by Attachment C, ninety nine percent (99 percent) of the cellular long distance calls that AT&T carries use LEC facilities to connect the cellular customers to AT&T's long distance network. Moreover, even if direct links to cellular systems were someday established, this would not diminish the local exchange monopoly. It would only create an alternative connection for the relatively few long distance calls that originate or terminate on cellular networks, as compared with the vast number of calls from corded or cordless phones over the local telephone network.[2] Nor would it impact the long distance market itself. In terms of long distance calling, calls originated on cellular phones account for only one tenth of one percent (0.1 percent) of all long distance calling—hardly a dent in the LECs' pervasive monopoly power over the facilities long distance carriers require to access their customers.

An appropriate set of metrics will make it clear when these facts have changed—as they plainly have not so far—and when, finally, it might credibly be concluded that the local exchanges have begun to experience competition. If the local exchange evolves into a genuinely competitive marketplace, moreover, there will be a further benefit: regulation can be reduced as the local market becomes competitive and the MFJ line of business injunctions could be lifted when that competition becomes effective. The contentious and distracting debates that arise in this forum and in the court and FCC over its continuance would end. If customers have a real choice of local service providers, there would be no longer any need to exclude the BOCs from the long distance and manufacturing markets. But just as experience with monopoly required the imposition of MFJ as a pro-competitive safeguard, only the experience of real competition can justify their removal.

The RBOCs' reactions to the broad issue of whether the local exchanges should be open to competition have been interesting—and not at all uniform. Of the seven RBOCs, only Ameritech has submitted a proposal to the FCC that it claims will have the effect of opening its local networks to competition. They are to be commended for that proposal. But, even that proposal is flawed, particularly by proposing MFJ and other changes prior to the existence of effective competition. Local exchange competition should be the first priority to build the telecommunications infrastructure and keep the communications capabilities of this country on the cutting edge.

Instead of working toward competitive local markets, many of the RBOCs have opposed competition and waged a campaign for long distance authority or other concessions. Five RBOCs have asked the FCC to declare outright that it is in the "public interest" for them to enter the long distance business, despite the continued monopoly over local exchange services.

Ameritech conditioned its "willingness" to become subject to rules for local exchange competition on the simultaneous grant of long distance authority by the court. This indeed highlights the degree of RBOC control over the local exchange, for only a monopoly has the luxury of "permitting" competition. It also underscores the fact that regulatory changes in rules should not be confused with the existence of real competition.

The RBOCs' only justification for their MFJ proposal is their claim that the long distance market is not truly competitive and that, somehow, only entry by the RBOCs can make it so. This is patently false, and they know it. AT&T has responded to the RBOCs' inaccurate and misleading claims before this Subcommittee in the July 14th hearing by letter to the Subcommittee dated August 2, 1993, submitted for the hearing record. A copy of that letter is attached to this testimony, as Attachment D.

---

1 Report of the Bell Companies on Competition in Wireless Telecommunications (p. 185), dated October 31, 1991 and filed with the Department of Justice December 13, 1991 in support of the RBOCs' wireless waiver request.

2 Customers placed an average 5.2 billion daily minutes of local calls from conventional phones (wired and cordless) in 1991, compared with only 26 million minutes from cellular phones. In terms of percentages, local calling initiated from a cellular set was only ½ of 1 percent of all local calling. Moreover, cellular originates only 1/10 of 1 percent of all long distance calls.

In fact, the long distance marketplace is robustly competitive. When the MFJ was announced in 1982, AT&T was the only truly national long distance network; and while it was the envy of the world, its network principally used the old analog technology. The networks of NCI and the other competing carriers were collectively 1/20 the size of the AT&T network. These other carriers, moreover, lacked the advantage of AT&T's integration with and access to the essential local exchange networks of the RBOCs.

After the break-up, AT&T's competitors took the Decree to the bank and the bond market—literally—and raised the money to expand the capacity of their networks on the strength of the Decree's promise that the RBOCs could no longer abuse their monopoly power and favor their long distance affiliate—because while they still had the power, they had no affiliate. As a result, FOUR major digital fiber networks have been built (AT&T's, MCI's, Sprint's and Wiltel's). EIGHT other large, regional fiber networks have been constructed. Today our competitors' networks have the capacity to readily absorb the nation's long distance demand—even without AT&T. Indeed, almost 500 long distance carriers are using this enhanced, competitive long distance infrastructure to introduce new services at breakneck speeds.

AT&T has itself responded, investing over $20 billion on upgrading its own network—to the point that the AT&T network today is essentially a new, fully digital network. We were not directed to do this by government regulators, nor did we seek government funding. We made this investment, and wrote off billions of dollars of analog facilities, for one simple reason: to remain competitive by meeting our customers' expectations.

During this same period since divestiture, tariffed consumer long distance rates of the three major carriers have dropped about 55 percent in real terms when inflation is factored in. This has been due in part to FCC required changes in access pricing, but it also reflects other price cutting by long distance carriers, and the increased usage of new lower priced services that competition has stimulated. The record in long distance demonstrates a clear downward trend.

In this regard, the access charge reductions are themselves a reflection of competition in the long distance market. Lower long distance prices have stimulated increased usage which, by lowering the unit cost of access to the local network. However, this has led to still lower access charges. These lower charges in turn have led to lower long distance prices, stimulating further calling, and so forth. The declines have not continued in 1993 because the local telephone company access increases exceeded access decreases for the first seven months of the year.[3] However, this beneficial cycle can resume if competition is able to develop for the access services the local carriers provide.

The bottom line on the RBOC claim that the long distance market is not competitive, however, is simply a matter of common sense. One need look no further than the declining prices, the vast array of real choices customers have in selecting long distance providers and services, and the extent to which they exercise these choices, to know that the market is vigorously competitive. These facts both contrast sharply with the current state of the local exchange, and attest eloquently to the infirmity of the RBOCs' claims.

The lower prices and the impressive degree of infrastructure growth of the competitive parts of the telecommunications sector in the past decade dramatically attest to the success of this nation's pro-competitive telecommunications policy. Even more importantly, the vitality of our nation's telecommunications industry is driving underlying rates of U.S. productivity and growth, increasing jobs and the strength of our domestic economy, and contributing to our nation's competitiveness worldwide.

At the outset, I referred to the need for government to recognize when its competitive policies have been successful, and the marketplace can take over. The FCC's continued pervasive economic regulation of AT&T is one good example of a failure to do this. The need to remove these burdensome, counter-productive rules was most recently underscored by British Telecom's announced investment of nearly $5 billion in MCI and its acquisition of veto-power over MCI's major business activities. Surely there is no longer any legitimate justification for applying regulations to AT&T that have been deemed unnecessary for our competitors. AT&T will soon ask the FCC to re-examine these thirteen year old classifications in light of current industry competitiveness.

This same theme—the need for government to know when competition is working on its own—guides my next comments on the legislation. AT&T opposes the provisions of S. 1086 that would require regulators to apply the same interconnection rules to the long distance industry that it adopts for the monopoly local telephone

[3] See Attachment D, letter dated August 2, 1993 from T.H. Norris of AT&T to Sen. Inouye.

industry. There is no justification for this. The long distance market is already competitive; the local market is not. These rules could require competitors to share proprietary network technology. Moreover, there are no structural or other barriers to entry into the long distance marketplace. Any firm has the option of building its own network—as many carriers already have—or of reselling services over any of the competing long distance carrier's facilities—hundreds of companies do. This is not true of the local exchange, where the LECs will continue to control the local loop for some time.

AT&T also objects to the CPNI rules in the bill—which have both privacy and competitive implications. We have several problems, not the least of which is that there are no reasons to apply these rules to competitive long distance carriers, even if there are reasons to apply them to the local exchange. We urge that they be severed from S. 1086 and dealt with separately. Even if the rules were revised to apply only to LECs, they would still create serious problems for AT&T and other long distance and cellular carriers with respect to security of our network and the provisioning and billing of services. It would also endanger proprietary marketing information, by requiring its disclosure to our direct competitors.

The Subcommittee has taken on a significant policy challenge. We at AT&T are ready to help any way we can, working with you to establish the rules that will guide the telecommunications industry—and the nation—to a more competitive, more productive, and more prosperous 21st century. Thank you for inviting me to testify. I would be pleased to answer any questions you may have now or for the record.

ATTACHMENT A—AT&T LONG DISTANCE REVENUES—PAID TO LECS FOR EXCHANGE ACCESS, 1992

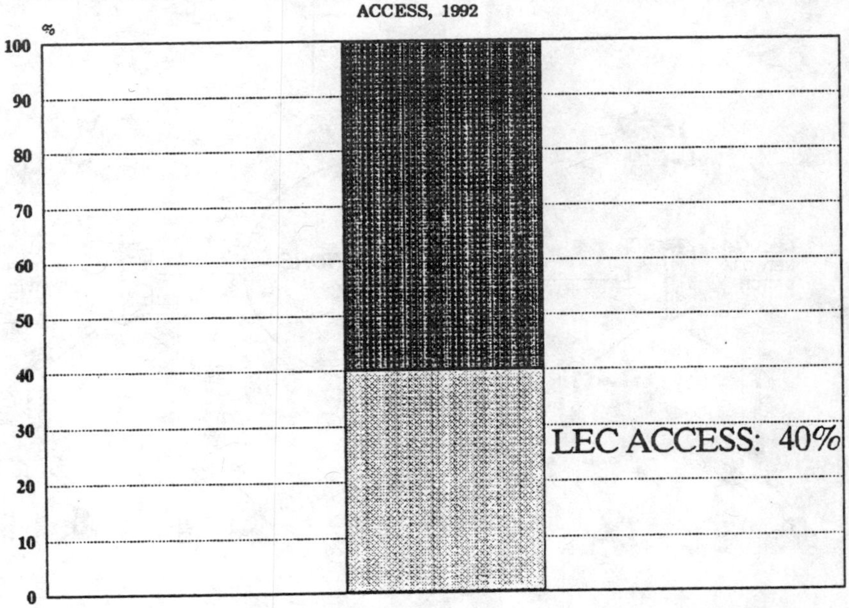

AT&T Long Distance Revenue: $35.5 Billion
LEC Access: $14.2 Billion
CAP Access: $ 0.019 Billion

ATTACHMENT B—1992 AT&T LOCAL ACCESS PAYMENTS—TOTAL: $14.2 BILLION

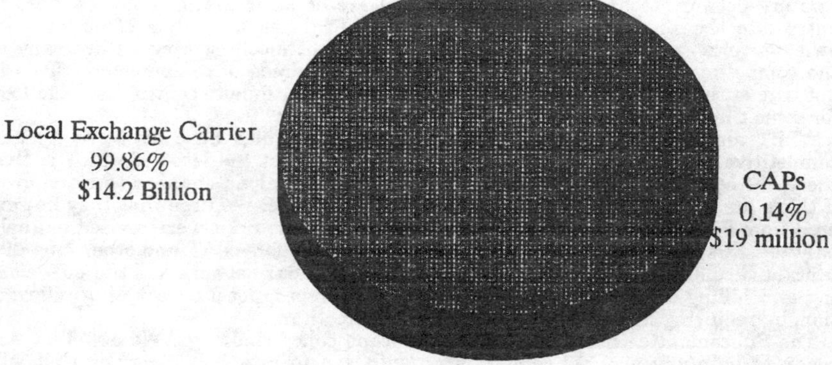

Approximately .40 of every long distance dollar is paid to local telephone companies or competitive access providers for local access. In 1992, AT&T's local access expense totalled $14.2 billion, with 99.86 percent to telcos and .14 percent paid to CAPs.

ATTACHMENT C—99 PERCENT OF ALL CELLULAR CALLS USE THE LOCAL TELEPHONE NETWORK

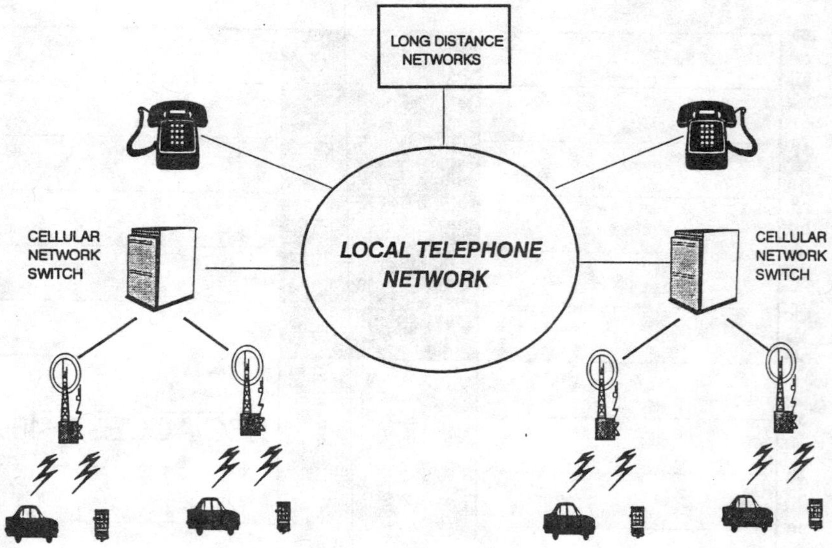

ATTACHMENT D—LETTER FROM TOM NORRIS, VICE PRESIDENT, FEDERAL GOVERNMENT AFFAIRS, AT&T

AUGUST 2, 1993.

The Honorable DANIEL K. INOUYE,
U.S. Senate,
Washington, DC 20510

DEAR MR. CHAIRMAN: During recent hearings before your Subcommittee on S. 1086, the Telecommunications Infrastructure Act of 1993, James G. Cullen, President of Bell Atlantic, made several highly inaccurate and misleading statements about AT&T's pricing. This attempt to shift the Subcommittee's attention away from local exchange competition and the monopoly that his company and the other Regional Bell Operating Companies (RBOCs) enjoy in local telecommunications must be corrected.

Mr. Cullen confused the issues raised in S. 1086 by selecting an issue that is inordinately complex, even to those in the industry. Specifically, he testified that the local exchange carriers (LECs) reduced access charges to long distance carriers, or IXCs, by approximately $250 million and that AT&T did not reduce its long distance prices by a comparable amount. In fact, over the recent period reflected in AT&T's price changes, LECs increased their switched access charges to IXCs by $20 million.[1] Mr. Cullen's statement is therefore not only a gross overstatement, it is also completely contrary to the actual direction of LEC access pricing.

Mr. Cullen's first mistake was to use a single access reduction to disguise multiple recent IXC access charge increases and then blaming long distance carriers for failing to pass through reductions we never received. He further misled Members of the Subcommittee by blaming AT&T for failing to pass through 100 percent of the fictional IXC access reduction. In fact, of course, AT&T pays only about 60 percent of IXC access charges and could hardly be expected to account for the entire reduction even if there had been one. The basic fact remains that the access charges in recent LEC filings have increased, not decreased, AT&T's access expense.

Mr. Cullen's testimony also included a chart from the February 1993 issue of Business Communications Review (BCR) which he claimed showed the IXC industry is engaged in oligopolistic pricing. He selectively used a chart but failed to include the entire article, which would have disproved the precise point he sought to make.

Had he done so, the Members of the Subcommittee would have seen that "basic long distance charges" as defined by the author, are the undiscounted per minute rates for AT&T Megacom, MCI Prism 1 or Sprint Ultra WATS for a "typical" customer with 1,000 hours of daytime traffic per month and include the cost of T1 access 24 voice grade lines) at one end.

The very point of the BCR article was to inform business telecommunications managers that there is substantial competition for such services and that they can and should negotiate discounts off these "basic" rates. The author wrote, "Tariffed or not, today's basic long distance service prices are just a *starting point* (my emphasis) in determining what your particular company might pay." This point is constantly reiterated throughout the article with further statements such as "base rates are just the 'sticker price'; pricing plans and special deals make the difference."

Nor are customers for the services whose base rates were charted representative of customers in general. But, in any case, every residential and business customer has a plethora of discount plans available, from AT&T's i plan to Sprint's The Most to MCI's Friends and Family to Tariff 12 and its clones. Customers have unique requirements, and Mr. Cullen's implication that long distance pricing can be described accurately in one chart shows either a deliberate attempt to confuse or a gross misunderstanding of how long distance competition works.

If Members of the Subcommittee and other policy makers are distracted by specious arguments such as Mr. Cullen's, they will fail to focus on the real problem and public policy need—the absence of competition in the local information infrastructure. It is noteworthy that nowhere in his testimony or rebuttal did Mr. Cullen refute the fact that the entire IXC industry is totally dependent on the LECs for access to our customers. Well over 99 per cent of AT&T's access expense in 1992 went to the LECs; only $19 million out of $14 billion, or 0.14 percent, went to their competitors while 99.86 percent stayed with the LECs.

The debate in Congress is, properly, how to ensure that competition flourishes in all segments of the telecommunications and information industry. Competitive conditions already exist in the interexchange industry and need to be established, if possible, in non-competitive segments—notably that of local exchange telephone companies.

AT&T looks forward to contributing to the fact-finding and deliberations of the Subcommittee and full Committee. You may rely on us to furnish facts as clearly as we know how.

Sincerely,

TOM NORRIS.

Attachment A—Below are LEC access expense changes that have an impact on the entire LEC access customer base. AT&T's Price Cap Regulated Basket 1 and 2

---

[1] A $250 million reduction in switched access charges to the IXC industry was made as part of the LECS' annual tariff filing effective July 1, 1993. But the $250 million number reduction in carrier switched access was more than offset by $270 million increases in access expenses.

Services are a subset of this total. LBC access charge changes affecting AT&T's Price Cap Regulated services are referred to as "Delta Y", and serve as input to AT&T's Price Cap Indices (PCT). AT&T's PCI is typically updated twice per year to include accumulated LEC access changes, inflation (6NP-PI), and FCC sanctioned exogenous cost adjustments. Delta Y is computed using 1992 Base Period Price Cap volumes and is relegated to AT&T's Price Casp Baskets using an "As Allocated" methodology consistent with FCC Docket 87-313 rules.

Switched access charge changes	Effective date	Estimated IXC industry impact in millions of dollars
LEC information data base	Fully deployed mid-1992	$100
GTE CCL increase (Trans. No. 778)	April 2, 1993	$33
LEC 800 number portability	May 1, 1993	$74
LEC annual access filing	July 1, 1993	($250)
General support costs reallocation and universal service fund adjustments.	July 1, 1993	$63
Sum of access charge changes		$20

LECs reduced total access expenses by $184 million in their annual filing. The net reductions to industry switched access services were $250 million, apportioned to switched access categories as follows: carrier common line charges reduced by $229.1 million, switched traffic sensitive charges reduced by $21.1 million. (Note: LBC Information Database charge changes were absent from AT&T's prices during 1992) Including the additional access charge changes, industry switched access expenses increased by $20M. AT&T has included approximately $9M worth of LBC switched access charge changes in its recent PCI filing.

Senator INOUYE. Thank you very much, Mr. Allen. May I now call on Mr. Weiss?

STATEMENT OF WILLIAM WEISS, CHAIRMAN, AMERITECH CORP.

Mr. WEISS. Thank you, Mr. Chairman. I am also pleased to be here to present the views of the seven regional companies on Senate bill 1086.

We commend the sponsors for their vision that greater infrastructure development is needed to bring advanced telecommunications services to broad segments of our society and to give consumers more choice.

We support the legislation's reliance on competition and private investment to do this. As much as we applaud these goals, we do not believe that the proposed bill in its current form will achieve those goals.

This industry is rapidly realigning. This year's proliferation of new coalitions and the court decision striking down the video programming ban confirm this. What is needed is a comprehensive revision of telecommunications policy to govern this dramatically changing industry.

We believe the solution is based on opening all parts of the telecommunications marketplace to even-handed competition.

Let me state that there is competition in the local exchange. To say otherwise is to perpetuate a myth. The local exchange business comprises many elements: business telephone systems, residential and business exchanges, cellular telephone systems, and interconnecting circuits and networks.

Now, some parties attempt to define the state of local exchange competition by who provides dial tone. This simply shifts the attention from the real situation that exists today.

Let me give some examples of how competition really works. Large businesses generate a tremendous volume of local traffic, mostly within their own leased or purchased systems, local exchange business systems.

And I must tell you that AT&T and other competitors dominate this portion of the local market. In providing the dedicated connecting links between the local exchange switching machines and the long-distance network, other companies already provide more than 3 out of every 10 circuits that are established.

Fifty percent of local business toll services are handled by the long-distance carriers. Seventy-five percent of local 800 toll free calling is carried by long-distance carriers.

Consumer choice is what this all about. In many elements of the local exchange wide consumer choice exists today. It is time to make choice available to consumers in all markets of this industry.

With AT&T's $12.6 billion acquisition of McCaw it will instantly become the largest cellular provider in the United States and will move further into the local exchange.

The acquisition enables the company's largest carrier to become the end-to-end service provider, not just to large businesses, but also to high-volume sophisticated local customers.

There is also the matter being discussed currently of AT&T cable alliances. This vision, as articulated by AT&T, is communications any where, any time, any place.

Clearly, the long-distance, cable, and local exchange markets are converging quickly, but they are still in limited hands.

Consumers want more choice in end-to-end providers. The proposed legislation must be modified if all consumers are to receive even-handed treatment and over time their service in all serving areas is not to be impaired.

This legislation provides no significant relief from a long-distance ban. Today, the restriction discourages us from bringing advanced services such as Ameritech's Wisconsin Health Information Network to less populated areas.

The long-distance carriers claim we will dominate them or the market if this ban is lifted. When we would start with zero percent of this market can anyone seriously believe that argument?

The sum of it is this: We too are investor-owned and we are competing with all other businesses for investment capital. We are not asking for handouts or for protected markets, but we do believe that we should be given the opportunity to compete for the resources or funds necessary to better serve our customers through an upgraded local exchange infrastructure.

Universally today analysts and commentators are calling for the restrictions to be overturned, to be eliminated. As Judge Ellis wrote in his decision striking down the video programming ban "there is no more draconian approach to solving the problems of potential anticompetitive practices by telephone companies than a complete bar on their entry into that industry."

That finding applies with equal force to the long-distance and manufacturing restrictions of the MFJ. We urge the subcommittee to modify S. 1086 to provide consumers the benefits of increased competition in all aspects of the telecommunications market.

Mr. Chairman, thank you.

[The prepared statement of Mr. Weiss follows:]

PREPARED STATEMENT OF WILLIAM L. WEISS

My name is William L. Weiss and I am the Chairman and Chief Executive Officer of Ameritech Corporation. I am here today to present the views of the seven regional telephone companies regarding S. 1086.

The regional companies commend the sponsors of S. 1086 for recognizing that the rapidly changing technology of telecommunications has outpaced the ability of the current regulatory framework to guarantee that the wide variety of services arriving in the marketplace will be available to all consumers. We agree that the world of telecommunications in the future should be focused on customers—customer choice, customer needs, and customer access to new products and services.

Developments in the marketplace, including the recently-announced acquisition by AT&T of McCaw, are leading to a society in which certain customers will have better access to new technologies than others. The Subcommittee recognizes this problem and is right to address it because, as S. 1086 states, "greater infrastructure development is needed to bring advanced telecommunications services to small business, disadvantaged, residential, low-income, educational, medical, and rural users."

My testimony today, on behalf of the seven regional Bell companies, is to reassure you that we have been, and will continue to be, concerned about customer access to quality service. But the solution currently proposed by S. 1086 will damage the long-term goal of quality, affordable service by placing the Bell companies at a severe competitive disadvantage.

We believe that the local exchange companies are well-positioned to build an infrastructure that provides these services, particularly to those consumers identified by the legislation as most in need. Unlike other providers in the marketplace, we have a tradition of providing universal service. However, if we are to continue to meet the challenge of improving services for those populations the legislation seeks to help, we must be able to fully serve the entire information marketplace.

The vision of the legislation is futuristic, but its solution—which relies on market distinctions that no longer exist—is a relic of the past. By imposing considerable new competitive risks on the regional companies without giving us the opportunity to compete, S. 1086 will have an effect precisely opposite that intended by the sponsors. It will retard investment in the information infrastructure and, as a consequence, virtually guarantee that the gap between the information haves and have-nots will continue to grow. The very citizens that this bill seeks to help may instead be relegated to the backwater of the Information Age.

The regional companies believe that the only way to achieve the Committee's objectives is through full and fair competition in all segments of the telecommunications marketplace, including long distance, manufacturing and cable television. As Senator Breaux said at the July 14 hearing: "Let everybody compete. If you can do it, do it. That is what America is about."

A recent study demonstrates dramatically the benefits that will be achieved if this Committee introduces equitable competition in the telecommunications industry. In a July 1992 study, A. D. Little had identified over $36 billion in annual cost reductions that could be achieved in the health care arena from the implementation of just four telecommunications applications. Long distance applications represent a significant part of the $36 billion savings.

As part of the process of implementing the MFJ, the U.S. was divided into 197 Local Access and Transport Areas, or LATAs. The Bell companies were prohibited from providing telecommunications services between LATAs and are therefore limited in their ability to achieve these large health care savings.

The cost reductions identified by A. D. Little are significant. Achieving them would dearly help control the spiraling costs of health care in this country. By permitting the Bell companies to compete for the long distance telecommunications business, you would permit them, working with health care providers, to assemble services needed to achieve these, and possibly greater, cost reductions in the health care that is so vital to us as a nation.

Also, the prestigious WEFA Group released an analysis entitled "The Economic Impact of Eliminating the Line of Business Restrictions on the Bell Companies." The study forecast the economic impact of entirely and permanently eliminating the MFJ restrictions from the Bell Companies and the video programming restriction in the Cable Act from all local telephone companies. It also assumed that all telecommunications providers would compete under equivalent terms and conditions.

WEFA projected that this relief would generate an additional 3.6 million jobs over the next ten years—jobs that are spread out across all states and all major industry groups. It also projected $247 billion in added real GDF by 2003 and $630 billion in gains to consumers over those ten years, freeing up a comparable amount of disposable income for purchases of other products and services. The study demonstrates that full and fair competition in the telecommunications market will create jobs and benefit consumers.

Based on widely respected economic models, the WEFA analysis found that competition resulting from line-of-business relief would bring these results:

- Long distance toll rates would fall 50 percent from baseline, saving customers over $490 billion over the next decade.
- Cellular rates would fall 15 percent from baseline over the first five years following relief, saving more than $25 billion by 2003.
- Cable prices would decrease almost 25 percent from WEFA's baseline forecast for the ten-year period, saving consumers nearly $75 billion by 2003.

For these reasons, S. 1086 should be recast to conform to the basic construct envisioned by the study.

Telecommunications providers, other than the Bell companies, already are free to compete outside their traditional businesses. Long distance carriers are offering local services in competition with the regional companies and other local exchange companies, and they have been quite successful. In the five states that we serve at Ameritech, long distance carriers now provide 50 percent of the local toll service provided to larger business customers. These carriers have also successfully targeted 800 toll-free services. In four years, we have lost 75 percent of our local 800 business, not because we are poor business people, but because the carriers could package local and long distance 800 services and we cannot. The MFJ long distance restriction permits the regional companies to provide only local 800 toll-free service, only a piece of what customers want.

Several national cable operators, most notably Time Warner, have stated their intention to offer local telephone service over their cable facilities. And AT&T has been holding talks with the nation's largest cable companies about linking customers into one big network, which would include allowing customers to phone one another through their cable lines.

The regional companies ask only for the opportunity to compete on an equal footing in the long distance, cable, and other markets from which we are currently barred. We are not asking for special privileges or wholesale deregulation in these businesses; we only want the chance to compete fully.

There are subtle but important regulatory practices that must be changed for there to be full and fair competition. Depreciation and capital recovery practices among cable companies, telephone companies and long distance companies vary widely today. This is difficult to justify where the companies use the same type of equipment to offer the same services. For example, the companies all use fiber optic transmission facilities, but each depreciates the fiber over different time frames.

As between carriers there is disparity. The FCC permits the long distance carriers to depreciate their plant over shorter periods than it permits the local companies to use. Such disparities among competitors simply is not proper in a fully competitive market. Thus, S. 1086 should require parity on all such items among telecommunications competitors.

The existing networks operated by the regional companies, with their tradition of universal service, are the most likely supplier of advanced services to the mass market. We want to upgrade our networks to provide voice, data and video service to all customers. But as competition intensifies, we need the opportunity to compete for new services that will provide the sources of funds for the necessary infrastructure upgrades.

Universal service has been achieved through a conscious and intricate system of subsidies. This has been critical in the past to bring services to economically disadvantaged customers and to millions of rural customers. The problem is that the existing system was created when AT&T had monopoly control of local and long distance services. That is no longer the case and, as competition spreads to all sectors of the telecommunications marketplace, sustaining subsidies from competitive services will be increasingly very problematic.

The issue of continued subsidies is significant. A recent study by the U.S. Telephone Association estimated that local telephone service is subsidized in the amount of $20 billion annually. It identified the sources of these subsidies as large and sophisticated business customers and toll rates. It is the large customer and the long distance carriers who have the competitive choice today to leave the public network. And they are making those choices.

S. 1086 must also deal with the remaining subsidy issues in ways that are specific and that spread the burden evenly over the entire telecommunications industry. Part of the answer is to free the Bell companies to enter any markets they seek to enter. This will allow them to maximize their efficiencies and to lower the costs of providing some currently subsidized services. Moreover, it will stimulate the market as shown by the WEFA study and lower consumers' overall telecommunications expenditures, as compared to retaining the restrictions.

Universal service is clearly in the national interest and is largely accomplished. It must be maintained so the promises of the Information Age can reach all Americans wherever they may choose to live and work. But, ultimately we must recognize that the burden currently borne by the local telephone companies cannot be sustained "as is" in an increasingly competitive local exchange business.

This legislation, by increasing competition in the local exchange, would only accelerate that trend just as occurred with 800 services. While some have argued that this threat to universal service should forestall further opening of the local business, we think the better view is simply to allow us the freedom to compete for new sources of revenue to replace those lost to local competition. As described more fully below, removing the entry barriers from long distance, manufacturing, and cable television will have benefits for consumers. If there are any lingering doubts about whether the regional companies could upset the terms of competition in the businesses from which we currently are excluded, these concerns are best dealt with through focused regulation and safeguards, not entry restriction.

Long distance relief. Our entry into long distance would bring immediate benefits to consumers in the form of greater choice and lower prices. No one can deny that the long distance business could stand a healthy dose of price competition. Recently, AT&T raised its prices by 4 percent, and MCI and Sprint quickly followed suit. These increases occurred at a time when local exchange companies had decreased access prices. Our entry would invigorate competition in this business.

In addition, elimination of the long distance ban is essential to achieving the Subcommittee's vision that advanced services be available to all customers, wherever they reside. The current restriction requires the regional companies to build separate facilities in each calling area if we seek to provide advanced services requiring data base storage and retrieval. These facilities could include data base applications for distance learning, remote health care monitoring and telecommuting. It is only economically feasible to build these separate facilities where enough people in the calling area will use the new service to cover the cost of buying, installing and maintaining the equipment. Less populous areas may lose out. If we could share facilities between calling areas, for example, by building a center in one city to serve an entire state or region, new services could be brought more quickly and cheaply to everyone.

AT&T and the other long distance carriers object to our entry into long distance before it has been demonstrated to their satisfaction that there is effective competition in the local exchange. In fact, Sprint has gone further and argued for a ten-year moratorium on our even asking for relief. These complaints are nothing more than self-serving attempts to prevent competition, not to serve consumers. Because market exclusions by definition harm consumers, the long distance restriction was designed to go away once there was no substantial possibility that we could use our market power in the local exchange to impede competition in the long distance business. Significantly, despite what the carriers may claim, the decree does not require that there be local competition in order for relief to be given. MFJ relief can be granted if there is a set of rules that requires us to compete fairly and not take advantage of our ownership of local facilities.

The regional companies now have 0 percent of the long distance business we seek to enter, and would be competing against dozens of companies, including three very prominent competitors. No one could seriously believe that we could somehow drive AT&T and the other carriers out of this business. Moreover, if the regional companies have so much power in the local exchange to dominate other services, how have the carriers taken half the business in toll? How could we have lost 75 percent of our local 800 business if our power in the local exchange was as great as the carriers claim? The fact is that the harms the carriers complain of are sheer speculation. Their dire predictions did not come true with customer premises equipment, with cellular, or with information services. It is time to remove the long distance restriction and let customers reap the benefits of increased competition in that business.

Manufacturing relief. In 1991, the Senate overwhelmingly passed Senator Hollings' bill to provide relief from the manufacturing ban of the MFJ. Manufacturing relief is even more important now than it was in 1991. The announced acquisition of McCaw by AT&T raises the specter that AT&T's manufacturing arm will favor McCaw over other cellular providers in supplying it first with new technology and features before they are made available to AT&T/McCaw's competitors.

Information services. The regional companies' ability to offer information services has been exhaustively litigated in the district and appellate courts. The regional companies have been offering information services on an unrestricted basis since the ban was removed in 1991. The speculative fears of domination and monopolization made by opponents of relief have not come to pass. The market is working and we urge the Subcommittee not to tamper with it.

Cable relief. Last month, a federal district court struck down the ban on telephone companies' provision of video programming in their service areas as an unconstitutional infringement on the companies' rights of free speech. The court's action was in harmony with the Subcommittee's intention to remove the restriction in this legislation. We would welcome an affirmation of the court's decision in any legislation in order to avoid the expense and delay of appeals from the court's order.

Judge Ellis wrote in his decision striking down the video programming ban that "there is no more draconian approach to solving the problem of potential anti-competitive practices by telephone companies * * * than a complete bar on their entry into that industry. " Although he was writing about cable television, his holding applies with equal force to the long distance and manufacturing restrictions of the MFJ. It is time for Congress to find less drastic means to protect against speculative harms to competitors that might occur from our entry into currently forbidden businesses, particularly given the very real benefits that are being denied consumers by our foreclosure from these businesses.

In conclusion, the regional companies want to work with the Committee to implement the vision of this legislation. But its terms must be modified if it is to be successful in realizing that vision. S. 1086 will only make it harder for the regional companies to bring advanced services to all consumers.

The solution is not the unbalanced competition set out in the draft. The answer is to free the marketplace to best serve all customers by opening up all markets to all competitors. The companies with the best products and the most efficient networks will be successful, and in the long run, all consumers will win. That is what the regional companies want. We think that is what the Subcommittee wants. We stand ready to work with the Subcommittee to achieve that objective.

Thank you.

Senator INOUYE. Thank you very much, sir. May I now call upon Mr. Kapor?

STATEMENT OF MITCHELL D. KAPOR, CHAIRMAN OF THE BOARD, ELECTRONIC FRONTIER FOUNDATION

Mr. KAPOR. Chairman Inouye, Senator Danforth, members of the committee, I want to thank you for the opportunity to come before you and commend you for taking such a strong interest in telecommunications infrastructure modernization.

I am the chairman of the board of the Electronic Frontier Foundation. The EFF is a nonprofit public interest group whose mission it is to ensure that new electronic highways enhanced free speech and privacy rights and are accessible to all segments in society.

The EFF actively builds coalition and consensus among public interest organizations, consumer groups, computer and communications firms, and cutting edge providers of new information services.

For those of you who do not know me, I am also the principal designer of the Lotus 1–2–3 spreadsheet program, and I served as the CEO of Lotus Development Corp. from its founding until 1986.

Properly implemented, the growing information and communications infrastructure is the basis for nothing less than a media revolution as important as the printing press, offering as yet unimagined opportunities for building and revitalizing communities and helping businesses maintain their competitive edge.

But we know to achieve these benefits we need more capacity than is available in today's telephone system and we need more than 500 channels of cable television offering just entertainment, news, and sports.

Today's phone systems does not have the capacity to enable us to exchange multimedia information. And 500 channels fail to meet our real needs if they only allow one-way distribution of information from the network operator.

And if that information is purely selected by the operator, the interactivity that is critical for educational services, the library access, the online medical assistance, for telecommuting or rural business connections, and for the next generation of multimedia entertainment it cannot be accommodated in a one way system or one with other limitations on its use.

So, our vision is one that promotes a rich diversity of information sources and new applications, that creates new entrepreneurial opportunities and is resistant to the monopolization of information channels.

We have given a name to the concept that ties together these goals. We call it "open platform service." It has four very simple points to it.

First, open platform services must offer widely available switched digital connections, point to point, anybody talking to anybody else, not just predetermined channels.

It must be affordably priced. It has to be open to everyone without discrimination as to the content of the message. And it has to have sufficient upstream capacity to let the users themselves originate as well as received good quality video multimedial services. Open platform services will provide the onramps for the Nation's growing information superhighway.

Congress has the responsibility to see that the national information infrastructure emerges from today's telephone and cable networks.

The Clinton administration has shown a great deal of interest in the NII, but we know that there is no way the Federal Government can or should build the national information infrastructure.

The task of actually deploying open platforms must be accomplished by the private sector. If not, the groundbreaking applications and systems which have already been developed through Government funding such as the Internet will never be accessible to the vast majority of the population.

So, what can Congress do? In my optimistic moments I am almost persuaded that policymakers do not have to do very much at all. If all goes well, telephone and cable industries will be persuaded by the logic of self-interest to build broadband networks which are consistent with these ideals of openness.

The moneymaking opportunities are there, I tell myself, even if they are unaware or uninterested in the full range of democratic goals.

On the other hand, a more cautious and pessimistic view suggests that providers may decide there is more safe money to be made by offering more of the same kinds of one-way video entertainment services we have today.

Or providers in some parts of the country may make the choice to retain full control over the content on the network rather than make them open. In that case, all of the potential for the national information infrastructure that we have been hoping for will be squandered.

Now, specifically with respect to S. 1086, there are two policy goals that I think apply equally in the optimistic and pessimistic scenarios. The first has to do with universal service for the Information Age.

The principle of equitable access to basic services is an integral part of our policy. We support the notion already in S. 1086 that all providers should share in the cost of maintaining universal service.

We also hope that a forward-looking definition of universal service, including multimedia information service access capabilities will also be incorporated in the legislation.

In other words, we have to move from plain old telephone service to plain old digital service. Equity also demands that these services meet the needs of people with disabilities, the elderly, and other groups with special needs.

Second, we need a new common carriage regime for the Information Age. We have to reshape common carriage duties for new medial environments.

The technology is converging, industry structure is changing. Telephone companies who were the traditional providers of common carriage services are moving closer to providing video.

Cable television, which would function as program providers, are showing great interest in offering telecommunications services.

Here is what we suggest: Any carrier that is willing to offer open platform services as I have described, on a nondiscriminatory basis, should be allowed to offer video programming as well.

By allowing any infrastructure provider to coexist in both regulatory categories, the provider would be encouraged to invest in both expanded entertainment services and at the same time make real contributions to the development of the NII.

EFF proposes that the price for participating as both a video programmer and telecommunications provider would be willingness to offer true open platform services on the telecommunications side. We would like to see S. 1086 amended to include this provision.

The recent judicial action casts some doubt on the constitutionality of absolute crossownership bars. We believe that strong statutory safeguards are constitutional and are certainly required where content and conduit services are provided by a single network owner.

The very existence of open platform services is itself the most important step to promoting diversity of information. Shaping the architecture of the new NII in a way that promotes easy access for all programmers is the most important safeguard of all, if the goal is diversity and fair access.

My third point has to do with the more pessimistic scenario. We believe competition is necessary, absolutely necessary, but not always sufficient to ensure adequate infrastructure for all.

S. 1086 wisely recognizes this. Policymakers must not assume that the open platform vision I have laid out is self-executing in all circumstances. Section 6 of the bill provides that the FCC has the authority to mandate infrastructure deployment to meet open platform standard if it is shown that competitive forces are not making such services available.

We would suggest a proactive regulatory process to have tariffs filed for affordable digital services. If the tariffs are inadequate or

70-334 O - 94 - 6

if there is some other market failure, then the FCC would be authorized to intervene under the authority granted in section 6.

We believe this is a less intrusive approach in that carriers are allowed to come forward with plans suggesting the ways in which they will meet the needs for open platform services.

Mandatory authority would only come into play if it is shown that the needs as outlined in section 6 have not been met.

In conclusion, the prescription here is not a heavy regulatory hand. Where the market is delivering the necessary services, intervention should be minimal. But merely lifting existing line of business restrictions will not guarantee that the market will delivery necessary new services.

I want to thank you again for the opportunity to appear before you on this very important issue.

[The prepared statement of Mr. Kapor follows:]

Prepared Statement of Mitchell D. Kapor

I. INTRODUCTION AND OVERVIEW

Chairman Inouye, Senator Danforth, and Members of the Committee: I want to thank you for the opportunity to come before this committee, and to commend your for taking such a strong interest in telecommunications infrastructure modernization policy. I am the Chairman of the Board of the Electronic Frontier Foundation (EFF). EFF is a non-profit, public interest organization whose public policy mission is to insure that the new electronic highways emerging from the convergence of telephone, cable, broadcast, and other communications technologies enhance free speech and privacy rights, and are open and accessible to all segments of society. To achieve these goals, EFF works actively to build coalition and consensus among public interest organizations, consumer groups, computer and communications firms, and cutting-edge providers of new information services. For those of you who do not know me, I am also the principal developer of the Lotus 1-2-3 spreadsheet program and served as the CEO of the Lotus Development Corporation between 1982 and 1986.

Over the last two years, EFF has been an active participant in the telecommunications infrastructure debate. Our chief concern in the past has been to identify economically, technically, and politically practical infrastructure improvements that can bring the benefits of new information technology to the American public without having to wait until the year 2015 for the deployment of a fiber4o-the-home network by local telephone companies. To this end, we have suggested that technologies which enable the delivery of video information services over existing copper telephone lines be implemented while fiber optic deployment proceeds. This effort is still vitally important to the development of the nation1s telecommunications infrastructure. In this testimony, we take the opportunity to offer a somewhat more long-term view of telecommunications regulation.

II. INFRASTRUCTURE POLICY GOALS

Before discussing the details of the legislation before us, I suggest that we all step back from the inter-industry political squabbles that have long characterized the telecommunications policy debate just long enough to reflect on what is at stake in the development of the national information infrastructure (NII). Technological advances now put in our reach a dazzling array of new information and communication capabilities which, if successfully deployed, can revolutionize the way that we all live, work, learn, play, and entertain ourselves. I know that this Committee, others in Congress, the Clinton Administration, state regulators, and private industry, all share the common goal of developing the NII.

The findings of S. 1086 carefully sets out these goals:

> (3) advancements of the Nation's telecommunications infrastructure will increase the public welfare by helping to speed the delivery of new services, such as distance learning, remote medical sensing, and distribution of health information;
>
> (20) access to switched, digital telecommunications service for all segments of the population promotes the core First Amendment goal of diverse information

sources by enabling individuals and organizations alike to publish and otherwise make information available in electronic form;

Today I would like to give you my thoughts on the importance of these two goals and ways that they might best be realized.

A. A Jeffersonian Vision of the National Information Infrastructure

Properly implemented, the growing infrastructure is nothing less than the new printing press, offering as yet unimagined opportunities for personal communications, for building and revitalizing communities, and for the development of vast markets for tomorrow's information entrepreneurs. The infrastructure can help business to work more efficiently, realizing global competitive advantage. And if the infrastructure is truly extended to all parts of the country, rural businesses can compete in markets previously closed to them. Workers, too, will benefit by being able to match their work demands to their lifestyle needs through telecommuting.

To achieve these and other benefits, we know that we need more capacity than is currently available in today's analog voice telephone system. We also need more than merely 500 channels of one-way cable television. Today's telephone system does not have enough capacity to enable us to exchange the multimedia information sources that will be the staple of our information diet in the near future. And the 500 channels which many look forward to fail to meet our real needs because they only allow one way distribution of information from the network operator down into each subscriber's home. The interactivity that is critical for educational services, for library access, for online medical assistance, for telecommuting or rural business connections and for the next generation of multimedia entertainment, cannot be accommodated in a one-way system or one with other limitations on its use.

This is our Jeffersonian vision of the information infrastructure: one that promotes a rich diversity of information sources and new applications, that creates new entrepreneurial opportunities, and that is resistant to the monopolization of information channels and promotes the free flow of ideas.

1. Promoting Diversity of information Sources and Equitable Access

Aside from the universal service guaranty, the driving communications policy value for the last fifty years has been promotion of the maximum diversity of information sources, with the greatest variety of view points. As we move into the multimedia information age, we have a new opportunity to shape communications policy in ways that promotes maximum diversity m a way not possible in earlier mass media such and broadcast and cable television.

Historically, the print medium has been the most successful at promoting a diversity of information sources because it allows easy entry as an information provider and easy access as an information consumer (a reader). Compared to both the broadcast and cable television arenas, print is the vehicle for the greatest diversity of viewpoints and has the lowest publication and distribution costs. Despite the regulatory steps taken to promote diversity in the mass media, the vexing problems of spectrum scarcity and limited channel capacity have always limited the variety of opinion and information.

The switched nature of advanced digital network technology offers to end the spectrum and channel scarcity problem altogether. Broadcast and current cable media have a built in distribution bottleneck because of the limited number of channels and the hierarchical nature of the distribution system. An independent content producer must always negotiate with the channel owner for the ability to communicate with others. In a switched, digital network, of the kind that phone companies and cable companies both speak of deploying in the near future, any user can communicate with any other user. The distribution bottleneck caused by having a small number of channel-holders is eliminated. Thus, anyone with content to distribute—whether to one, one hundred, or one hundred thousand users—can do so without the permission or advance approval of the carrier.

2. The Need For Open Platform Services

To achieve the full potential of new digital media, we need to make available what we call Open Platform services, which reach all American homes, businesses, schools, libraries, and government institutions. Open Platform service will enable children at home to tie into their school library (or libraries all around the world) to do their homework. It will make it possible for a parent who makes a video of the local elementary school soccer game to share it with parents and students throughout the community. Open Platform will make it as easy to be an information provider as it is to be an information consumer.

Specifically, Open Platform service must meet the following criteria:

- widely available, switched digital connections;

- affordable prices;
- open access to all without discrimination as the content of the message;
- sufficient "up-stream" capacity to enable users to originate, as well as receive, good quality video, multimedia services.

Open Platform service itself will be provided by a variety of providers over interconnected networks, using a variety of wires, fiber optics, coax cable, and wireless transmission services. But however it is provided, if it is affordable and widely available, it will be the on-ramp for the nation's growing information superhighway.

B. Public and Private Roles in Infrastructure Development

Congressional action on the issues raised by S. 1086 is absolutely critical to the growth of the National Information Infrastructure as a whole. This year there are a number of legislative and executive branch actions contemplated in support of the NH. Senator Hollings, through 5A, and Congressman Boucher, through H.R. 1757, are both working to promote the development of new network-based applications for the Internet and the NREN. Chairman Markey, on the House side, has been hard at work on an infrastructure bill which we hope will be introduced very soon. Last year, Senator Burns, in S. 1200, also showed a desire to work for infrastructure improvement The Administration has indicated that the National Telecommunications and Information Administration will also play a role in funding network access for local community institutions such as schools and libraries. These funding initiatives are important steps toward making electronic networks more useful to individuals.

But those funding efforts will only bear fruit for Americans if appropriate public policy and private sector investment succeeds at extending the on- and off-ramps of the data superhighway into Americans' homes, schools, and libraries. This task of actually deploying Open Platforms must be accomplished or the groundbreaking applications developed through government and private initiative will never be accessible to the vast majority of the population. Any suggestion that the government might be able to build an NII through public funding is misguided. The government has neither the money, as you know, nor the expertise, to undertake such an effort. But EFF firmly believes that given the proper regulatory environment, the private sector will build the kind of infrastructure that the country needs. However, we should recognize that if the infrastructure goals of S. 1086 will not be achieved without comprehensive, legislative restructuring of telecommunications policy.

III. PUBLIC POLICY RECOMMENDATIONS

What can Congress do to ensure that the potential of this technology is realized? In my optimistic moments, I am nearly persuaded that policy-makers don't have to do much at all. The many announcements of new technology experiments and increased infrastructure investments by both the cable and telephone industry certainly suggest that the private sector is rising to the challenge of building new information infrastructure. If all goes well these industries will be persuaded by the logic of self-interest to build broadband networks consistent with Jeffersonian ideals of openness, even if they are unaware of or uninterested in the full range of-democratic goals.

Optimism does not always carry the day, however. If competing cable and telephone companies have too narrow of vision of what will be successful telecommunications services, the country will never realize the great potential that Open Platform services offer. For example, providers may decide that there is more safe money to be made simply be offering more and more one-way video entertainment services. Or, some infrastructure providers in some parts of the country may make the choice to retain full control over the content on their networks, rather than make their networks open to a diversity of information, consistent with Open Platform principles. If these or other less optimistic scenarios are realized, we would simply go from 57 channels and nothing on, to 570 channels and still nothing on. All of the potential for the NH that we have been hoping for would be squandered.

Sadly, it does not take much to fail in building the NII. Islands of progressive infrastructure investment surrounded by a few market failures would lead to infrastructure balkanization and of disenfranchising individuals and organizations in underserved areas. Moreover, the whole country would be left with an incomplete infrastructure. Therefore, if this committee and the Congress in fact find that infrastructure growth and full access to the information age is indeed important, then active steps must be taken to ensure that these goals are realized.

A. Optimistic View: Achieve Social Policy Goals By Reliance on Market Forces, Fairly-Shared Universal Service Subsidy, and Appropriate Safeguards

To the greatest extent possible, it is wise to rely on market forces to deliver the services and functionalities that the public needs as part of the National Information Infrastructure. Federal policy makers should not put themselves in the position of mandating a particular technology or of anointing a particular industry as the infrastructure provider of choice. Rather, Congress should concentrate on ensuring that basic information and communications needs are met, and on creating conditions for fair competition which protect consumers and providers alike.

In many areas, vigorous competition among enlightened providers in local exchange markets may well lead to the deployment of Open Platform services that we believe are essential to a flourishing NII. Where such deployment occurs, regulation can be limited to the interconnection requirements and universal service guarantees outlined in S. 1086, as well as comprehensive anti-competitive safeguards.

Dramatic advances in network technology over the last few years promise to render obsolete traditional communications industry categories. Technology exists today which enables telephone companies to offer video programming service in competition with cable television providers even without waiting until the year 2015 for the deployment of a full fiber-to-the-home network. By the same token, cable companies are already experimenting with systems to offer both voice and date telecommunications services over their networks. EFF believes that this competitive environment can be a major positive force in bringing affordable Open Platform services to the American public. The competition in video programming could also, of course, improve the cable consumer's lot.

In order to be sure that this competitive opportunity produces real infrastructure improvements for the American public, regulation should be structured to meet two criteria.

1. Universal Service Goals for the Information Age

The principle of equitable access to basic services is an integral part of nation's telecommunications policy. From the early history of the telephone network, both government and commercial actors have taken steps to ensure that access to basic voice telephone services is affordable and accessible to all segments of society. Since the divestiture of AT&T, many of the internal cross-subsidies that supported the "social contract" of universal service have fallen away. Re-creation of old patterns of subsidy is no longer be possible, but serious thought must be given to sources of funds that will guaranty that the economically disadvantaged will still have access to basic communications services.

The universal service guaranty in the Communications Act of 1934 has, until now, been interpreted to mean access to "plain old telephone service" (POTS). In the Information Age, we must extend this guaranty to include "plain old digital service." Extending this guaranty means ensuring that new basic digital services are affordable and ubiquitously available. Equity and the democratic imperative also demand that these services meet the needs of people with disabilities, the elderly, and other groups with special needs. Failure to do so is sure to create a society of "information haves and have nots."

In a competitive telecommunications environment, regulatory paradigms must be industry neutral and treat all similarly situated providers equally. S. 1086 sets out just this kind of framework by defining interconnection and universal service fund obligations for all entities that provide telecommunications service, regardless of which traditional industry category they are associated with. So, a cable television company that provides voice or data telecommunications service, would have the same obligations as any other telecommunications provider, such as a local phone company or a wireless service provider. The scope of these obligations should certainly be proportionate to the companies market presence, but otherwise, all who chose to provide telecommunications services should be subject to the same requirements. We hope that a more thorough consideration of the universal service funding mechanisms is included before S. 1086 is reported from this committee. Also, a forward-looking definition of the universal service level, including multimedia information service access capabilities, must be incorporated in legislation.

2. A New Common Carriage Regime For The Information Age

In a society which relies more and more on electronic communications media as its primary conduit for expression, full support for First Amendment values requires extension of the common carrier nondiscrimination principle to all of these new media. Common carriage platforms will be critical as the new electronic public fora for politics, culture, and personal communications. They are the soap box, the local op-ed page, and the printing presses of the Information Age. If all carriers were to limit access to their networks based on the content of messages sent, the opportunity for free expression in society would be dramatically limited.

Re-shaping common carriage duties for new media environments will be necessary as mass media and telecommunications services converge and recombine in new forms. Telephone companies, the traditional providers of common carriage communications services, are moving closer and closer to providing video and other content-based services. By the same token, cable television companies, which have functioned as program providers, are showing great interest in offering telecommunications services. The desire of these industries to cross over into new businesses can be a source of great opportunity to consumers, if proper regulatory safeguards are put into place.

Any carrier that is willing to offer Open Platform services on a nondiscriminatory basis should be allowed to offer video programming as well. EFF believes that it will be possible to structure a regulatory regime in which infrastructure providers can provide both video programming, and common carrier-like telecommunications services on the same network. In framing common requirements for all telecommunications carriers, S. 1086 already suggests this direction (see Section 5). By allowing any infrastructure provider to co-exist in both regulatory categories, the provider will be encouraged to invest in both expanded entertainment services and, at the same time, make real contributions to the development of the NII. EFF proposes that the "price" for participating as both a video programmer and a telecommunications provider, would be willingness to offer true Open Platform services on the telecommunications side. We would like to see S. 1086 amended to include this provision.

A venerable regulatory tradition exists which argues that content and conduit providers must be separated in order to guard against anti-competitive behavior which could stifle, not enhance, diversity. Recent judicial action does cast some doubt on the constitutionality of such absolute cross-ownership bars. However, strong statutory safeguards are certainly required where content and conduit services are provided by a single network owner. We would support safeguards in the tradition of antitrust law, that allow victims of discrimination to seek remedies directly from carriers.

The very existence of affordable Open Platform services will be the most important step toward promoting diversity of information in the new multimedia environment. Shaping the architecture of the new NII in a way that promotes easy access for all programmers is the most important safeguard of all, if the goal is diversity and fair access.

B. Competition is Necessary, But Not Always Sufficient, To Ensure Adequate Infrastructure for All

S. 1086 wisely recognizes that competition is necessary, but not always sufficient to promote the widespread availability of advanced infrastructure (See Section 6). Despite the virtues of marketplace competition, policy makers must not assume that the Open Platform vision that I have laid out is self-executing in all circumstances. The importance of having an information infrastructure that promotes diversity, and that is accessible to all Americans is simply too great to be left to the vagaries of as yet undeveloped markets.

Section Six of S. 1086 provides that the FCC has the authority to mandate infrastructure deployment to meet Open Platform standards if it is shown that competitive forces are not making such services available:

> If State regulatory authorities fail to achieve the goal of ensuring that telecommunications carriers provide consumers in rural markets and noncompetitive markets with access to high quality telecommunications network facilities and capabilities which—
>
> "(1) provide subscribers with sufficient network capacity to access information services that provide a combination of voice, data, image, and video; and
>
> "(2) are available at nondiscriminatory rates that are based on reasonably identifiable costs of providing such services, then the Commission may take any action necessary to achieve that goal.
>
> "(b) FULL EFFECTUATION.—The Commission shall have the authority to preempt any State or local statute or regulation, or other State or local legal requirement, that prevents the full effectuation of the goal embodied in subsection (a)."

We would suggest, instead, a pro-active regulatory process to have tariffs for affordable digital services. If these tariffs are inadequate, or if there is some other market failure, then the FCC would be authorized to intervene, under the authority granted by Section Six. We believe this is a less intrusive approach, in that carriers are allowed to come forward with plans suggesting the ways in which they will meet the need for Open Platform services. Mandatory authority would only come into play if it is shown that the needs, as outlined in Section Six, have not been met. We believe that this proposal is fully consistent with the spirit of this bill, and that it will ensure that all Americans have access to Open Platforms services in the near term.

In order for the NII to truly benefit Americans, we need more than just transport services which will carry bits of data around the country. We need to spur the development of applications—the hardware, software, and services—which will make the NII truly useful to the country. This will not be done merely by lifting regulatory restrictions on the current carriers. Rather, we have to create opportunities for small, independent entrepreneurs to develop the Information Age equivalents of the PC industries word processors, spread sheets, and data bases systems. In the few years that I have been investigating the telecommunications industry, I have learned that it is different in many respects from the personal computer industry which I came from and helped to found. Perhaps the most important difference, from a public policy perspective, is that you cannot build a whole National Information Infrastructure in a garage. Steve Jobs and Steve Wozniak could build the Apple II in their garage, on their own initiative. And it was that kind of entrepreneurial initiative that gave me and others in the software industry the platform on which to build the innovative products and services which enabled the personal computer industry to grow from zero to $150 Billion dollars in just ten years.

We need that same kind of innovation in the development of new information and communications services, but it won't happen unless Open Platforms are in place. The development of this kind of platform, however, requires more than just a Silicon Valley garage, or a small factory in Boca Raton (the birthplace of the IBM PC). The development of the NII requires coordination of many industries—from telephone companies, to cable companies, to wireless providers—which are operating in a great variety of urban, rural, suburban, business and residential markets. Failure to achieve consistent infrastructure deployment across the country will seriously impair the usefulness of the new services, and more importantly, will have the effect of disenfranchising those communities and individuals who do not achieve adequate service.

The prescription here is not a heavy regulatory hand. Where the market is delivering the necessary services, intervention should be minimal. But merely lifting existing line-of-business restrictions will not guaranty that the market will deliver necessary new services. As Section 6 of the bill notes, regulatory action should be authorized to see that a minimum level of connection to the infrastructure is available to all areas of the country in the event the market does not provide such connection. We encourage the Committee to be sure that the terms of this section give regulators a clear mandate for action when infrastructure deficiencies are identified.

IV. CONCLUSION

I want to thank you again for the opportunity to appear before the committee on this very important issue. The Electronic Frontier Foundation is committed to working with you, as well as a range of public interest and industry groups to reach consensus on a new regulatory paradigm that will bring the benefits of new information infrastructure to the American public.

Senator INOUYE. Thank you very much, Mr. Kapor. May I now call on Dr. Duke?

STATEMENT OF DR. DAVID DUKE, VICE CHAIRMAN, CORNING, INC.

Dr. DUKE. Thank you, Mr. Chairman. My name is Dave Duke and I am the vice chairman of Corning, Inc. I am currently responsible for managing Corning's research, development, and engineering resources.

As you may know, Corning invented low-loss optical fiber. I was personally responsible for bringing this revolutionary product from the laboratory to the marketplace.

I am here today as a technologist, not as a legal expert, to discuss this experience, because I believe in this case the past is prolog.

I believe that many of the obstacles that we faced over a decade ago in deploying fiber optic technology in the long-distance market are similar to the ones that we face today in deploying the technology in the local loop. I believe the solutions are similar as well.

We do not need to mandate investment in any particular technology, whether it be fiber optics or ISDN. Before I get into the details of S. 1086, I would like to provide a little history.

Corning, along with Bell Labs and others, began research into optical fibers in the mid-1960's. Most were ready to give up by 1970, but Corning scientists persisted and made the primary optical invention later that year.

Then the commercial fun began. Corning had to convince reluctant customers to use glass fiber instead of copper wire in telephone systems. Our customers just did not buy it.

AT&T, which owned most of the telephone line in America at that time, said it would be 30 years before its telephone system would be ready for optical fiber.

Finally, in 1982, after Government action clearly established competition in the long-distance market, the commercial breakthrough happened. MCI took the risk and placed a 100,000-kilometer order for a new generation of fiber.

The moral of this story is that a Government decision to create a competitive long-distance market stimulated demand for new technology, including fiber optics, to better serve subscribers.

Not only did Corning benefit from this decision, but the Nation did. Annual investment in long-distance networks immediately expanded by more than 50 percent. Long-distance usage increased in excess of 12 percent annually and long-distance rates fell by almost 10 percent annually.

As an added advantage, the entire long-distance infrastructure was converted to fiber while rates dropped by over 40 percent in nominal terms.

The situation today at the local loop is not dissimilar to the conditions that existed for the long-distance market in the 1970's.

Local telephone service is a regulated monopoly, just like long distance was two decades ago. Local telephone companies are feeling the threat of competition, just like AT&T first felt the heat of competition by MCI.

Some companies like Ameritech and Rochester Telephone recognize this threat and have made some innovative regulatory proposals to deal with it and I applaud their foresight.

As was the case with AT&T in the 1970's, the local monopoly is reluctant to invest in advanced technology, particularly fiber in the loop, in the distribution and drop part of the loop particularly.

Overall investment by local exchange companies has remained relatively flat in nominal terms over the last 10 years, at about $20 billion annually, and has declined in real terms.

And about $3.5 billion is spent annually on copper wire rather than more advanced broadband technology.

While I believe these monopolies would like to invest at a faster pace in advanced technology, a regulatory mine field discourages them from moving forward. They are restrained by State regulatory authorities that disallow investment.

Subsidies from businesses to residential service necessitate minimal investment to keep rates low. And they are still largely precluded from providing the one video service with demonstrated demand, video programming, which requires investment in fiber optics and other broadband infrastructure.

So, the bottom line is, if we are satisfied with plain old telephone service over copper wire, then there is no need to act. If we want Americans to have higher value services necessary to remain internationally competitive, then the local infrastructure must be upgraded. This can be achieved without increasing the rates for telephone service, just like it was achieved in the case of long distance.

Do not be threatened into inaction by outlandish estimates from technology opponents that broadband technology will cost $450 billion to deploy. We estimate the deployment of an interactive broadband service over the public network can be accelerated by 20 years for a minimal investment of only $23 billion in 1992 dollars over that 20-year period.

I believe that S. 1086 builds on our experience from the past. As was the case with long distance, it is based on the fundamental realization that competition for local communication service can become a reality and bring enormous benefits to subscribers.

It is very possible with today's technology for two or more providers to serve most local areas. A similar precondition existed for long distance over a decade ago.

In recognition of this fact, the bill opens up the local loop to competition. It gives the local exchange carriers more flexibility on pricing and freedom to enter into new businesses.

Whether the bill goes far enough in some areas is debatable. But it does include all of the fundamental features to encourage rather than discourage investment in advanced technology, as is presently the case.

I believe an enactment of legislation based on the fundamentals of S. 1086 would result in the accelerated deployment of an interactive broadband network of networks to serve all subscribers in your lifetime and mine.

And this network of networks would include local telephone companies, cable companies, wireless carriers, long-distance carriers and others all interconnected into a seamless web.

This is not to say that S. 1086 is perfect as it is. I think there will be and must be refinements. But please remember, as refinements are made, there is a sensitive balance between necessary regulation to ensure fairness and over regulation that will squelch investment.

And please keep in mind that narrowly drawn statutes may quickly become obsolete, or worse yet, dampen the dynamic nature of this industry.

Thank you for the opportunity to appear before this panel and present Corning's views.

[The prepared statement of Dr. Duke follows:]

Prepared Statement of Dr. David A. Duke

Mr. Chairman, I'm delighted to be here today to present Corning's views on S.1086, a historic piece of legislation.

My name is Dave Duke. I'm the Vice Chairman of Corning Incorporated. I'm currently responsible for guiding and managing Corning's research, development, and engineering resources.

As you may know, Corning invented low-loss optical fiber. Earlier in my career, I was responsible for bringing this revolutionary product from the laboratory to the market place. I'm here today to discuss this experience because I believe in this case the past is prologue. My views are those of a technologist/businessman, not a lawyer/regulator.

I believe that many of the obstacles we faced over a decade ago in deploying fiber optic technology in the long distance market are similar to the ones that we face today in deploying the technology in the local loop. And, I believe the solutions are similar as well.

I'm not here to suggest that you adopt legislation that will mandate the universal deployment of fiber optics to every home, school, and small business in America. You may find this strange.

But, I believe that requiring by government mandate the deployment of a specific technology, be it ISDN, fiber optics, or something else that may come along, would be a huge public disservice. The pace of technological change is too fast; the demand for new services is too uncertain; and the industry structure is too dynamic to lock communications carriers into specific technological solutions.

Rather, I think we need to focus on creating a new balance between competition and monopoly regulation to ensure an optimal environment for the deployment of technology in the local loop to serve the subscribers' interests. S. 1086 provides lust the framework for such an approach.

THE IMPACT OF COMPETITION ON TECHNOLOGY DEPLOYMENT IN LONG DISTANCE SERVICE

Before getting into the details of S. 1086, I'd like to provide a little history. Specifically, I'd like to discuss Corning's experience with fiber optics in the early years. I think it will provide an excellent object lesson on the importance of competition as a condition precedent for technology deployment.[1]

Corning, along with Bell Labs and several foreign companies, began research into optical fiber about the same time in the mid-1960's. Most were ready to give up by 1970, when a well-known scientist in the field announced at a prestigious technical conference that glass fibers were decades away.[2]

But Corning scientists persisted and made the pioneering optical fiber invention later that year. In fact, it was in August 1970, 23 years ago.

Once the breakthrough was made, the commercial fun began. Corning entrepreneurs had to convince reluctant customers to use glass fiber instead of copper wire in telephone systems. Imagine, 23 years ago, burying cables underground containing glass fibers that could carry thousands of telephone calls simultaneously. You must admit, it sounded strange.

Well, our customers didn't buy it. AT&T, which owned most of the telephone lines in America at the time, said it would be 30 years before its telephone system would be ready for optical fiber. And when it was, AT&T planned to make its own fiber.[3]

At that point, we began to think that we had a technical success, but a commercial failure.

Then we took our technology overseas. Frankly, we had a more friendly reception. We negotiated some technology-sharing arrangements and formed joint ventures to help support further research.

The research paid off technically. By 1975, Corning scientists had developed a fiber which was an 8,000 percent improvement over their original creation. But still, we had no serious customers in the United States. Nevertheless, we built a pilot facility in 1976.

Finally, in 1982, after government action clearly established competition in the long distance market, the commercial breakthrough happened. MCI took the risk and placed a 100,000 kilometer order for a new generation of fiber, single-mode fiber. We took the MCI order, built a full scale plant, and started a technological revolution.

The moral of this story is that a government decision to create a competitive long distance market stimulated demand for new technology, including fiber optics, to better serve subscribers. Our scientific, manufacturing, and marketing expertise alone weren't enough. It was a government decision to move toward competition and away from monopoly regulation that made the difference in the end. Had this decision not been made, optical fiber may still be in the lab.

Not only did Corning benefit from this decision, the nation did. Annual investment in the long distance networks immediately expanded by more than 50 per-

---

[1] For a complete historical account, see Ira Magaziner and Mark Patinkin, The Silent War: Inside the Global Business Battles Shaping America's Future, Random House, New York (1989), Chapter 9.

[2] Op. cit., p. 274.

[3] Op. cit., p. 275.

cent.[4] AT&T alone has invested over $20 billion modernizing its own network since divestiture.[5] Long distance usage increased in excess of 12 percent annually.[6] And, long distance rates fell by almost 10 percent annually.[7] [8]

In short, as a result of a decision to move toward competition in long distance service, Americans got better quality service, more innovative service, and far cheaper service than they would have otherwise.

And, as an added advantage, the entire long distance infrastructure was rebuilt. When the first order was placed in 1982 by MCI, almost no fiber existed in the long distance network of this country.[9] Today, over 90 percent of the long distance network is fiber optics. In eleven short years, the network was entirely converted to fiber, while rates dropped by over 40 percent in nominal terms.[10]

Our experience with long distance demonstrates beyond any reasonable doubt, that competition can drive the deployment of advanced technology while simultaneously reducing prices and improving service.

CURRENT SITUATION

The situation today with the local loop is not dissimilar to the conditions that existed for the long distance market in the 1970's. Local telephone service is a regulated monopoly market, just like long distance was two decades ago.

Another parallel to the long distance experience is the rapid onslaught of competition. Just like AT&T felt the heat of competition by MCI in the 1970's and early 1980's, the local telephone companies are today witnessing the threat of competition from alternative access carriers, cable companies, and wireless technology.

Some local companies have recognized this threat and are aggressively embracing it. Ameritech, for example, has made a bold proposal under its Advanced Universal Access Plan to completely open its network functions to competitors in exchange for regulatory relief. Rochester Telephone has made a similar proposal in New York State.

Finally, as it was the case with AT&T in the 1970's, the local monopoly is reluctant to invest in advanced technology, particularly fiber in the loop. Let me give you some statistics. Overall investment by local exchange companies has remained relatively flat in nominal terms over the last 10 years at about $20 billion annually and has declined in real terms.[11] And, about $3.5 billion is spent annually on copper wire rather than more advanced broadband technology.[12] Some argue that the local exchange carriers are investing at their depreciation rate, choosing to invest excess retained earnings in other more profitable endeavors.[13]

---

[4] Immediately after divestiture in 1982, investment by inter-exchange carriers increased from $3 billion in 1982 to $4.8 billion in 1983. It continued to increase to $6.3 billion in 1986 and since then has trailed off to about $5 billion annually. These numbers were derived from two sources: Statistics of the Local Exchange carriers (1993), United States Telephone Association (to be released in Fall 1993); and After the Breakup: U.S. Telecommunications in a More Competitive Era, Robert W. Crandall, The Brookings Institution (1991), Table 3-3, p. 47.

[5] Robert E. Allen, Chairman, AT&T, Statement before the House Energy and commerce Subcommittee on Telecommunications and Finance, March 24, 1993, p. 7.

[6] Real output of interstate switched-message-transfer services ("MTS") and interstate wide-area telephone service ("WATS") increased by an average annual rate of 12.1 percent during 1983 to 1988, the time period immediately following the AT&T divestiture. See After the Breakup: U.S. Telecommunications in a More Competitive Era, Robert W. Crandall, The Brookings Institution (1991), Table 3-1, p. 44.

[7] The real Consumer Price Index for interstate toll service fell by an annual average rate of 9.8 percent during 1983 to 1989, the period immediately following AT&T divestiture. Op. cit., Table 3-9, p. 61.

[8] In testimony before the House Energy and Commerce Subcommittee on Telecommunications and Finance on March 24,1993, AT&T Chairman, Robert E. Allen, elaborated on benefits of competition in long distance by stating: Particularly in the eleven years since the divestiture decree was entered, the competitive parts of the telecommunications industry have experienced extraordinarily dynamic growth. Hundreds of new carriers have begun providing long distance service. Private business communications networks have proliferated; specialized value-added networks have become virtually ubiquitous; thousands of databases provided by hundreds of competing suppliers are now accessible over the public switched network; and a sophisticated array of both customer premises and network telecommunications equipment is now available to customers.

[9] AT&T had deployed some multi-mode fiber in the Northeast Corridor. But, virtually no single-mode fiber, the technology of choice for long distance applications today, had been deployed.

[10] Supra, Note 5, p. 8.

[11] See Table 1 attached.

[12] Telephony, June 25, 1993, p. 36.

[13] Oral statement of James Q. Crowe, Chairman of MSF Communications Company, before the Senate Committee on commerce, Science and Transportation, Subcommittee on Communications made on July 14, 1993.

While I believe these monopolies would like to invest at a faster pace in advanced technology, a regulatory mine field discourages them from moving forward. They are restrained in their investment behavior by state regulatory authorities that disallow investment. They can't charge market clearing prices for new services made possible by investment in advanced technology. Subsidies from business to residential service necessitate minimal investment to keep rates low. And, they are still largely precluded from providing the one video service with demonstrated demand, video programming, which requires investment in fiber optics and other broadband infrastructure.[14]

So, the bottom line is, if you're satisfied with plain old telephone service over copper wire, then there's no need to act. The existing infrastructure can do the job.

But, if you want Americans to have higher value-added services necessary to remain internationally competitive, then the local infrastructure must be upgraded. And, this can be achieved without increasing the rates for telephone service, just like it was achieved in the case of long distance.

In essence, our experience in long distance has proven that Americans can have access to advanced services over a modern infrastructure without paying higher prices for basic service.

THE RED HERRING

Before discussing how to translate the experience in long distance into the local loop, I'd like to spend a minute talking about the big "red herring"; that is, the argument that numerous groups have made that we can't afford to build this new infrastructure because it's going to be outlandishly expensive.

The argument goes like this. Building a fiber optic network to every home in the country will cost $450 billion. Absorbing this increased investment will raise rates particularly for subscribers who want to use the telephone system just to make telephone calls. The country is bankrupt, and we can't afford to pay the price.

This argument is a gross exaggeration intended for the sole purpose of scaring policy makers away from broadband technology and into inaction. A number this big is enough to scare anybody.

But, the reality is, the local exchange carriers will invest $400 billion over the next 20 years anyway. As I've already demonstrated, the local telephone companies are investing at an annual rate of about $20 billion per year. Telephone networks need to be constantly added to and rehabilitated to maintain reliable service.

At some point, when the cost of new technology falls below the cost of existing technology, telephone companies will invest in more advanced technology in the normal course of building and maintaining their networks. In other words, the deployment of advanced technology could very well become simply a cost driven phenomenon.

If it is, we expect that the entire network could be rebuilt for no additional cost whatsoever. However, universal deployment in this situation would take a long time. We estimate that universal interactive broadband service over the public switched network would not be available until sometime around the year 2035, if deployment is solely a cost driven phenomenon.

Fortunately, we have a choice. If policy makers would like to advance the pace of deployment, they can do so at a relatively small incremental investment. We estimate that the incremental cost of accelerating universal deployment from the year 2035 to the year 2015 is only $23 billion in 1992 dollars, or about 70 cents per subscriber, per month.

While $23 billion is certainly a considerable sum, it is nothing like $450 billion. And, even if our estimate is off by a factor of 2, a $50 billion incremental investment at the national level spread over 20 years is not something to be frightened about, especially when measured against the benefits. The Economic Strategy Institute measures these benefits at $321 billion in additional GNP.[15]

So please, do not be threatened into inaction by outlandish estimates from technology opponents. Certainly cost is an issue. But, it isn't such a huge obstacle that it should halt the process or press you to pass legislation that will further discourage investment in broadband technology.

---

[14] The August 24, 1993 U.S. District Court Order to enjoin the enforcement of the cable-telco cross-ownership prohibition applies only to Bell Atlantic. The application of this decision to other RBOCs remains uncertain at this time.

[15] Economic Impact of Broadband Communications on the U.S. Economy and on U.S. Competitiveness, Robert B. Cohen, Economic Strategy Institute (1992), pgs. 1 and 13.

THE S. 1086 FRAMEWORK

Now I'd like to turn my attention to the bill in question, S. 1086. I believe that the bill builds on our experience from the past. As was the case of long distance, it is based on the fundamental realization that competition for local communications service can become a reality and bring enormous benefits to subscribers. It is clearly possible with today's technology for two or more providers to serve most local areas. A similar precondition existed for long distance over a decade ago.

In recognition of this fact, the bill opens up the local loop to competition. It does this by requiring that the local exchange monopoly unbundle its network functions and make them available to competitors. In exchange for such openness, the bill gives the local exchange carriers more freedom on pricing and freedom to enter into new currently prohibited businesses.

In addition, the bill includes a number of important features:

- its provisions are technology-neutral;
- its provisions are intended to be provider-neutral;
- it includes safeguards to protect telephone subscribers against cross-subsidization;
- it includes safeguards to protect competitors against anti-competitive practices;
- it includes provisions to guarantee equal access to competitive providers; and
- it recognizes the need to ensure universal service to subscribers in areas not benefited from competition.

In short, the bill includes all of the fundamental features to establish a new regulatory framework necessary to encourage rather than discourage investment in advanced technology, as is currently the case. I believe that enactment of legislation based on the fundamentals in S. 1086 would result in the accelerated deployment of an interactive broadband 'network of networks" to serve all subscribers in your lifetime and mine, not by the year 2035 as we currently project. And this "network of networks" would include local telephone companies, cable companies, wireless carriers, long distance carriers, and others, all interconnected into a seamless web.

This is not to say that S. 1086 is perfect as it is. It is not. It will need refinements as the legislative process progresses.

But please remember, as these refinements are made, that there is a sensitive balance between necessary regulation to ensure fairness and over-regulation that will squelch investment. Every change that is made in S. 1086 must be measured against this balance. After all, the intent of this bill is to encourage local exchange carriers and others to invest in local networks.

Finally, please keep in mind as this bill is refined, that the telecommunications industry is rapidly changing. Its dynamism is its strength. Hence, narrowly drawn statutes may quickly become obsolete or, worse yet, dampen the industry's dynamism.

CONCLUSION

As you can see, the parallels between the long distance market in the 1970's and the local market in the 1990's are striking. The problems and the solutions are similar. We need to build on our experience from the past in order to build a better future.

We have this opportunity today with S. 1086. It provides us with the framework to move forward. Armed with that framework and with the experience of the past, we can refine S. 1086 and make it a better piece of legislation to serve the public's interest and to serve the interests of all the players in the telecommunications field.

Mr. Chairman, thank you for the opportunity to appear before this Subcommittee.

TABLE 1—Stagnant LEC Network Investment

[In billions of dollars]

Year	Construction expenditures in current dollars	Construction expenditures in constant 1987 dollars[1]	Year	Construction expenditures in current dollars	Construction expenditures in constant 1987 dollars[1]
1983	17.5	18.4	1988	20.0	19.6
1984	17.9	18.8	1989	18.9	18.3
1985	20.3	21.2	1990	19.3	18.4
1986	20.0	20.3	1991	18.4	17.4
1987	21.9	21.9	1992	20.0	N/A

[1] Deflated by implicit price deflator for nonresidential fixed investment, producers' durable equipment (see 1993 Economic Report of the President, Table B–3, p. 352).

Source: U.S. Telephone Association.

Senator INOUYE. Thank you very much, Dr. Duke. I would like to begin, if I may, by asking a few questions. Mr. Allen, I am certain my colleagues will be touching upon the AT&T-McCaw merger, so I will stay away from that.

I think we all agree that there is increasing competition for local telephone services at this time. And an important result of this competition is that telephone companies are losing some of the revenues that they have traditionally used to subsidize high-cost areas and low-income customers.

What do you think is the best way to ensure universal service in this new competitive environment?

Mr. ALLEN. Well, I think, Senator, I would probably take issue with your only assumption with respect to whether competition may exist in the local exchange.

I, along with many others, have commended Bill Weiss on the bold move they took, or have suggested that be taken, which would open up their local exchange for competition, although I think they have the cart before horse; that is, seeking entry into long distance before competition really exists.

I think there is very little evidence that in the bottleneck local exchange that competition really exists. As I suggested in my own testimony and my open remarks, customers who are asked the question today, do you have a choice, by and large have to say no, they do not have a choice.

Having said that, I hope that the competition will exist in the local exchange at some point in the future. I think it needs to be tested.

I think we need to have methods to determine not only whether competition is available, but whether indeed customers are choosing when competition really exists, that is, whether customers are making a choice.

And as I asserted in my earlier testimony, I think that is one of the changes that needs to be made to this legislation, so that we will know competition when we see it.

So, I would take issue with your premise, having said that, and hoping that competition will come. And recognizing that adjustments will have to be made as a result of that, I am very concerned about universal service and believe that proper safeguards need to be taken at the regulatory level or through direct subsidies, perhaps to poor people or others who do not have a choice, who therefore might have their rates potentially raised beyond their affordability. I think, however, that is on the margin.

I also believe that it is not axiomatic, that permitting competition into the local exchange automatically means that somebody's rates have to go up.

In fact, I believe that is monopolistic thinking and is a return to the rate base rate of return era that we knew before divestiture, which unfortunately still exists in most states, that is to say, competition forces costs out of the business, and not necessarily a shift in cost.

You may have a shift in cost, in your booked costs, but it did not necessarily mean a shift in pricing, and therefore higher rates. What competition does is drive out unnecessary costs. And so I do not automatically assume that universal service will be in jeopardy, because some people's rates have to be increased.

Senator INOUYE. Thank you very much, Mr. Allen.

Mr. WEISS. Mr. Chairman, may I comment at this point?

Senator INOUYE. Please do.

Mr. WEISS. It seems to me—to just respond to some of the points that Mr. Allen made—having zero percent market share in his business, I have pointed out that he has some market share in my business. It seems to me that we would not harm them very much if we are given freedom to compete on a wholesale basis.

I would like to point out the issue of our attempting to become competitive, as have the carriers such as AT&T. There has been no rate increase in my territory for local telephone rates for a period of 6 to 7 years.

As a matter of fact, over that period of time we have reduced rates or refunded to customers amounts up to, last year, a total of $800 million. So, this is a significant effort on our part to also work in the new world that exists.

Do customers have a choice? I must tell you, today the price of providing a local dial tone line in the basic exchange is about $1,000.

It is about $1,100 to provide a line for cellular service. They are very close and very competitive, and customers are making those choices, even at the local residential level. That will increase.

There are cases in our industry where people are using cellular service as a substitute for a normal wire-base local exchange service simply because it is the most economically efficient way to provide that service to customers. So, this is happening.

You know also, I think, that the regional companies have maintained, in everything they have said or written, a commitment to universal service. We think that is a very great public interest matter, and we do not intend to step away from that. Thank you.

Senator INOUYE. Thank you very much. I would like to ask you, Mr. Weiss, a question, if I may. I think all of us here have suggested that the goal of this bill is to develop infrastructure to speed the delivery of new services, libraries, schools, medicine, et cetera, et cetera.

Which of the following would result in the greater investment in the network to attain goals of this bill, allowing the Bell Co. to compete for long distance or to compete for cable television and video programming services?

Mr. WEISS. Mr. Chairman, as I tried to point out in my comments, these markets are all converging. What we are really talking about is all of the players in this increasingly competitive market having an opportunity to compete for the revenue of the customer, the consumer, so that they can continue their investment pattern in the future.

It is just about that simple. I think with the long-distance carrier being able to move in at will to the local exchange, with the cable company being able to move in at will to the local exchange, granted there will be regulatory discussions in the process, but when that is occurring, they can become full-service provider to consumers, which is something we cannot do.

And it seriously disadvantages us in the marketplace. So, again, what we are talking about is, as Mr. Kapor stated, I think, the broadest level or broadest framework that will permit regulators to deal with these situations as they develop on a local basis.

Both of these are essential for us. Both of these are essential to the other players, the long-distance players and the cable players in the industry. We are just asking that the rules be equal.

Senator INOUYE. Before I proceed any further, I would like to invite all of the witnesses on this panel and other panels to submit to the staff, if they so wish, suggested amendments with explanations.

And the sooner we get that I think the better we can serve all of you. So, we would appreciate such assistance and cooperation from your side.

Senator PRESSLER. Mr. Chairman, could those be shared with other members of the committee as they come in?

Senator INOUYE. Oh, absolutely.

Senator PRESSLER. Thank you.

Mr. ALLEN. Senator, could I make a clarifying point for the record?

Senator INOUYE. Yes, sir.

Mr. ALLEN. Mr. Weiss and I have been colleagues for many years in this business, and our behavior here is not any different than it normally is outside the room. And we are good friends and good golfing friends.

But I do have to clarify one point that I do not think is generally understood. And that is that the RBOC's today—in fact, all the local exchange companies—are in the long-distance business. They are essential partners.

In fact, 40 cents of every dollar AT&T collects, and any other interexchange carrier collects on a long-distance call goes to the local exchange company as a partner in accessing those local customers for long-distance calls.

Our analysis indicates that in many cases that of 40 percent, or 40 cents out of every dollar, which aggregates to $25 billion a year, constitutes many times all and sometimes more than the total profits of some of the entities among the RBOCS, among the telephone companies.

So, they are in the long-distance business. They are essential partners—$25 billion is paid to them as essential partners by the interexchange carriers. And last year $250 million, or something on the order of 1 percent if my arithmetic is correct, were paid to other providers to access our customers, our local exchange customers.

And therefore, 99 percent of our business is with their customers, our joint customers, is done with them. And to suggest that we have the right and the capability and are behaving in a sense that going direct to their customers, to serve them without the local exchange company, just violates the facts that exist. Thank you.

Mr. WEISS. Mr. Chairman, I would like to respond.

In this analogy that Mr. Allen makes, he is absolutely right about the percentage of revenues paid to AT&T, in terms of these connecting links, but let me tell you how the game is played.

AT&T customers in Chicago, for example, Sears and Roebuck or the First Chicago Bank, are provided service in their basic on-premise systems, their large business systems by AT&T.

That service is collected and delivered directly to AT&T's connecting point in Chicago, so that the long-distance traffic goes directly to wherever without touching the local exchange.

Senator INOUYE. What is the method of connection?

Mr. WEISS. Pardon?

Senator INOUYE. How do they connect?

Mr. WEISS. By a competitive access provider. And the issue here is that that is not counted in the revenues AT&T pays us, because Sears and Roebuck or the First Chicago Bank pays directly to the competitive access provider for providing that linkage.

So, the statistics we are talking about here have to be understood in their full context. And I can tell you, in Chicago, we, today, have 60 percent market share of that special access connecting business.

So, you can make numbers do anything, but it is the way the game is played that we are talking about. And that is why, at the local exchange, it is truly a competitive environment today.

Senator INOUYE. Now, we will get out of our box and AT&T. We will go to Mr. Kapor.

We have received suggestions from many sources that, by legislation, we require phone companies to deploy certain types of advanced technology, such as the integrated services digital networks or isometrical digital subscriber line, in order to deliver video over today's network.

Do you think that the Government should promote one technology over another or should we leave that up to the open market place?

Mr. KAPOR. Quite clearly, we do not think it makes sense for the Government to mandate particular technologies. Our emphasis on the importance of ISDN, which is a point that EFF has been very much involved with, has actually produced some benefits—helped to produce some benefits—in the voluntary undertaking of wider deployment of ISDN by Ameritech and other regional companies.

I won't say that battle is completely over, but many of the goals for ISDN as an interim step in infrastructure deployment appear to be in the process of being accomplished through voluntary private sector action, which makes us perfectly happy.

Given in today's environment, which is changing rapidly, technologically, and in the changing regulatory framework—we are also not suggesting mandating a technology.

We have suggested, though, that there is a kind of bargain which can be struck, which is that to the extent that we have entities, carriers, providing video programming—and being in the entertainment business—they undertake an obligation to provide open platform service.

We envision this applying in a nondiscriminatory way to many different carriers and also across the full range of whatever technologies they choose to deploy these services on. No mandates.

Senator INOUYE. Dr. Duke, you indicated that you are appearing as a technician more than a policymaker here. And you spend much time on telling us about the value of fiber optic. And we agree with you.

I think we were late in appreciating that, but do you think that this will be the major answer to our debate at this time, the use of fiber optics?

Dr. DUKE. Mr. Chairman, I do not think any specific technology is the answer to the debate. In my statement, I said we should not mandate, you should not mandate, any particular technology—fiber optics or ISDN or anything else. Things move too fast. There are still many applications for satellites and for coax and for fiber and different types of fiber.

So, I do not think any technology is the answer. I think it is a policy issue to enhance competition. I think this is what it is all about.

And then, the technology, the innovation, the market, will put in whatever is most cost effective; whatever brings the best service to improve the quality of life at the best cost.

Senator INOUYE. Thank you very much. I will be calling upon the author of the measure, but before I do, if you look at the makeup of the subcommittee, you will notice, there is an over abundance of Senators representing rural America. And I am certain several of them will have questions on rural America. So, you should prepare yourselves.

Senator Danforth.

Senator DANFORTH. I would like to thank the panel. I think this has really been a very excellent panel. And I appreciate everybody being here. And, Dr. Duke, I especially appreciate your testimony.

I think you are exactly right. The idea is not to change the way in which Government gets in the way, but to get us out of the path of progress and allow competition to prevail. And I appreciate your comments.

I would like to raise a question to the two golfers and see if it is possible to at least narrow what the issue is, so that we can precisely understand what the issue is. Tell me if I am wrong.

The rationale—and Mr. Weiss might contest whether there is any justifiable rationale, but the rationale for keeping the Bell companies out of long distance is that the Bell companies constitute monopolies within the regions with respect to local telephone service.

Those monopolies, in turn, constitute bottlenecks and, therefore, if the Bell companies, which, if they have monopolies for local service, get into long distance, that is an unfair situation.

That is the rationale, is it not, Mr. Allen?

Mr. ALLEN. That is the rationale. And that was the basis on which, as you know, the divestiture or the breakup of the Bell System occurred. And that was the linchpin. And, obviously, that has worked well in separating the competitive from monopoly businesses.

Senator DANFORTH. Now, the converse of that would be that if there was competition within a region—if there was competition for local telephone service—the rationale for keeping the Bell companies out of long-distance service would no longer exist. They would be comparable to anybody else who wanted to get into long distance.

Mr. ALLEN. Customers, first of all, would have choices. And second, they would make choices, when there is some point at which it is clearly demonstrable that competition exists. Yes.

Senator Danforth. Yes. Now, you, I think, Mr. Allen, said something which is very, very interesting and very important. And you said what we need is methods of determining whether competition exists and that you recommended changes in the bill so that we know competition when we see it.

Now, what I am wondering is: Would it be possible to focus on the definition or to focus on the test? What does constitute competition? This is, really, how we approached the cable legislation, trying to determine what it was that created the threshold of competition.

Is it possible to agree on what the problem is, namely, how to define competition? And second, is there any possibility that we, you and those of us on the committee and others, could try to come up with some definition, some test of what constitutes competition and what does not?

You, Mr. Allen, said, "There really is no competition now."

And Mr. Weiss says, "Yes. There is competition."

Obviously, that is a very, very strong difference of opinion, but is there a possibility, without getting into what your difference of opinion is right now, to try to concentrate on how do we define what the issue is?

Mr. ALLEN. Well, I do not have a corner on the market on this subject. And I am not sure that anyone else does. I would suggest that what you did with respect to cable is probably not—or some structure like that, some framework—is not the question in my mind.

And, indeed, to the extent that we disagree on whether competition exists in small or large segments of Ameritech's market or somebody else's, I think, can be measured by—in some terms, on the basis of do customers—can they answer "Yes" to the question: I have a choice of some other carrier to provide my local exchange service, whether it is the Sears in the world or the local exchange customer who lives in Peoria?

And there ought to be some point at which choice exists. That would be the first guidepost in the framework. That is to say, I think, in the cable bill legislation you talked about 50 percent, as I recall.

The 50 percent—at least, 50 percent in a given territory or area or market have a choice. And that some percent are actually exercising and have exercised that choice.

And perhaps, I do not know what that number ought to be, but maybe it is 25 percent. And I think if you could segregate markets, territories, regions, areas or some relevant markets, that some test like that might be applicable and would be some matrix, at least.

Now, I am sure there are other people who have different ideas about matrix, but that would be one way.

Senator DANFORTH. Mr. Weiss.

Mr. WEISS. May I respond, Senator? To Mr. Allen's point of choice, if you ask Sears the question, they will say "Yes." If you ask somebody in Peoria, they will say "Maybe." And it totally defines—

it gets to the point of how do you define the markets that exist in the local exchange.

There are, as I tried to point out in my opening comments, many markets in the local exchange. And the interesting fact is that monopolies exist today only where somebody else has chosen not to enter the market.

Now, any businessman thinking about that understands why they have chosen not to enter the market. It is because they cannot make the highest margins there.

And generally, that remains the last vestige of what people would call a monopoly. But with alternative technologies available, like local exchange compared to cellular service—to do the same thing, to make telephone calls, either local or long distance, customers have a choice today, not among just two, but three providers in most cases.

With the FCC moving forward with further frequency allocations, we will have multiple providers in every local exchange in the United States. And I would suggest that AT&T will probably follow their business strategy, as I understand it looking from afar, and will continue to move into the wireless business as a direct contact, an end-to-end kind of contact with the customer.

So, I think we could get to a broad definition of "competition," but then I think we have to get down to what is a local exchange versus what is the intercity business or the toll business?

And what are the elements? And where does competition exist? And I think, rationally, it exists in far more of the industry than we have ever really been willing to talk about to date.

Senator DANFORTH. Well, my point is that it would be possible—in fact, I think this should be the policy, but it would be possible for Congress to create a policy which said that if there is competition, then the Bell operating companies should be able to get into the long-distance market.

Mr. WEISS. I would agree.

Senator DANFORTH. Now, if we created such a policy, then the question—the next question—it's a hard question—is, all right, what do we mean by "competition"? That is a definitional issue. And obviously, it would be a contested definitional issue.

It just seems to me that that is where the contests should be taking place, now, between the Bell operating companies and, you know, others who are—AT&T and whoever is concerned about the Bell operating companies getting into long distance.

And what I am asking is: Would it be fruitful? In the few days that exist, I hope, before this bill goes to markup, would it be possible and would it be fruitful to try our hand at defining just what this—what the meaning of "competition" is? Is that a worthwhile pursuit and would your two companies participate in that pursuit?

Mr. WEISS. I think, yes, but I have some concerns about how this would be legislated here at the national level. If you look at the progress made a number of State regulatory agencies today, they are making judgments about each element of the local exchange which is going competitive.

And then they change the rules for that portion, compared to some other portion, where they still see the local provider, the Bell

Co. or whatever, being—having monopoly control. And that is unfolding.

They do not tend to use any system of matrix, because technology will move around matrix so quickly that they just cannot stay in place.

What they tend to do is look at the exchange, look at all of the evidence—you know, the growth of the exchange, the revenues and so forth, and how they are divided—and they make a judgment.

That kind of broad supervision is certainly possible, but to get beyond that, when it is so difficult at the Federal level for us to change some kind of a matrix, once established——

Senator DANFORTH. I am not talking about, you know, making a lot, but, I mean, what we did in cable was to say that, you know, the question is the existence of multichannel providers.

And it was not—in other words, it was not in any effort to be in any way specific or detailed. It was a very general kind of a definition.

Mr. WEISS. That is possible, I think.

Mr. ALLEN. We would certainly be willing to cooperate in such an effort. In deference to your disdain for a judge making communications policy, I would, apologetically, at least, remind you that there is a provision in the decree for triennial review to determine whether market conditions have changed, so that the restrictions can be lifted.

That is another avenue here for pursuit, but we certainly would be pleased to work with you in trying to find what do we mean by "competition"? How do we know it when we see it? And at what point does it make sense to alleviate all restrictions?

Mr. WEISS. And, Senator, in the 10 years since divestiture, we have had one triennial review completed, which tells you something about the speed and efficiency of that process.

Senator DANFORTH. We do not like courts either.

Mr. WEISS. For a regional company to get a waiver, the average timespan of a waiver being completed and granted is 5 years. This industry will move past that so quickly that that is not an adequate way in which this industry can be governed.

Mr. ALLEN. I was not recommending. And I am just reminded that we do have another alternative. And it is timely.

Senator INOUYE. Thanks. Senator Pressler.

Senator PRESSLER. Thank you very much, Mr. Chairman.

I am just following up. Indeed, I would not say that this is necessarily a rural State question. I think it could be an inner city question or it could be a question in New York State, where there are smaller communities. The local telephone companies currently have universal service obligations.

They argue that opening the local loop will allow competitors to cream, skim, or cherry-pick—that is, serve the most lucrative customers, while ignoring the others.

S. 1086 attempts to address this problem by requiring all carriers to contribute to the universal availability of affordable phone service.

Now, Mr. Allen, you suggested any subsidies should generally be provided directly to the end user. Would subsidies to end users ensure that new services are brought to higher cost service areas? These may not be rural areas in all cases.

A United States Telephone Association study estimated that local telephone service is subsidizing universal service at $20 billion annually. Are you in agreement with the findings of that study?

After you have responded, anybody else who wants to make a comment is welcome to.

Mr. ALLEN. Well, first of all, I think there's adequate history that demonstrates there are adequate resources at the State level—public utility commissions and otherwise—to protect those who potentially cannot afford telephone service or other services for that matter.

And, therefore, while I think it is a very important issue and needs to be dealt with and it is one that we commit to, even through the subsidy that now flows from our long-distance rates to the local exchange and, indeed, to those of low-income and in rural areas, I do not believe that is a concern that—it is a concern for this committee and this legislation, but I think that can be dealt with, perhaps, at the State level.

One of the difficult questions, obviously, is determining who should be subsidized. And I do believe that we must face the fact that modernization of the local exchange—and I am not in that business and so this is an opinion—will logically come, as it has in the past, to the larger metropolitan areas first, because that is where more customers are.

That is where you get the most efficient investment. And, typically, if you look at the history in this industry, the newer technology has come to the larger areas first and then it has followed along.

That does not mean they have to be denied. It is a question of timing, a question of affordability on the part of the telephone companies to make the huge investments that are required to upgrade all of the facilities.

So, I do not think we have to sacrifice that. I think that we have to be realistic about how capital investments are made.

I am not familiar with the study, which you cite. So, I have no comment about that.

Mr. WEISS. Senator Pressler, technology today is permitting us to do a much better job of providing advanced capabilities in the rural or the less populous areas of any given territory.

For example, in the provision of ISDN service, we will have, in my territory—and this is not too different in the other companies—about 80 percent of the businesses or residences. They will have this service available to them by 1995.

Now, when I say 80 percent will have it available, what I am suggesting is we have equipped the central offices and related technology with the frames in which the plugging units are placed when a customer orders service.

What it does is keep our investment low for the moment, but makes it available on an instant response for the customer. That is what is happening with the ability of technology today. And it is a very forthright way of going about it.

The subsidy question, though, must be dealt with. I would only add to what Mr. Allen said, I think, by saying it is both a State and Federal Communications Commission issue. It has got to be dealt with at both levels.

Senator PRESSLER. I would address this to Mr. Allen and anyone else.

S. 1086 is designed to open the local telephone loop to competition. You have stated that competitive forces are the best way to spread delivery of advanced telecommunications services.

For example, when AT&T faced stiff competition in the long-distance market, you accelerated investment, as I understand, in digital fiber technology and wrote off billions in older analog, but still serviceable plants.

Why would local telephone competition encourage similar investment and who will pay for the investment—ratepayers or shareholders?

Mr. ALLEN. Well, in final analysis, ratepayers pay for everything, even though you may book a shareowners loss at a given point in time to update your technology, which is what we did when we wrote off $6.7 billion of analog equipment.

And so our shareowners took the hit on that, but it was mandated by a competitive marketplace, in which we operated at that point in time.

I believe, while competition may not be applicable in every aspect of the local exchange company, for reasons that you and others have cited about concern for rural areas and for poor people who cannot afford higher rates and so forth, I believe that incentives in a competitive marketplace will drive the local exchange carrier or any other provider to put forth the best possible technology, in terms of price and performance, service and so forth.

And so—otherwise, you have to say, "Well, I cede a certain portion of my market to competitors, because I am unwilling to make the investment." And so I think the natural forces of competition provide those incentives.

Senator PRESSLER. Mr. Weiss, I know you are testifying on behalf of all seven regional Bell companies. At the same time, I understand your company, Ameritech, has taken a somewhat different approach to local competition than the other Bell companies.

Would you briefly describe the Ameritech plan pending before the FCC and is congressional action needed for Ameritech to implement this plan?

Mr. WEISS. Yes. I would be happy to do that and then make a couple of comments about it.

The Ameritech plan is simply a plan to totally open the local exchange network to any provider using any component of that network that that provider chooses to complete his or her service offering to the customer.

So, what it does is it is fully open to local exchange in its pieceparts, if you will, and permits access to those parts on the cost-base pricing basis to competitors who would enter the market.

Now, our philosophy is fairly simple. Over time, it is going to happen anyhow. And No. 2, we believe that if we make ourselves more available for others, our network will grow or the components of our network will grow. So, we do not see this, in my case, as a necessarily losing business proposition.

On the other hand, the conditions across this country served by the seven regional companies, do differ. Each of the regional companies has a plan that will move toward these same goals they are talking about and working with at either the State or Federal level. So, it is an issue very much in the minds of the regional companies.

We have simply said, in our case, that we think it is so important to get these policies cleared up for the country that you need a place to try.

You need a place to go as far as you can go and see if it works and learn from it, and then create policies based on those findings. That does not need to be done everywhere, but it does say that all of the companies stand ready to move ahead in this competitive provision.

Senator PRESSLER. Mr. Kapor, how would you define "universal service" in the Information Age? I know that the Communication Act's guarantee of universal service must mean more than access to plain old telephone service.

In a digital world, universal service should ensure access to digital technology. Advanced communications capabilities are a vital ingredient for economic development initiatives.

That is what my community leaders seem to want, up-to-date digital switches and advanced telecommunications services. How would you define "universal service" in the Information Age that we are in?

Mr. KAPOR. Let me emphasize two aspects of universal service for the Information Age. The first is that it must have the character of a switched network.

The way today's telephone network is, anybody who is on the network can talk to anybody else. That is in contrast to today's cable television networks that are one-way networks in which all programming originates at a head-end and comes out to subscribers. To promote diversity, the network has to be switched.

Second, there has to be enough capacity in the network to let people originate—this is users, at the community level—schools, hospitals, towns—their own sources of programming.

In other words, something which is more symmetric. It is not sufficient to have a system in which subscribers can simply receive video programming, but it has to be easy for somebody who is on the network to originate some form of interactive video. That is the way that town governments will be able to talk to their citizens and school districts and so on.

So, there has to be enough bandwidth in the network, in the upstream direction, to permit people to originate programming.

Senator PRESSLER. This is my final question. I will submit the rest of my questions for the record. Mr. Duke, you tell a compelling story about how the AT&T divestiture and emergence of MCI spurred investment in fiber optic networks.

You note that overall investment by local exchange companies has remained flat over the past 10 years and change is needed to stimulate similar investment in the local loop.

You specifically mentioned that State regulatory bodies may discourage such investment. How do current rules discourage investment and what effect would the changes you recommend have on universal service?

Dr. DUKE. Well, Senator, again, I'm not a regulatory expert, but my experience is that the public utility commissions look at what is the cost of—at the moment, telephone service, plain old telephone service.

And if any investment comes in which would enhance the service, like Mr. Kapor talks about, since the local telephone exchange or the local telephone company may not carry that enhanced service, then the PUC's do not agree with an additional tariff.

I would like to add to what Mr. Kapor said. I think if we are going to have the broadband information highways and the quality of life we all talk about, we must have a switched network.

It has got to be point-to-point. It has got to be broadband. And it has got to be two-way. And at the moment, if you have the PUC's which are looking at that in terms of telephone service, then the cost of this network will be higher. We must enhance the services and put more things on the network.

Senator PRESSLER. Thank you, Mr. Chairman. I have several additional questions, which I will submit for the record.

I thank the panel. I think it is an excellent panel.

Senator INOUYE. Thank you. Senator Burns.

Senator BURNS. Thank you, Mr. Chairman. I am sorry I missed out on the statements, but I will read and I have read most of them. And we will get into the questions.

First of all, I want to congratulate Mr. Weiss and Mr. Allen, both. They have made some extraordinary moves with a little vision and boldness to advance not only their own companies and corporations, but also would tell the rest of America that there are things that are going to happen in this great industry whether Congress moves or not.

And I congratulate both of you. And I do not know what the negotiations would be like on the first T-box. I cannot even imagine what lunch would be like, but I would imagine that the negotiations starts much before we arrive at the parking lot.

With regard to the reports and the editorials in the Wall Street Journal—and I would ask of both of these gentlemen about safeguards and what safeguards that we can develop in the legislation to protect residential and rural communities in this connection.

I'm recalling our experience under the airline deregulation. Even those who believe in airline deregulation, they believe it was unwise, but it is impossible to return to the old system. In other words, you cannot put Humpty Dumpty back together again.

And so I would just like to get some type of comment from both of you that would ensure that States, such like Montana, and the inner city areas, where there are limited resources for investment, and those areas are protected and will be—these services will be made available to those people. And I guess I am saying about universal access.

And Mr. Weiss, if you want to respond to that, and then Mr. Allen, I would appreciate that.

Mr. WEISS. Well, one of the safeguards that has to be clearly in mind is this issue of the subsidies. It is a carryover from the old monopolistic industry. And it still exists.

And it has got to be dealt with, because that, in fact, will make a difference if it is not dealt with, particularly between the rural areas and the metropolitan areas.

It seems to me that in the broad area of safeguards, what we are really talking about here is looking at the various elements of who provides what within this industry and establishing the rules of the game in such a way that no party is disadvantaged.

We have never opted for anybody to be disadvantaged in any kind of rulemaking. And I think it is important for legislation at this level to start there and craft this set of governance issues on that basis.

Mr. ALLEN. I do not have any disagreement with Mr. Weiss. I think my words would be superfluous on the issue.

Senator BURNS. Well, is it true that AT&T and other long-distance carriers are being prevented to compete against Bell companies and other telephone companies in the provision of inter-LATA toll service? Mr. Allen.

Mr. ALLEN. Inter-LATA toll?

Senator BURNS. Yes. Well, I mean, what does inter-LATA competition mean for universal service goals? In other words, that is a goal that we have seen to have embraced and latched onto.

Mr. ALLEN. Well, first of all, there is no competition today between the local exchange companies and interexchange carriers and inter-LATA services. That is one of the restrictions in the decree.

There have been some States in which competition has been open for intra-LATA services by local exchange carriers and interexchange carriers. And I do not think that number is very large, but perhaps half of a dozen at the outside.

And I do not think we have a lot of experience there, but it certainly will drive—to the extent that prices, precompetition, were higher than costs, it will tend to drive margins down—prices down, margins down.

And therefore, to the extent that that was a subsidy for other services, it must be dealt with by the local exchange carrier. And you can either take the costs out or you can get your regulatory body to shift the costs some place else and perhaps the pricing, but it does, potentially, present a problem for protecting universal service in rural areas where, typically, customers do not pay for the cost of providing the service. And so it is something that has to be watched very carefully.

Mr. WEISS. To your point on competition, Senator, there is no competition, as Mr. Allen said, in the inter-LATA business, which is the intercity or between the areas served by local exchange companies.

Now, within the local exchange——

Mr. ALLEN. There is competition, but not between us.

Mr. WEISS. I am sorry. What did I say, Bob?

Mr. ALLEN. You said there is no competition.

Mr. WEISS. There is no competition between us and AT&T or MCI for that service, between cities. Now, within the LATA or within the local exchange territory, there is what some regulatory commissions still call toll service and others call it different things.

For example, Chicago, which is a local exchange area that is about 80 miles long—it goes from Indiana to Wisconsin in terms of its broad dimension—there is no toll service. It is all local measured service.

And a call made from one point to another is an accumulation of local message units as the call is made.

In Indiana, between Indianapolis and Muncie, which are in one LATA or in one serving area, so to speak, it is a different rate structure. And there is a toll route between Indianapolis and Muncie that is within the LATA and is not within the intercity or the inter-LATA definition.

So, when I talk about trying to design some kind of a matrix that would permit sanity to prevail here on market entry, that is the difficulty we have. There just are so many variations on the theme that it is not possible to do that, in our opinion.

Senator BURNS. Well, I would ask Mr. Weiss, we have heard that the R-boxes want to get into the long-distance business. And I contend that there is enough competition in that area right now to provide customers with a competitive market with regard to access and price.

Then, on the other hand, with the move that AT&T has made with cellular or wireless, and they are asking for some protection to keep the Bells out of the long-distance business, I would say that I do not know how to fashion or craft legislation to promote placement of all of the work that Dr. Duke has done.

By the way, Dr. Duke, I want to congratulate you. You are an American hero, as far as the information service is concerned, in developing this broadband stuff.

But to develop this technology and deploy it—how to fashion that and to allow competition in both long-distance and in local loop competition.

Mr. WEISS. What I am talking about is we have a responsibility to the industry systems. We are free to maintain that. There is example after example of market share, because we are going up against competition that has structural advantage over us. That is the issue. There is adequate capacity in the inner city network.

Somebody would have to be foolish with their investment to want to build another inner city network, conditions being what they are today.

Senator BURNS. Mr. Kapor, in your testimony before the House Telecommunications Subcommittee in January you said, and I quote, "The challenge for public policy is to promote widespread affordable deployment of ISDN by local telephone companies."

In an August article in Wired magazine, you indicate that the Nation needs a system of high bandwidth, open architecture, and distributed two-way switching. It seems as though your position has evolved? Am I correct in that?

Mr. KAPOR. I think you are quite correct. If my memory serves me well, when I testified in the House, in that testimony, I indicated that the move toward ISDN should be seen not as an alternative to broadband, but as an interim step toward broadband.

In the context of those remarks, it made a great deal of sense, because it is only within the last few months that we have seen this dramatic acceleration toward broadband deployment, particularly in cable television, but also among the telephone companies.

Our feeling is that ISDN is actually well on the way to becoming, if not ubiquitous, widespread, because of moves by Ameritech and Pacbell and other regional companies and the rapid advancement of the technology of broadband.

The conclusion that broadband was economic to put in place today by companies, because they did not need to do full fiber to the home, but fiber to the curb and use coax the rest of the way, and the opportunities that the provider sees for getting revenue through entertainment, all suggested that broadband is coming, one way or another.

And we think that is terrific. We, therefore, turned our efforts, as I did in the Wired magazine article, to consider, since we are getting broadband and we will have hundreds of channels, how is the public interest going to be served in the broadband regime?

And so we are, indeed, focusing our attention very much on that issue.

Senator BURNS. Mr. Duke, you testified that, incrementally, deployment of a broadband or fiber optic technology probably at a cost around of only around $23 billion.

We have heard much higher estimates of that. Can you tell us how the reason or how you come to that conclusion?

Dr. DUKE. Well, yes, Senator. We have done some studies and there have been several others—Delloite Touche and others—which we could provide you with some of the details. If you look at the investment now in local distribution, it is about $20 billion a year.

When people talk about $400 billion or something like that, they include all of what would be done anyway.

The incremental costs of running fiber to the curb or the final drop to the home is, we estimate, about $23 billion over about a 20-year period of time, or about 70 cents per subscriber per year. And some of the independent studies that we saw were about 66 cents. So, they are very close to what we calculated.

Senator BURNS. Mr. Chairman, in the essence of time, I have a couple of other questions for both Mr. Weiss and Mr. Allen, but we have a tremendously long day. And I know some of my colleagues have questions, too.

And I appreciate both your time and being very open with us here and what we are trying to do in this tremendous industry. And I thank you for that.

Thank you very much.

Senator INOUYE. Thank you. Senator Stevens.

Senator STEVENS. I am conscious of the time, too, but let me say, Dr. Duke, I admire your background and only wish we had some way to turn loose the technology that deal with some of the problems I face, which I will be discussing with your colleagues.

Mr. Allen, I spent the last part—a part of the last month discussing matters with your representatives, with GCI, which is like MCI, as you know, in our State, and Alascom people, discussing this bill.

And as I got into it, I found out that I was dealing with the wrong subject. We really ought to be talking about the survival of telecommunications in my State, in view of the fact that your company has indicated that it will soon replicate the existing delivery system in the State, because it is cheaper to do that, because of new technology, than to buy out Alascom.

Now, that taught me a lesson, I'll tell you, very quickly, that the tumbling technology does bring about what Dr. Duke mentioned, a total reduction in cost of new systems.

And we are now in a situation of trying to figure out what is going to happen. I do not want to color that with you. I do not think that is fair here today, because we are dealing with another subject.

I want to get back to what I mentioned at the beginning. I am of the opinion that this bill gives the FCC a tremendous amount of new authority. We have all endorsed it overwhelmingly.

Yet, I think the FCC are the busiest paperhangers in the State—in the Nation's Capital right now. They have so much to do. I am not sure that they can perform what they would have to perform here, particularly in view of the fact that I think that the authority of this bill completely will eliminate the State regulatory authorities.

And I would call your attention to section 6. One of you mentioned section 6. If the Commission finds that the existing carriers fail to provide service under two categories, sufficient network capacity, basically, and nondiscriminatory rates that are reasonably—based on reasonably identifiable costs, then the Commission can take over or take any action that is necessary to achieve the goal of the bill.

Now, in my State, that is worrisome to me, because of the identifiable costs concept. We have not had rate regulation based upon identifiable costs, particularly rural areas. My State is one-fifth the size of the United States.

I do not need to tell you. It is 3,000 miles from east to west and 2,500 miles north to south. I would urge you gentlemen to give up golf and take up a little marine research and come up my way and try the rivers of Alaska and learn how big it is, and to try to make a call from there.

In the article I mentioned, that the lessons of AT&T cellular move, it is pointed out that McCaw already offers toll-free calling on its cellular network in what it calls the city of Florida.

I am interested in the city of Alaska. And the problem with it is right now, is trying to decide what this bill is going to do to our system.

As I said, when I came to the Senate, Mr. Allen, the Army provided our communications system. They sold it to RCA, then PTI came into Alascom. And now you are coming in.

I am trying to figure out, if you are coming under this bill, what happens to us? And that is my question to you. That is not unfair, I do not think.

It does seem to me that there is very little protection in here for a State that is interested in the carrier of last resort concept. And there is very little protection here for a State that still believes that telephone service ought to be equated to postal service in the sense of having a rate base that provides a commonality, whether you are in Kotzebue or Ketchikan or Anchorage. Your cost to a family and to a business should not be that much different.

Now, is it possible that we are going to be able survive under this bill, Mr. Allen, if you come in? And I think you are coming.

Mr. ALLEN. Well, Senator, first of all, you are to be commended for your long history of being concerned about and protecting the people of Alaska in this particular area that I am familiar with and I am sure in other areas, as well.

And of late, I believe you have joined the belief that competition will be good for your customers in Alaska. And that is one of the reasons why I hope you welcome us there.

We are, indeed, today, every place we serve, whether we wish to be or not, the long-distance carrier of last resort. It is a part of our service history.

It is part of our belief about universal service. It is a part of our culture. And, indeed, it is a part of the result of divestiture, when we were the only interexchange carrier in many locations.

And the competition did not choose to serve many of those areas, because they tend to be low-margin areas and questionable in terms of their investment.

I am proud to say that we have not abandoned any area in this United States. And if we come into Alaska, I can give you the same kind of assurances there.

We will come in there on a business proposition that we hope is successful. And we will have to align our costs appropriately and our rates appropriately to do so, but if we come into Alaska, you can bet that we will serve your customers. And while we will not identify ourselves as the carrier of the last resort; that is, in fact, what we will be.

Senator STEVENS. Well, I appreciate that. As a matter of fact, your former associate, Charlie McCord and I used to argue at length about whether AT&T should come in when we became a State. And I still—if it would be about—you know, about the price of McCaw, better off if you had.

When I was there this last month, I ran into a situation where there is a small village on the west coast, where it costs $70,000 to put in a new switch.

It costs about $10,000 a year to maintain it. It has about an $8,000 revenue—an $8,000 annual revenue. And I understood what was told me at the time, that the reason that that switch was put in there was that a company likes to be able to tell its consumers, nationally, that it can deliver its calls wherever there is a telephone in the country.

Now, maybe that explanation I've simplified it too much, but then I come back and I read this bill. And it says that nondiscriminatory rates will be based upon reasonably identified costs of providing the services.

If those 1,000 people in that village are going to have costs based upon identifiable costs of providing services, they are going to be billed on the $70,000 plus the $10,000 a year. Today, those rates are not anywhere near compensatory to costs.

Now, how can you provide services, say, in rural areas like mine and throughout the country, under this concept of nondiscriminatory rates based upon reasonably identified costs, when you are really talking about competitive situations that were described to me, and that is that you want to be able to deliver your service on a universal basis?

Today you would absorb those costs in your system. And our people would get service based upon what is a reasonably expected cost of service, rather than one based upon reasonably identifiable costs.

Am I wrong to pick out those three words and say that means bad news for rural States?

Mr. ALLEN. Well, I am not sure. I do think, Senator, that you ought to be wary about any legislation that puts all of the power or authority at the FCC level or in any central point in this country.

It is not my understanding—maybe I have misread the legislation—that the elimination of State regulatory bodies is inevitable, if this legislation is passed. I do think you ought to be concerned about that.

Our offerings for a long-distance service are generally uniform around the country, irrespective of the particular location in which we operate, whether it is Alaska or in a high metropolitan area like New York, so that our—one of the advantages we offer our customers is that they know that the rate will be comparably the same any place they go.

So, we will certainly not enter any market, as an institution, hoping to lose money, but we would have to go into Alaska or any other market on the basis of the relevant costs and what our nationwide rates structure would produce in the way of profits and margins.

And it is not likely that the early entry would cause us to realize any profits for a time, but it is a longer term investment in the marketplace that would be of interest to us. But I do not know what specific cost numbers you are dealing with and, therefore, I cannot comment.

Senator STEVENS. I do not mean to be unfair about that. And I am not going to prolong it. You know, the great problem I have is having seen the progress we have made in our State and still witness the terrible Third World poverty in areas of our State, I know that we cannot have service based upon identifiable costs.

Now, we are either going to have service and be part of the network—the telecommunications network of the United States or we are going to go back to the point where we are going to have to have some Federal system serve us.

When we do that, then, and you have mothers and fathers in Florida that are trying to call their sons or daughters that are stationed at some far off point in my State, I think you are going to bear the burden in the long run. You do hold yourself out to be able to deliver those calls wherever an individual is.

Now, I do not know whether the cellular system is going to provide the breakthrough or not, but I just tell you, I have real reservations, based upon the specifics of this bill that deal with, on one hand, the pledge of universal service; on the other hand, a concept of giving the FCC the authority to allocate the responsibility for regulation based upon whether or not the State agency provides a system of nondiscriminatory rates based upon reasonably identifiable costs. I would urge you to study that.

And in my opinion, unless there is some way that the industry can explain to those of us from rural States what happens to those people thousands of miles from any other—1,000 miles from a metropolitan area, what happens to them in terms of basic cost of telecommunications services, then, I think, this bill is going to have to wait for a long time in being enacted, as far as I am concerned.

Mr. WEISS. Senator, if I may comment just a moment.

Senator STEVENS. Yes, sir. I apologize to all of you gentlemen for passing you by, but I think Mr. Allen is my current new found pen pal. [Laughter.]

Yes, sir, Mr. Weiss.

Mr. WEISS. I agree with you——

Mr. ALLEN. I was hoping you played golf. [Laughter.]

Mr. WEISS. I agree with you that the regulatory process must be reformed. The historic investment-based regulation absolutely will not deal with the fast moving marketplace or the rural areas or whatever it is you are talking about in this industry. And that is my point.

We argue to move to price regulation, because that keeps the customer clearly in focus and does have an interest in what they get for what they spend, but rate of return regulation or investment-based regulation gets so detailed and tedious that there is not a regulatory body in the world today that can keep up with this industry and regulate it properly.

So, one of the precepts of this bill, I think, has to be new forms of regulation to deal with the industry as it is moving forward.

The final point, to change the subject a minute, your comment about McCaw and the city of Florida and service that they provide—toll-free calling throughout the State.

One of the points I would make in spades here is if, in fact, AT&T acquires McCaw it is really mandatory that all of the other cellular providers be able to offer similar service offerings, which means, at the minimum, intercity permission ought to be granted to the cellular providers, because when those cellular systems are run by regional holding companies, they cannot do what McCaw does. And they are clearly disadvantaged in the marketplace.

Senator STEVENS. Well, I understand the difference between the investment based concepts and the cost based concepts that we are coming into now.

And that is one of the great feuds we have, however, in the rural areas, because I think that we are at the tentacles of this system as it tumbles. And we do not want to be left out. I want my people to come into the 21st century with everyone else.

And this bill indicates to me that we will stay back in the investment base, while everyone else is going to the cost base system, unless we are part of the national. And we will have to work that one out.

But thank you very much, Mr. Chairman. Thank you, gentlemen.

Senator INOUYE. Senator Lott.

Senator LOTT. Thank you, Mr. Chairman. Now, in retrospect, I think I should have perhaps made my statement at the beginning of the hearing, but now I am going to acknowledge the good work that you and the Senator from Missouri have done on this.

You certainly have stirred up a lot of interest. And I think that it is very healthy. And I hope some good results will come out of these hearings in what finally is devised as legislation.

I want to thank the panel. And I apologize for the length of time you have had to wait to get through these questions. And I just really want to get into one area, since time has just completely gotten away from us.

I think, Mr. Allen, that you should be congratulated for what you have—are trying to accomplish there with McCaw. It is certainly an outstanding company, but there has been, again, a lot of interest and, I guess, some questions about how that is going to work, trying to understand the short- and the long-term implications of that merger and what it will mean both in local and long-distance competition. So, toward that end, I have just got two or three questions I would like to ask at this time.

Over the next few years, could AT&T use McCaw and wireless technology to bypass the local exchange? And if not, why not? Could that not be one of the results?

Mr. ALLEN. Well, there is actually no incentive for us to do that. If you remember the chart I put up there, in order to build a totally wireless connection between wireless callers and anybody else, would cause us to commit untold capital dollars, which we have no incentive to do.

That is to say, less than 1 percent of long-distance calls today are placed by cellular users. Now, hopefully, that will increase over time, but it is less than 1 percent today.

And so we have no financial incentive or other incentive to build something around the local exchange that now connects those customers to the long-distance network, either at AT&T or one of our competitors.

So, it is not our intention. It does not make strategic or financial sense. And we believe that the local exchange carriers can and should be the best providers of access. They have the embedded plant.

They have—in a competitive environment, they certainly would have higher incentive to be the most efficient, the best supplier of access. And why would we want to commit our capital to building something around them, when it is unnecessary, when we have so many other capital needs?

Senator LOTT. So, your answer is you could do it, but at this time, the financial incentive to do it would not be there; is that right?

Mr. ALLEN. Technically, it is possible, because I asked the president of the Bell Laboratories that last week, because this has become such an issue. And I had never asked the question before, which, obviously, means it is not a strategy of AT&T's.

And he said, "Yes. Technically, anything is possible." But it does not make any sense.

Senator LOTT. As far as the number of people using those cellular systems for long-distance calls, I assume that as they are used in greater numbers and, you know, we are such a mobile society. I would say one-half the calls I make are long distance.

Mr. ALLEN. Well, I hope they are AT&T calls. And I thank you for that, but you do not represent the average user of cellular serv-

70-334 O - 94 - 7

ice, which tends to be in the local exchange or in the cellular calling area.

Mr. WEISS. I would just, again, call your attention to the exceptions, which will become the rule before too long. If you look at the city of—Florida service for McCaw, it is a network lashed up to provide statewide free calling within the cellular system.

And so, quickly, at every turn, we have passed what currently seem to be normal conditions and technology will just change the world for us.

Mr. ALLEN. I do believe that Bell-South, who is the other cellular provider there, has the same option to provide the same service.

Senator LOTT. Well, the followup question to this, the Bell companies now have cellular or wireless subsidiaries, and under—but under the current MFT restrictions, they are prohibited from providing the same range of service, I believe, that AT&T and McCaw could provide a customer, and that is both local and long-distance service.

Does that not cause a possible disparity between the two?

Mr. ALLEN. Well, they have the same option to provide local calling that McCaw now has and that, presumably, we would have if this deal is consummated, through cellular calls for local purposes. So there is no disparity there.

The disparity, perhaps, exists with respect to the restrictions of the MFJ that you talk about on connecting long-distance calls outside of the particular serving RBOC territory.

And as I said, that is less than 1 percent of the calls today, but whether it was 1 or 50 percent, the fundamental restrictions that were imposed by—in the MFJ were based on the fact that until and as long as the monopoly bottleneck exists, which it does here and which I have demonstrated we have to use even as a cellular provider, then there still remains that incentive and tendency, if you will, although perhaps not now intended, to subsidize, to cross-subsidize, to favor your own supplier, et cetera.

So, competition, which this bill calls for in the local exchange, eliminates all of those problems. And then the market is open to everybody on an equal basis.

Senator LOTT. Is this a disparity that makes any difference, Mr. Weiss?

Mr. WEISS. Well, Senator, I simply disagree with Mr. Allen on the premise he has laid out. And I think my facts are correct. A regional company is required to provide equal access opportunity for his customers in both local exchange and cellular.

And what that means is we cannot lash up with an intercity carrier to provide a phone network that goes beyond any one of our tightly confined LATA's. You just cannot do it, by law.

Senator LOTT. Just one last question, Mr. Allen, because we are out of time here. Could a consumer subscribing to McCaw select a long-distance carrier other than AT&T?

Mr. ALLEN. They can in some territories. And it is our intention, if the merger consummates, to provide equal access to all of our customers.

They do today in several markets, particularly where they are joint partners with Pacific in San Francisco and with Bell-South in Los Angeles and Houston. And it would be our intent to give all of our cellular subscribers equal access to any interexchange carrier they wish.

Senator LOTT. Well, this is certainly an interesting development. And I am sure we are going to be watching it very closely as it goes forward.

Thank you, Mr. Chairman.

Senator INOUYE. OK. Senator Mathews.

Senator MATHEWS. Thank you, Mr. Chairman.

Mr. Weiss, we indicated a few moments ago that perhaps there was not an area of computation and of rates in some areas that both you and Mr. Allen had jurisdiction over, but there seems to be a pretty good healthy computation for the service in those areas.

It seems to me that we are getting down to the point of what happens in a LATA, the things that you can do in one, the things that Mr. Allen can do and the things that neither of you can do or what.

Would you share with me as a newcomer on this, you know, what is a LATA? What do you do in it? What does Mr. Allen do in it? What does MCI or somebody else do in it?

Mr. WEISS. I will try to keep it as brief and simple as I can. At the time of divestiture, the Department of Justice and the AT&T Co. were seeking a means of separating the local exchange from the intercity business. Intercity being the toll call between major cities or major areas of the country.

The term, LATA—or local access and transport area—is a defined area in which the local exchange can operate fully. It does not preclude competition from entering it, but that is the area in which the local exchange company can operate fully.

Now, as that was worked out across the country, there are dramatic differences in the way LATA's were devised. For example, Wisconsin has four LATA's. The State has four LATA's—northeast, south, southeast, and so forth.

In Illinois there are more than a dozen LATA's, defined more around cities or communities—communities of interest. And any traffic between those, any calling between those, then goes to one of the MCI's, AT&T's, Sprints, or other carriers that provide that service.

So, it was simply a way of defining and dividing the industry on divestiture.

Senator MATHEWS. Could you—do you have the capability or do the local companies have the capability of crossing lines and legislation restricts you or is it a matter that you are just set up within the LATA to serve that area?

Mr. WEISS. No. We used to, but then with divestiture, we separated those facilities, so that the things that went across that boundary went to AT&T, in effect, on divestiture. And we retained the facilities in the local exchange or the LATA, if you will.

Now, could we do it again? Yes. In nearby arrangements, where, for example, the city of Rockford, IL, might want to reach up into southern Wisconsin. And we might be the full service provider, if we had full freedom to do that. Yes. We could do that relatively easily.

Could we become a competitor on a national basis? Absolutely not. We do not have the investment capability to even think about that.

So, our approach is to try to become full service providers to customers by being able to be their account manager, if you will, and work with AT&T or MCI or whoever the customer chooses to set up the total service.

Today we cannot even do that. We cannot resell. We cannot touch that. We cannot manage a customer's business or his intercity component. We are totally prohibited from any part of that.

Senator MATHEWS. I have one other question. And then I want to ask Mr. Allen to respond to this, basically the same one. To what extent does cellular service enter into this equation of within a LATA and outside a LATA? Are you—can you offer cellular service in your LATA's?

Mr. WEISS. Yes. We can offer cellular service within the LATA, but we are restricted to the same intercity prohibitions that we are for local exchange service.

Senator MATHEWS. In Florida, if you are—does a Florida—cellular programs, other than McCaw, now, can they offer statewide service just as McCaw?

Mr. WEISS. Not on the basis that McCaw is doing it. That is my understanding.

Senator MATHEWS. All right. Mr. Allen, on the news last week, I guess it was, a McCaw and AT&T merger, I think, sent some shockwaves through the country, but also sent a lot of hope through the country, too, because of the things that are developing.

I am thinking we are probably at a point where the technology is way ahead of where we are, in terms of allowing that technology to serve the people.

And I am hoping that we are not behind, as a result of the question who is going to serve these people. And I hope we all have the same hope in mind and the same goals in mind on that.

In terms of talking about an acquisition of McCaw, you seemed to say or suggest that it is sort of a routine-type acquisition.

I mean, it is a natural business step in a progression of one step up the line. What other acquisitions has AT&T made since the breakup? What are we moving toward? What are you moving toward?

Mr. ALLEN. Well, regardless of what I may have said, a $12.8 billion acquisition is not routine, even in a company the size of AT&T. And the other major acquisition that we made a few years ago was the NCR Corp., which has now become the computer division of AT&T.

Except for an interest that we have in new technologies and new companies that bring different ideas on the scene—mostly Silicone Valley kinds of interests—we have no further plans in terms of major acquisitions.

We have stomached—if we complete this one successfully, we have done about all that I think we are capable of doing. And, in fact, the division of our business does not suggest that we need to move into any other areas, as long as we can execute on the strategy.

Senator MATHEWS. Have you made any—do you own any cable companies or is this—or is that a next step which would tie all of these operations together?

Mr. ALLEN. No. We do not own cable companies and have no interest in owning cable companies. We would like to be a provider of technology to both the telephone companies and the cable companies, as we see both of those industries change, and hopefully seek a competitive environment stimulated there between the two of them in their own markets.

And we believe we have the technology which can enable both the telephone companies and the cable companies to realize their vision. And there could be other suppliers of that, obviously, but our interest is in partnerships, as opposed to ownership.

Senator MATHEWS. Just one other. Would you care to comment on any of the questions that I asked Mr. Weiss about service within a LATA area?

Mr. ALLEN. No. No. Thank you.

Senator MATHEWS. Thank you, Mr. Chairman.

Senator INOUYE. Thank you. Senator Breaux.

Senator BREAUX. Thank you, Mr. Chairman. And I thank all of the panel members for being here all morning and providing your testimony.

Let me continue with my general theme or premise, and that is, that all communication companies that come to Congress, basically, come to the theme of let me in your business, but stay out of my business.

And I hear that from just about everybody. I mean, some want to do manufacturing. Some want to do local service. Some want to do long-distance service. Some of them want to be in the cable business. Some want to be in the broadcasting business, but it does not leave anybody back in their business.

Now, Mr. Allen, you and Mr. Weiss are incredibly successful and outstanding businessmen, captains of industry, supporters of the free enterprise system.

What is wrong with the Congress just saying, "We are out of here, folks. If you can do, go do it. If you want to manufacture, Mr. Weiss, manufacture. Mr. Allen, if you want to do long-distance service, go to it. If you want to provide local service, if you can do it better than Mr. Weiss' companies, go to it."

I mean, we are trying to determine what competition is here. Should not the free market determine what competition is? What's wrong with us just getting out here? Mr. Allen, do you buy that?

Mr. ALLEN. I think if we could assure the American public that competition—there were no barriers to competition in all the markets you have described and, in fact, customers had choice and the true competitive environment existed——

Senator BREAUX. Well, would that not happen if Congress got out of it? Would that not happen if Congress got out of it? I mean, it is not happening now.

I mean, your argument is that you do not have competition in the local systems, because we have got a monopoly, which was created by the Government. Suppose we did not have that anymore. Would not the competition exist?

Mr. ALLEN. Well, I would hope it would, but, Senator, I came from that environment. I was never malintended. I never intended to cross-subsidize.

I never intended to disadvantage someone who might be a competitor against my business, but in retrospect, the whole Bell System did so.

That was the basis on which that business was broken apart, because as competition entered parts of the business, we could not coexist with competitive businesses and monopoly businesses, because the temptation is there for mischief.

I was a part of it. I did not do it to hurt anybody, but I know that the collective efforts of all of us—well intended—set back the development of competition in this industry by a number of years.

We all did a good job in building a telecommunications system for this country. It is the best in the world. The old system worked. It is time for a new one.

Competition has been demonstrated to work. Let us get competition in every marketplace, except those areas where we need be concerned—in the rural areas or areas where competition may not work. I am willing to concede that.

And then let us all go at each other. And we do not need this committee. We do not need the FCC. We do not need anybody.

Mr. WEISS. I already voted in favor of this proposal, Senator, but we cannot even get started. I could ask Mr. Allen, if he acquires McCaw, would he not at least agree that we ought to have intercity or long-distance parity for cellular providers?

Senator BREAUX. You all have been very nice to each other. I mean, you are trying to bury each other economically. I mean, that is what it is all about. [Laughter.]

Mr. WEISS. Well, would it not be reasonable——

Senator BREAUX. That's right. These gentleman are just being very nice. When they leave the room, you are trying to bury each other economically. That is what it is all about. And I am glad.

Mr. WEISS. Would it not at least be reasonable for parity to exist in those parts of the industry which are similar, like cellular, if they acquire McCaw?

Senator BREAUX. Mr. Allen, why did you all buy McCaw?

Mr. ALLEN. Well, we invented the technology. It seems natural to our business. It is a growing business. And it is something that AT&T needs to do. It needs to grow. It is not growing in its core business. It is a relatively new and developing technology.

Senator BREAUX. What does it allow you to do that you cannot do now?

Mr. ALLEN. It allows us to be in a business that is growing, changing; one that is dynamic; one that we are not currently in, except as a provider of——

Senator BREAUX. Local service?

Mr. ALLEN. No. I do not know how many times I have to say it. I do not want to be in the local service business.

Senator BREAUX. Is AT&T experimenting with facilities of our using cellular or their personal communications system to bypass the local exchanges?

Mr. ALLEN. To my knowledge, the answer to that is "No." We have no intent to bypass, as long as the access providers give us good service, cost base pricing and something—why would we want to rebuild an infrastructure, when we have got so many other capital needs?

Senator BREAUX. Has AT&T not experimented with the question of bypassing the local exchange by the use of——

Mr. ALLEN. I am sure people in our organization have looked at every possibility. Every time we have even thought about bypass as any conscious strategy, we have rejected it on the basis that we have seven RBOC's out there and a whole lot of other local exchange companies who ought to be in the best position of anybody in the world to provide the best access in the world. Why do we want to bypass?

Senator BREAUX. Mr. Weiss, why——

Mr. ALLEN. Occasionally—may I just add one thing?

Senator BREAUX. Sure.

Mr. ALLEN. Occasionally, some customer will come to us, like Sears, he talks about, and say, "Look, I am going to put my service out for bid to a long-distance carrier. Would you like to bid on it?"

Well, my inclination is not to turn away, because I have a loyalty to Bill Weiss, who is an access provider.

Senator BREAUX. Mr. Weiss, why are you worried about Mr. Allen getting into the long-distance service? He says that when he goes to the local service, that it has to come through you anyway, that 99 percent of all of his use of local service, that you are going to benefit from it, because he has to buy it from you?

Mr. WEISS. Well, that simply denies the movement of technology, Senator. I mean, I tried to describe in my opening comments the fact that the local exchange is the multifaceted operation.

And in many elements of that, AT&T is a fairly significant player. So, I mean, I think—and I also tried to say that if you get down to simply trying to define this on the basis of which market AT&T wants to enter or does not want to enter, and therefore, defining only those that they do not want to enter as monopolistic, denies the reasonableness of this whole scheme.

The final point, simply, is this: I have got to take Mr. Allen's word. He has the law on his side. It seems to me this just ought to be changed. That is all.

Senator BREAUX. Mr. Allen, Mr. Weiss, in his testimony, says that your company's long-distance carriers are already in their business.

He says that long-distance carriers are offering local services in competition with the regional companies and local exchange companies. And they have been quite successful.

In the five states that Ameritech serves, long-distance carriers now provide 50 percent of the local toll service provided to larger business companies.

These carriers have also successfully targeted 800 toll-free services. In 4 years, we have lost 75 percent of our local 800 business.

I mean, what he is saying is that you are, collectively, in his business, but he cannot get in your business.

Mr. ALLEN. Well, Senator, we can dispute the relevance of the small segments of the market where Sears is not in the world where we, in fact, compete.

Senator BREAUX. You said it does not happen. It is not that much.

Mr. ALLEN. Well, it is not relevant to the question of whether there is competition existing in the larger marketplace.

It is of concern to Mr. Weiss and should be, but if one complains about losing 75 percent of the 800 service in a LATA, I think one has to go examine its offering in the marketplace with respect to price and service and features, not complain about whether somebody else is in the market or not.

If the telephone companies who have owned the LATA's for all of their existence cannot be the best service in those LATA's on any basis, having them as their captive customers, if you will, then I do not know who can be. It's a fair marketplace with some exceptions, the possibility to walk away from this and let the marketplace decide, because there are forces who will eventually decide these issues in favor of almost all consumers, and we need to worry about those who are on the fringes.

Mr. WEISS. And I agree with that.

Thank you, Mr. Chairman.

Senator INOUYE. Thank you.

Senator Packwood.

Senator PACKWOOD. Mr. Kapor, in your draft statement, under a Jeffersonian vision of national information—this is not your statement today—but your draft statement we had, in your first paragraph you indicate what the benefits are that you want to achieve.

And then you say, "To achieve these and other benefits, the infrastructure must be open, accessible, and affordable to all citizens throughout the country."

In your statement today you say, "To achieve these and other benefits we know that we need more capacity than is currently available in today's analog voice telephone system."

I cannot find this open, accessible, and affordable anyplace else. Have you changed your policy?

Mr. KAPOR. No, not at all. In the interest of keeping my oral testimony to a reasonable length, we simply shortened the full text of remarks. But I certainly stand firm on open, accessible, and affordable.

Senator PACKWOOD. Then on page 4 of your testimony today, you use the argument about printing, and that being, perhaps, the most open, affordable, and accessible, and it is easy to get into, and easy to access. But to an average citizen, how is a DNA tax notice affordable at a couple of thousand bucks a year?

Mr. KAPOR. I am not familiar with the particular—is this an online service that——

Senator PACKWOOD. It used to be a print service. I mean, you can get them electronically now. But these are expensive, usually overnight, daily services of what is going on in some industries or Congress, and they are very expensive, comparatively speaking. But how are those affordable to an average citizen?

Mr. KAPOR. Well, there is an interesting dynamic that happens that we have seen with expensive information services, in the context of the Internet.

The same service which formerly might have been $100 an hour to access is made available commercially on the Internet for $5 to $10 an hour, profitably. How can that be possible?

Well, it turns out that when we have a big infrastructure that is capable of reaching a lot of people you do not have to charge the high prices. To the extent that the infrastructure underlying the service is limited, the providers are going to have to charge more.

Senator PACKWOOD. Well, let me give you another one then. "Open platform services," the infrastructure which you are talking about, "will enable children at home to tie into their school library to do their homework, or make it possible for a parent who makes a video of the local elementary school soccer game to share it with parents and students throughout the community."

What if the parent cannot afford the video camera?

Mr. KAPOR. Well, the interesting things to look at are the cost curves for all of this technology. For instance, the cost of the equipment that you would need to receive this open platform service would be extremely small, especially at the point when it can be integrated into every single television set, just as we have seen captioned decoders being integrated in so that every TV will have that capability.

Receiving information, interactive multimedia, it is not going to be a cost barrier, as long as you assume that people will be able to buy TV's. The last time I heard, more people had TV's than either telephones or indoor plumbing.

Senator PACKWOOD. OK.

Mr. KAPOR. But on the other side——

Senator PACKWOOD. Well, wait. Wait. This is what I want to get to. In other words, affordable does not mean that the Government has to provide it. If people can afford to buy a television, that is adequate.

Mr. KAPOR. Absolutely.

Senator PACKWOOD. OK.

Mr. KAPOR. The whole idea is to ramp up the technology by making it available on a mass scale. That is what is going to drive down costs.

Why do personal computers come down in cost from $5,000 to $500? By the way, the $500 base machine you buy today is about 50 times more powerful than the one you paid $5,000 for 15 years ago.

Senator PACKWOOD. I just want to make sure I understand the goals of your organization. You are not saying at the start that we have to make new technology affordable to everybody in the country as it becomes available.

And it is initially expensive. I realize that it will come down, but we do not have to make it affordable to everybody.

Mr. KAPOR. Absolutely. When we say affordable, we think that the economics of deploying open platforms will cost—that is the single thing that will cause the costs of the technology to come down in the most rapid fashion.

We do believe that at the margin there are still going to be some class of users out there who are going to require some sort of subsidy, but we really think that is the marginal case, just as it is the marginal case today.

Senator PACKWOOD. And there will always be some class of service that is quite expensive, and only a few people can—there are only a few people who want it, quite frankly.

Mr. KAPOR. That is right. If the market is allowed to operate, I mean, it will sort itself out, and clearly there are going to be some very high-end services.

Senator PACKWOOD. My hunch is that the average citizen does not want DNA tax notes. I seldom want it.

Mr. KAPOR. You may also find, though, that there are sources of information today that are very expensive, that are assumed not to be of interest, but, in fact, turn out to be of interest. And there is a barrier, in terms of an artificial scarcity that was created by its price or its technological unavailability.

Senator PACKWOOD. Let me ask you this question, and then I want to go to the two telephone men. In terms of cable television, should there be any charge for premium channels at all, or should that be available to everybody?

Mr. KAPOR. The distinction that we make in the first instance is between the entire range of kinds of services that are offered today on cable—and that includes basic and premium, and so on—which are one-way, noninteractive, and where the content is controlled and selected by the cable provider.

We contrast with that with open platform services which will be interactive, two-way switched, and where the content will be determined by the users. What we are saying is——

Senator PACKWOOD. No. No. You are not following my question.

Mr. KAPOR. OK.

Senator PACKWOOD. I understand there is one-way versus two-way. What I am saying is: Should that benefit be free to the user, whether as one-way or two-way, or are there some things for which you will be allowed to charge which some people cannot afford? I do not care if it is two-way. It is input, rather than receiving.

Mr. KAPOR. I do not have an answer to that question, and I am not going to tell you that I do. But I would say that we would have to look at specific instances, not formulate the general rule.

We have to decide whether something like HBO is important enough to interfere with the working of the market, and demand that it be offered in a certain way. That is outside my area of expertise, but I agree that it is a very legitimate question.

Mr. WEISS. Senator, if you would get back to the ISDN service——

Senator PACKWOOD. The basic what?

Mr. WEISS. The basic ISDN service, the rates for that service in our five States range from $20 to roughly $48. The differences at the high end are because of a good deal of usage added into the rate, where at the low end it is basic service. Now, that is very comparable to cable television rates in most States right now.

So, if you are talking about, "Is it affordable?" broadly, I think so.

And we have had no investment help in making that happen.

Senator PACKWOOD. Eventually, I am going to get back to the old argument of subsidy. I go back far enough on this committee that I can remember the advertisements that Mr. DeButts used to cut on television, and Charlie Brown. I have not seen Mr. Allen a lot on television.

Mr. ALLEN. You will not see me enticed into that arena.

Senator PACKWOOD. OK. [Laughter.]

I would pay to see you. There might be a small base of support.

This statement that Senator Breaux read of yours, Mr. Weiss, "Telecommunication providers, other than Bell companies, already are free to compete outside their traditional businesses. Long-distance carriers are offering local services in competition with the regional companies and other local exchange companies, and they have been quite successful."

Who are the long-distance carriers? Mr. Allen says he is not doing it. Who are the long-distance—or is Mr. Allen doing it?

Mr. WEISS. Yes. He is. And so——

Senator PACKWOOD. OK. Mr. Allen, are you doing it?

Mr. WEISS. So is everybody.

Senator PACKWOOD. All right. Are you doing it, Mr. Allen?

Mr. ALLEN. Are you talking—are you back to Sears? [Laughter.]

Senator PACKWOOD. I am talking about long-distance carriers offering local service in competition with regional companies.

Mr. ALLEN. No. No. We do not offer local service anyplace in competition with local——

Senator PACKWOOD. Now, Mr. Weiss, you said they do.

Mr. WEISS. Yes. They do. [Laughter.]

Senator PACKWOOD. Can you——

Mr. ALLEN. We are prohibited from owning an RBOC.

Senator PACKWOOD. Well, Mr. Weiss, can you respond then?

Mr. WEISS. Mr. Allen very tightly defines a local exchange in his response to that part where they are not involved.

But I have tried to lay out in my oral comments here this morning that there are many elements at the local exchange. And they participate in a number of those markets, as do others. And that is absolutely the way it ought to be.

Senator PACKWOOD. What is Mr. Allen, in your mind, defining very tightly so that it excludes the definition of local competition when he says he is not in it?

Mr. WEISS. He has to define local residential telephone service as the basis of choice.

Senator PACKWOOD. So, he is saying, in your judgment, "I am not in the local residential telephone service; therefore, I am not involved in local"——

Mr. WEISS. Essentially.

Senator PACKWOOD. Is that what you are saying, Mr. Allen?

Mr. ALLEN. I am not in the local residential telephone service business. I am not in the local business telephone service. We provide long-distance service to customers who are both residential- and business-oriented.

And in a few instances, we do direct connect with large businesses who want to aggregate their usage, and bypass the local exchange, which may be——

Senator PACKWOOD. For purposes of long-distance calls.

Mr. ALLEN. For purposes of long-distance calls, and presumably for efficiency reasons, and are managed under their own networks, which is not something that we——

Mr. WEISS. They provide PBX systems in many cases.

Senator PACKWOOD. They provide what?

Mr. WEISS. They provide PBX systems, if that is——

Senator PACKWOOD. You mean equipment.

Mr. WEISS. Yes. And they provide key systems——

Senator PACKWOOD. Well, I do not mean equipment. I am still confused by local competition, as to what you are saying and what Mr. Allen just said.

Mr. WEISS. I am saying of all the revenue streams available to the local exchange companies, to make it a vital company, and to permit it to move forward with the investments that this committee is very concerned about, Mr. Allen is in many of those elements of my business.

Senator PACKWOOD. What are they?

Mr. WEISS. A large business.

Senator PACKWOOD. A large business making local phone calls.

Mr. WEISS. The equipment business.

Senator PACKWOOD. No. No. No. Let us go one at a time. He is a large business making local phone calls. He is in that business.

Mr. WEISS. Yes.

Senator PACKWOOD. Mr. Allen, are you in that business——

Mr. ALLEN. Well, I do not understand. I do not know what his definition of local is. It is possible, I suppose, for a business customer to make long-distance calls that comes back into his phone exchange. It is not a very efficient way of doing it.

Senator PACKWOOD. No. What I essentially mean—although, I understand you are not the Sears provider; somebody else is, from Chicago.

Mr. ALLEN. Even if we were.

Senator PACKWOOD. All right. Even if you were. I think what he means is that the local Sears employee calls the garage to see if their car is ready. He got it repaired, and somehow you are handling that call.

Is that what you mean, Mr. Weiss?

Mr. WEISS. If we sell the PBX, and the call is made——

Senator PACKWOOD. No. I do not mean the equipment. I understand the equipment part of it.

Mr. WEISS. These are major systems that are just as large as our local exchange switches. When you take the First Chicago Co.—and since Mr. Allen lost Sears, why, I will talk about First Chicago.

The issue here is that that is a multistructured, major business account, provided in large measure by AT&T, for the equipment and the service within that system.

Senator PACKWOOD. OK.

Mr. WEISS. And those calls are occurring within the local exchange just the same as two calls between business systems. But by definition, AT&T would very much like to have those excluded. Now, they are part of my revenue structure, if I can be competitive for them.

Senator PACKWOOD. I still do not understand. I understand AT&T provides equipment. And what does Allied—to use whatever example you said.

Mr. WEISS. The First Chicago Corp.

Senator PACKWOOD. All right. The First Chicago Corp. AT&T provides a lot of their equipment, maybe all of their equipment, for all I know.

Now, the secretary at First Chicago is going to call her husband. She picks up the phone. Does this completely bypass your network? And this is a local call, not only put on AT&T's equipment, but it goes over lines that are not yours to the secretary's home in Chicago.

Mr. WEISS. Not normally, if he is not located in the same general business configuration that she is.

Senator PACKWOOD. In other words, if he is in the bank——

Mr. WEISS. If he is in the bank, she would be calling totally outside of our service.

Senator PACKWOOD. I cannot get the answer.

Mr. WEISS. Because you see——

Senator PACKWOOD. Senator Gorton says the answer is intrabusiness calls. He says Boeing-Everett to Boeing-Seattle might be handled by somebody other than the local exchange; is that right?

Senator GORTON. I think they are the same. That is right.

Senator PACKWOOD. So, that is what you mean by local competition, that they are handling on their own lines, or not with your lines, or do they interconnect——

Mr. WEISS. Either way.

Senator PACKWOOD [continuing]. With some intermediary carrier?

Mr. WEISS. Either way. If the First Chicago Corp. has a series of business locations around town, they may connect those locations with somebody that has nothing to do with us, like Metropolitan Fiber Systems, or Teleport, that provides connecting links and dedicated channels between those. So, it is a totally internal system, in various locations in the city, that bypasses the local exchange totally.

Senator PACKWOOD. Basically, it is an in-house communications system for the company.

Mr. WEISS. Yes. But what is the difference between that and any other call that is being made within the local exchange?

Senator PACKWOOD. Sure. You may have a point. I am just trying to get down to how the physics of this works. Now, the company is all hooked up intracompany. To the extent that you make a call on the company phone that is outside the company, it goes to your home, and you normally carry that——

Mr. ALLEN. Yes.

Senator PACKWOOD. OK. So, the competition is not what we would normally call, or what the average citizen would think of as local competition, because we do not think of it as intrabusiness initially. We think of it as residential.

Mr. WEISS. That is right.

Senator PACKWOOD. And if one business were to use this intraphone system to call another business, you would handle that call.

Mr. WEISS. We probably would. We would not necessarily, but we probably would if it were within Chicago.

Senator PACKWOOD. OK. Now, what would happen if we were to simply say to the two of you, and Senator Breaux was pursuing this, "We would preempt the States' competition, and why do not the two of you just go at it for long-distance and local?"

What would happen?

Mr. WEISS. There would be a readjustment of the industry. In other words, my business would change, and so would Mr. Allen's, but in ways that would not—for example, if I did not have a nationwide intercity network, as he does, I cannot provide that service.

So, what I would do with this freedom is work with customers, so that I would become their agent to negotiate and handle all of the services they want, even though some of it might be provided by AT&T, or some of it might be provided by MCI or others. I cannot do that today. So, I really am very limited when I even get in the negotiations.

Senator PACKWOOD. I understand that you cannot do it today. I am wondering what would happen if both of you could go at each other in your respective businesses, let us say within a deadline of January 1, 1995, and at that stage we would preempt State regulations and local regulations, and we say to you and Mr. Allen, "Get ready."

Now, what would happen?

Mr. WEISS. What would happen, in my opinion, is that all elements of the industry would grow faster than they are today. To Mr. Allen's testimony, that is what competition generally does——

Senator PACKWOOD. Mr. Allen, what do you think would happen?

Mr. WEISS [continuing]. Expand the marketplace.

Senator PACKWOOD. You are given a reasonable time, I do not know if that is a reasonable time, but a reasonable time where you are all to start together.

Mr. ALLEN. Well, I certainly agree with Mr. Weiss that there would be radical changes in the industry.

Senator PACKWOOD. Would it be a desirable change?

Mr. ALLEN. Under the current circumstances of that bottleneck, called the local telephone network, which all of those calls have to move through, it would not be a workable situation, in terms of real competition.

If we had all this competition, I am all for it. I would like to think that we could do it by then, but it seems unrealistic.

Senator PACKWOOD. You were apparently able to eliminate the local bottleneck intrabusiness calls, using a wire that is not theirs, within one business.

Mr. ALLEN. Well, that is the nature of the way most PBX systems are built. In fact, that is a competitive marketplace which many other companies could have won the account for Chicago, including Illinois Bell.

Senator PACKWOOD. But you managed to get around the bottleneck.

Mr. ALLEN. No. It is not around the bottleneck, because calling within most companies is done through some switching device on the premises, or in the central office, if it happens to be served by what is generally called a Centrex service or some variation on that by the local exchange company.

Senator PACKWOOD. Does it use Illinois Bell's lines at all to do it?

Mr. WEISS. Yes. It does.

Senator PACKWOOD. Oh, it does.

Mr. WEISS. Yes. Just as——

Senator PACKWOOD. Does it even for the intrabusiness?

Mr. WEISS. I believe that is the way it is.

Mr. ALLEN. No. No. Normally the traffic would not flow back into the central office and back out, although, I think in earlier days that was probably the way it worked, which was not particularly efficient.

Senator PACKWOOD. If you were given—again, I do not know what a reasonable time is—a reasonable time to prepare, you would still not get around the bottleneck.

Mr. ALLEN. I know of no way to do that. But I can do it. As I said earlier, it is technically feasible. It is possible. It is not realistic or practical from a financial point of view. And they ought to be the best providers of access.

Senator PACKWOOD. Thank you, Mr. Chairman.

Senator INOUYE. Senator Gorton.

OPEING STATEMENT OF SENATOR GORTON

Senator GORTON. I would like to go back to the fundamental questions which Senator Danforth was asking, and followup on them.

I take it that, Mr. Allen, you said that this competition about which you spoke, and Senator Packwood spoke, even with respect to this local bottleneck, will be perfectly valid and acceptable, as far as you are concerned, as long as you were not going through a monopoly provider, or as long as there is real competition in the provision of the local services, is that correct——

Mr. ALLEN. That is correct.

Senator GORTON [continuing]. But that such competition does not now exist. Mr. Weiss, now, on the other hand, feels that that competition does now exist, and so he should be permitted into a business from which he is now prohibited.

You have asked in your written testimony that we come up with an objective definition of "competition," to show when that competition exists.

I, in turn, would ask that both you and Mr. Weiss submit to us a definition of what you believe real competition is, so that we could determine—Mr. Weiss, evidently, the definition that you would provide us would be one which is now, in fact, in existence.

Nevertheless, from the point of view of policymakers setting broad policy, when you are at the heart of the question of competition, we ought to have some concept of what each of you believes that competition is.

Are you simply asking us, Mr. Allen, to come up with that ourselves, or do you have a proposed statutory definition of the form of competition which is necessary, from your perspective, before we should allow Mr. Weiss to be in your business?

Mr. ALLEN. We do not have a statutory definition. We would be glad to work with you and the committee to try and develop one.

In retrospect, I think Senator Danforth's, to follow the golfing analogy, was asking us to establish a handicap system.

Until we can be sure that we have a handicap system which properly defines the marketplace and its competitors, neither of us can have the proper handicap to go head to head on the golf course.

Now, that assumes that I want in the local exchange business to begin with, which I do not. But if——

Senator GORTON. You want a competitive local system.

Mr. ALLEN. I want a competitive local system, because I think that will benefit consumers, it will benefit the Nation, and, yes, it will benefit AT&T, because you pay 40 percent of every dollar to them.

Senator GORTON. Since that is so vital to you, it would seem to me that you would have a high degree of incentive that is presenting to us a statutory definition of AFTA conditions, so others can critique it.

Mr. Weiss, from your perspective, while I think I followed the distinctions between you and Mr. Allen, in connection with Senator Packwood questions, I would like to see in writing why it is that you think that that competitive situation exists at the present time, and not just the statement that competition exists.

But what is the definition of "competition," which the present situation fits, and, therefore, in your view, qualifies you to enter into these other businesses? Are you capable of giving us an answer to that question?

Mr. WEISS. Well, we will certainly try, Senator. As I said before, it is a complicated issue, but I will try to do that.

Senator GORTON. I have one ancillary question for you, and I think, by extension, to Mr. Allen, who very graciously admitted that unconsciously, competitors were inhibited by the previous situation, when AT&T, in effect, did it all.

We have—I am not going to be here when we get to the next panel. But we have one example, someone from Hello, Inc., who I think is quite typical of many small and struggling companies, in one aspect of the communications business or another, who says, "Well, maybe theoretically this competition is available now, but since I have to go through the RBOC's with all of my customers, my customers are told, 'No, my service really cannot be provided, and besides we do it better.'"

And, in effect, in the real world, whatever the theoretical competition is, the small person is greatly disadvantaged, and prohibited from competing effectively.

It is one thing for us here to talk about whether we can create a situation in which the two of you, huge corporations, can compete.

But how can we also assure that these smaller companies that are just starting into the business, and perhaps engaged in only one part of it, are treated fairly, I think particularly by your company, and I suspect by Mr. Allen's as well?

Mr. WEISS. Well, if you look at their growth characteristics today, they are very successful companies. They have good cost characteristics. They have good quality, and they are formidable competitors.

Not all, but the good ones are. I think that really takes care of itself. They have found areas of our business that they can enter, where they can do that business very well, and they are being very successful. And none of the rules that I am proposing will change that capability on their part.

Senator GORTON. So, you do not think, in fact, that there is discrimination or unfair competition on the part of RBOC's with most or many of the struggling companies; is that right?

Mr. WEISS. They would not have gotten started if there were. Now, there are continuously developing rules and policies at both the State and Federal level, and interconnection in other matters, on a full and fair basis.

I mean, that is happening. Most of those things are worked out among the parties and the local regulator. But they are growing companies, and they are vital companies.

Senator GORTON. Finally, for you, Mr. Weiss, how much do you require before you will be perfectly comfortable with the AT&T purchase of McCaw? Do you require only what I heard you say at one point, the ability for the RBOC's cellular companies to do exactly the same things that McCaw can do?

Do you require being able to get back into the long-distance business on a fairly unlimited basis, or do you require not only getting back into the long-distance business on a fair basis, but being in the manufacturing business, and the information services business as well?

Do you have to have the whole kit and caboodle before you will agree that AT&T should be allowed to take over McCaw?

Mr. WEISS. If AT&T acquires McCaw, which is their first major entrance into the cellular business, all I ask is parity. Make the rules equal for the players of that game.

Senator GORTON. What will that require? That does not have anything to do with manufacturing, or does it, because AT&T can manufacture——

Mr. WEISS. No. There is a manufacturing issue, because AT&T is the provider of both the cellular equipment as well as the intercity service. But that is not what we are talking about here.

We have asked AT&T, and at least we have Mr. Allen's word that says they will provide us the same technology in the same terms and timeframe that they do anything they use themselves.

I will accept that, if, in fact, that policy is carried forward. So, what I am really asking for is the same parity on intercity capability, that the cellular systems owned by regional companies, which are hindered by the MFJ, are given the same kind of privileges that AT&T/McCaw would have.

Senator GORTON. So, information services and manufacturing are separate.

Mr. WEISS. Those are other issues. They are broader issues.

Senator GORTON. All right. Thank you.

Thank you, Mr. Chairman.

Senator INOUYE. Thank you very much.

Senator Burns.

Senator BURNS. I just want to followup, and then we can all go to lunch here, and I have some other things to do, too.

For fear of oversimplifying what we are facing here, I want to draw an analogy. I spent about 20 years refereeing football.

I can work a game in Montana, or Texas, or Florida, or, yes, even here in Washington, DC, and have very few problems, or hardly any problems, maybe in judgment, but not with regard to the rulebook, because the rulebook is the same in every State. It is the same.

I can draw an analogy, when we start forming up legislation, and drawing up rules about the RBOC's, and then I think we can go over and do the same thing as far as the long-distance carriers are concerned, because there is competition there, and maybe one rulebook over here, and one rulebook over there. But whenever we start making those rulebooks, that is where we start running into trouble.

Now, you, Mr. Allen, say that you do not have any incentive to go into the local business, in other words, the local business, or the universal business—$25 billion dollars is a lot of incentive.

And I look at that, and I have a feeling: I get very nervous about that, but the problem we are having is trying to meld these rulebooks, and do it fairly. Is that an oversimplification?

Mr. WEISS. I am sorry. I thought you were speaking to Mr.——

Senator BURNS. Well, whomever.

Mr. ALLEN. I am not sure it is an oversimplification. It is pretty hard to play a game with a bat when you are football referee. You are mixing up the two games.

I think, to some extent, that is the challenge we face, in that we have a competitive game, and we have a monopoly game, if you will, and trying to find rules that apply commonly to both groups is very difficult.

And that is why they need rules from one area through the transition period, and then the same rules would apply to everybody. But it is a difficult challenge.

Senator BURNS. Well, I just think that, I just happen to believe that the 40 cents on every dollar that you send to the local universal service provider provides quite a lot of incentive to circumvent or bypass the local provider, and I will be willing to watch that very closely.

Mr. ALLEN. That is precisely why I would like to have competition at the local level, so that 40 cents out of every dollar moves down to some lower number.

Mr. WEISS. Senator, while I was waiting for Mr. Allen—my apologies—I was noting two things. One is that, as Senator Stevens pointed out, the game is different in Chicago and Alaska.

So, the rulebook has to be modified to deal with those circumstances, and that is why I still appeal that the local regulator has to be brought into the process, as these things unfold, and our industry moves forward.

I also would say to this subcommittee, that the last legislation that effectively governs the industry was passed in 1934; therefore, I expect that whatever legislation occurs now would last a long time.

The pace of change in this industry is such that if it is not very broad, and permits a lot of discretion, it will be outdated within 2 or 3 years. And that is my concern. It is very hard to stay up with what is happening. There is a land mine every week in our industry.

Senator BURNS. Thank you very much.

Mr. Chairman, it is all yours.

Senator INOUYE. OK.

Senator DANFORTH. Mr. Chairman, could I just add one thing?

I appreciate everybody's patience. I do not think anybody on this panel or anybody in the audience guessed that we would have been here this long today, and we really appreciate your help to the committee.

I would like Mr. Weiss and Mr. Allen to take you up on what we talked about earlier, and my hope would be that each one of you would designate somebody who has your personal ear to meet with us to try to do our best to define what does constitute competition in the regional market. If you could do that, we would very much appreciate it.

Mr. WEISS. Senator, I am here representing seven regional companies, so I must say that I feel obligated to include all parties in that request.

Senator DANFORTH. Sure.

Mr. WEISS. Thank you.

Senator DANFORTH. I mean, that would be my understanding, that—I do not think we want a huge room full of people, because if you do that, you just get bogged down immediately. But just for the sake of discussion, or edification of us, if we could have one person from each.

Mr. WEISS. I will try to do that.

Senator DANFORTH. Thank you. I mean, from each of you, not from—thank you.

Senator INOUYE. I have a little problem here. I am prepared to stay here all afternoon, if necessary, but as Senator Danforth indicated, we had no idea that the first panel would be sitting here this long.

It has never happened in the past. Therefore, I feel rather bad for the second and third panel, because in all likelihood, because of other commitments, you would be lucky if you had two Senators sitting here.

I am willing to have additional hearings, so that you will have a broader panel here of Senators to listen to your testimony, but I also realize that some of you have traveled long distances to be with us, and it may be difficult for you to get back again.

So, may I call upon the remaining eight witnesses—we will have a short recess. If you will, come up here and we will discuss this. If you wish to continue, we will continue.

Thank you very much at this time.

[A brief recess was taken.]

Senator INOUYE. We have just decided that panel three, consisting of Mr. Dean J. Miller and Mr. William F. Squadron, will return to the committee at our next hearing. We will continue this morning's hearing with the second panel.

The panel consists of the president of West Publishing Co., Mr. Vance P. Opperman; the legislative director of the Consumer Federation of America, Mr. Gene Kimmelman; the superintendent of Glasgow Electric Plant Board, Mr. William J. Ray; president and chief executive officer of Eastern TeleLogics Corp., and chairman of the board of the Association for Local Telecommunications Services,

Mr. Gary E. Lasher; the secretary-treasurer of Communications Workers of America, Ms. Barbara J. Easterling; and the vice president of Administration of Hello, Inc., Ms. Paula Smith Preston.

On behalf of the committee I thank you ladies and gentlemen for your patience and your cooperation. It is very important to us. May I now call upon Mr. Opperman?

STATEMENT OF VANCE K. OPPERMAN, PRESIDENT, WEST PUBLISHING CO.

Mr. OPPERMAN. Mr. Chairman, Senator Danforth, my name is Vance Opperman. I am the president of West Publishing Co.

We are here to support all of the thrust and much of the language of S. 1086. In the interest of time, let me just briefly outline why we believe our experience, and the experience of the information industry, speaks eloquently for why a law such as 1086 should be passed.

I will speak first for my company, and then for the industry. In 1982, when the Bell monopoly was broken up—and I will speak in general numbers—our company had less than 100 people employed full time in our electronic publishing products.

Today, and using round numbers, to give you an idea of the order of magnitude, we employ more than 1,000 people in that type of activity.

Our fine competitors, such as the Mead Corp., barely existed prior to the breakup of the Bell monopoly, and today is a very large and very successful organization doing electronic publishing out of the State of Ohio.

That experience is mirrored in the entire industry. For example, according to Department of Commerce studies for the industrial outlook for the last 10 years, prior to the breakup of the Bell monopoly, less than 10,000 people in this country were employed in electronic publishing and the information services industry. And today, that number is over 1 million people employed.

We have a positive balance of trade. We lead the world. The increase of information services in the world, in the United States, 56 percent of that information is originated and comes from the United States, 32 percent in Europe, and 2 percent in Japan. And that gap has widened since the breakup of the Bell monopolies.

I would be belaboring something that I think this committee knows well, and that everyone else in this room I think would agree with, and that is, competition is good. Competition created this industry.

Competition gives us new products, new innovation, lower prices, and greater access. And the thrust of 1086, the goal is toward breaking up now the second type of monopoly, and that is the bottleneck monopoly.

And the local exchange, if our experience is any kind of guide in this country, will accomplish the same kind of beneficial results that we have already seen in information industries.

I thought for a minute—and with all due respect to Mr. Allen, a man I admire—I thought for a moment I might have attended a meeting of monopolists anonymous. [Laughter.]

Mr. Allen said that, "Yes, it was true, there was a time when, even though I tried not to cross subsidize, I tried not to discriminate, yes, indeed, as a monopolist, I did."

I do not mean to make fun of that, because that is, of course, what the experience has been. And that will be the experience, and has been the experience, if we allow the local bottleneck monopoly to go without the kinds of protections that we have had to date.

And this bill addresses that experience, and addresses it in two significant ways. First, there is a recognition that there has to be structural safeguards.

And if those structural safeguards are not put in the local exchange, and if they are not required to be followed, we will have exactly that cross-subsidization problem, that problem that we know existed, that problem that we observed, that problem that was broken up, and once it was broken up, that led to the tremendous American success story that the information industry is today.

And second, the legislation talks to the second problem, and that is to protect Americans' privacy, the so-called CPNI issue, the issue that when you have a monopolist, who gets all of your signal—and I use some of the colloquy that was here again—yes, it is true, we are a customer of the gentlemen who were up here, and we have to use the local exchange today.

Our signals and the signals of our customers go through the local exchange. The local exchange knows who those customers are. They know the customers of our competitors. They know what they order.

They know how much they pay. They know where they are. And they can know what they want, the content of that type of signal. Now, I recognize the people who will take the pledge: "I do not want to do that. I will not discriminate. I will not do any of those things that monopolists do." It is not a matter of individual inclination; it is a matter of economic incentive.

This legislation recognizes that we are to have the advantages of competition now, to move in on the local exchange and break that up, so that we can enjoy that advantage, that we have to have some transition rules.

And those rules and protections are found exactly where this bill finds them, first, in the area of structural safeguards, to prevent that cross-subsidization, and second of all, in the CPNI area.

We have suggested some ways in which we think both of those can be strengthened, and they are contained in our testimony. I thank you very much for the time. I enjoyed the discussion as a customer.

I certainly want to see the local exchange more efficient. We hire them a lot. We want a better signal and a more efficient signal. We would like to pay less and have more, which would be the exact results to this company and this country, if we pass 1086.

Thank you once again, Mr. Chairman.

And thank you, Senator Danforth.

Senator INOUYE. Mr. Opperman, your prepared statement is going to be made part of the record, as all of the others. But your statement is much better than the prepared one. [Laughter.]

Senator IIT WAS VERY CONVINCING.

Mr. OPPERMAN. Thank you, Mr. Chairman.

[The prepared statement of Mr. Opperman follows:]

PREPARED STATEMENT OF VANCE K. OPPERMAN

Good morning, Chairman Inouye and distinguished members of the Subcommittee. Thank you for the opportunity to present my views on S. 1086, the "Telecommunications Infrastructure Act of 1993." For reasons that I shall amplify shortly, I sincerely commend you for conducting what are path-breaking and potentially momentous hearings that strike at the heart of the shibboleth that local exchange telephone service is a "natural" monopoly. These hearings are particularly timely because pieces of the country's telecommunications structure are rapidly falling into place. How well the pieces fall together depends in part on whether Congress overhauls the current system of telecommunications regulation.

I am Vance K. Opperman. Since August 1, 1993, I have served as President of the West Publishing Company of Eagan, Minnesota, a leading publisher of legal materials. For the past few years, West has actively worked with a coalition of other electronic publishers, the Electronic Publishers Group ("EPG"). Although EPG members publish diverse materials, we are all concerned about the potential threat that the local bottleneck monopoly poses to our businesses.

Prior to assuming my current position at West, I was a lawyer in private practice for 23 years, serving as counsel to West and specializing primarily in antitrust and regulated industries. I have handled numerous cases involving allegations of monopolization and bottleneck abuse. I have also had an opportunity to witness first hand how some of the most well intentioned federal and state regulatory schemes foundered in their implementation.

As you may know, West is an employee-owned information service provider, specializing in legal and statutory material. West occupies a distinct place in our Nation's legal and intellectual history. As the preeminent publisher of legal materials—including case reporters, digests, annotated statutes, legal treatises, casebooks, textbooks, dictionaries, and encyclopedias—West has pioneered a unique headnote and key numbering system that in large part organized the American legal system. In addition, West created and publishes the United States Code Annotated and annotated statutes for many states.

West publishes its materials in a number of formats and distributes them in various ways. We have printed books for nearly 117 years. Recently we began offering CD-ROM products. We continue to ship both to our customers through traditional air and surface transportation systems, such as the U.S. mails, U.P.S. and other similar services.

In 1975, we started WESTLAW, our on-line computer-assisted legal research database, which has become an increasingly important component of our overall business. Through WESTLAW, our customers dial up our computer data banks in Minnesota and enter queries or search terms. Through technological advances, we have now been able to provide simplified search and retrieval capability using "natural language." Our computers retrieve pertinent legal materials and then pass them through the public telephone network to the 'customer's computer screen. In certain circumstances, WESTLAW can provide lawyers with the full text of a judicial decision within minutes of the court's rendering of that decision.

WESTLAW is entirely dependent on the speed and reliability of a delivery system—the public telephone system—that we do not and could not own. Any discrimination or disruption in telephone service could be extremely injurious to WESTLAW.

Everyone familiar with the legal research process knows that many lawyers—particularly younger lawyers—have a strong preference for computer-assisted legal research over the traditional book system. Everyone familiar with the legal system also knows that lawyers win and lose cases on the basis of the thoroughness, accuracy and speed of their research. Our customers have important reasons for demanding complete, accurate and timely service. Under these circumstances, no one could dispute that inferior telephone service of any type poses a substantial threat to WESTLAW.

That is why West, and 1 believe most of the members of the Electronic Publishers Group, have been so vocal in court and in Congress about the lifting of the 1982 Modification of Final Judgment's restriction concerning information services. Simply put, our concern is with unfair competition from the monopoly supplier of one leg of our distribution system—the local exchange bottleneck.

In the past, we have favored legislative measures to prevent monopoly abuses and to ensure that electronic publishers, captive to the local telephone monopoly, are not unfairly disadvantaged. As long as the local exchange carriers have monopolies, we will continue to favor those measures. At the same time, S. 1086 is more than a good first step. It is a bold, direct strike at the heart of the problem—the local telephone monopoly. For the past 100 years, it has been almost an article of faith in the telephone industry that the "last mile" of telephonic communication is a "natural monopoly." As recently as three years ago, the idea of viable local competition was so remote as to be almost the stuff of science fiction writers. S. 1086's approach is to say, in the manner of Gershwin, that "It Ain't Necessarily So."

S. 1086 would create a new national policy discouraging monopolies and favoring infrastructure development through private investment. With private investment and without monopolies, there should be—and will be—plenty of progress with no stifling of competition. S. 1086 would emphasize market mechanisms instead of government regulation as a national telecommunications policy, which will tend to maximize output and quality, and minimize price. The genius of the approach is that, if properly implemented, it would obviate the need for FCC regulation of competition, leaving only the flexible and dynamic antitrust laws in place. In short, S. 1086 presents an historic opportunity to make competitive rules that will promote progress.

Even a casual glimpse at the headlines shows that the country is in the middle of a technological revolution in computing and telecommunications, complete with huge realignments of industry participants and capital. I have set forth, for your review, only a few of the articles about the telecommunications and computing revolution that have appeared in the national press from the last Subcommittee hearing on July 14, 1993, through the end of August:

July 16, 1993, Wall Street Journal—"PictureTel to Introduce $6,000 System to Make PCs Work as Video Telephones."

July 16, 1993, New York Times—"MCI and Northern Telecom to Test Cellular System."

July 19, 1993, Wall Street Journal—"TV Broadcasters Itch to Deliver Data, Too." (WavePhore develops TVTI which will allow broadcasters to transmit 1.5 megabits of digital data per second.)

July 26, 1993, Wall Street Journal—"Time Warner, In Turnaround, to Join Test of Cable TV Service with NYNEX."

July 29, 1993,Washington Post—"Game for a Multimedia Opportunity, AT&T's Plan to Buy a Stake in On-Line Firm Reflects a Bet on Interactivity." (AT&T acquires a 20% stake in the Sierra Network.)

July 30, 1993, Wall Street Journal—"MCI, in Bold Move, Wins Allies in Wireless Communications War."

August 3, 1993, New York Times—"Sprint Tying into Wireless Global Net: Investing ($40 million] in Motorola's Satellite System."

August 16, 1993, Business Week—"The Parallel Universe Grows: Unisys, Intel and IBM Are Moving Into 'Alternative Mainframes.'"

August 22, 1993, Washington Post—"AT&T's Deal: A Giant steps Into New Arena: Deal May Make AT&T a One-Stop Phone Source." (AT&T acquires McCaw Cellular.)

August 24, 1993, New York Times—"New British Pact for NYNEX." (NYNEX to invest $2.25 billion in Mercury Communications, Britain's second largest long distance company.)

August 25, 1993, Wall Street Journal—"Novell, SynOptics to Link Computer Networking Efforts."

August 25, 1993, New York Times—"New Competitor for NYNEX." (Locate Telephone Company to offer local telephone service- in -competition with New York Telephone Company.)

August 25, 1993, Washington Times—"Major TV Cable Firm to Link with Internet." (Continental Cablevision's system will be a pipeline for Internet into the home.)

August 25, 1993, Wall Street Journal—"Sprint is Expected to Offer New Transmission Service." (System will use new traffic switching technology called ATM or asynchronous transfer mode.)

August 26, 1993, Wall Street Journal—"Chip Maker, Pacific Bell Team Up on PC Applications." (Intel and Pacific Bell will work together on integrating new audio, video and data applications for personal computers.)

August 27, 1993, Washington Post—"The Baby Bells Go Hollywood, And We're Talking Deals, Baby."

August 31, 1993, New York Times—"Basic Gear Gives Access to Network of Networks." (Anyone with a computer and modem can access the Internet.)

I think it is a fair conclusion that, given this level of activity in the industry, it is clear that the pieces of the telecommunications infrastructure will fall into place with or without Congressional guidance. How well the pieces fall together depends in part on whether Congress overhauls the current system of telecommunications regulation. S. 1086 recognizes that now is the time to make sensible national policy based on competition and private investment.

The information services industry is a textbook example of how intelligent regulation can produce vigorous competition. Virtually nonexistent during the years of the Bell monopoly, the information industry has flourished since—and we say because of—the decree. Explosive growth, fantastic diversity and broad participation characterize information services in America today. Legal materials are but a small part of this electronic information cornucopia. Indeed, the 1993 Industrial Outlook Report states:

> U.S. information services are expected to expand faster than GDP in 1993 to remain among the most active sectors of the economy. Global demand for information services to advance economic, business, and social development will continue unabated. Revenues of the electronic information services industry are projected to grow more than 16 percent; data processing and network services providers by almost 14 percent; and computer professional services by about 10 percent.
>
> * * * * * * *
>
> The dynamic U.S. information services industry comprises about 26,000 establishments competing in the marketplace. In 1992, the industry was estimated to have more than one million employees for the first time. (1993 Report at 25-1).

The 1991 U.S. Industrial Outlook makes clear that the United States is by far the largest producer of information services, providing 56 percent of the world's databases, compared to 32 percent for all of Western Europe and only 2 percent for Japan. (1991 Report at 27-2). The 1993 U.S. Industrial Outlook shows that the trend of U.S. leadership is continuing: "[I]n 1991, U.S. controlled companies earned 35 percent of all European computer services and systems integration revenues * * * Only U.S. and French-controlled computer services gained market share; companies from all other countries lost market share." (1993 Report at 25-1). The 1993 report goes on to say that U.S. companies lead the market overseas because "they are acknowledged to provide the world's most advanced and diversified information services." (Id. at 25-2).

Growth, innovative products, vigorous competition, trade surpluses, American dominance. Not bad. Prohibiting the Bells from monopolizing information services has produced powerful benefits. If you can expeditiously and successfully usher in an age of competition within the local exchange, you will produce additional powerful benefits to generations of Americans—without huge expenditures of public funds and without an elaborate bureaucratic regulatory apparatus.

However, the care that must be taken in moving from a monopoly environment to a competitive environment cannot be overstated. Failure to nurture competition can jeopardize the entire process and skew markets for decades. While we can all rejoice at the new economic freedoms of the formerly Communist world, I am sure none of us envies the disruptions those economies have experienced in moving from state monopolies overnight. Closer to home, Congress chose to deregulate fare structures in the airline industry by abolishing the Civil Aeronautics Board in the late 1970's. Thereafter, the antitrust proscriptions ensuring full competition were at best enforced with laxity. The result, of course, is the current drastically altered landscape in the airline industry. Whether or not that landscape appeals to you, it is indisputably the product of federal policies adopted to spur the move from cartelization to competition.

My point is that the transition rules necessary to promote competition have effects that far outlast the transition itself. They will determine for generations to come whether consumers will have the benefits of a competitive market. While I do not purport to know exactly what the right balance is, I do know that laissez faire is not the correct approach.

West and the Electronic Publishers Group believe that a separate subsidiary requirement—such as that contained in section 233 of S. 1086—is absolutely essential. Requiring separate subsidiaries is one form of prophylactic regulation that is clear and entails a minimum of government activity. In fact, West and the Electronic Publishers Group support changes that require the RBOCs' separate subsidiaries to be truly separate and proposals that' mandate that the RBOCs deal with their subsidiaries at arms-length and in a non-discriminatory fashion. Such proposals should: (1) prohibit the sharing and ownership of all facilities; (2) require the separate subsidiaries to have separate directors, officers and employees; (3) require separate subsidiaries to obtain transmission capacity from their affiliated BOCs solely pursuant to generally available tariffs; (4) prohibit joint activities of any kind, including operations, production and research and development; (5) require more than de minimus outside ownership of separate susidiaries; and (6) require separate subsidiaries to have separate financing and to prepare and file financial -statements as if they were publicly held corporations.

Why not continue to rely on Computer III-type regulation by the Federal Communications Commission? Because pricing manipulation, withholding of access to new innovations in the network, and strategic timing in the release of technological developments are all practices that the FCC has been unable to detect adequately in the past.

As George Santayana said, "Those who are ignorant of history are bound to repeat it." The AT&T case was largely about the shortcomings of FCC regulation in preventing the Bell system from stifling progress and crushing competitors. Even now there are grave warning signs about the inadequacy of FCC regulation. In a February 3, 1993, Report to Congress, entitled "Telecommunications, FCC's Oversight Efforts to Control Cross-Subsidization," the General Accounting Office reported that:

> In 1987, we reported that the FCC had insufficient staff to ensure that consumers were protected from cross subsidization. Since that time, FCC's responsibility for overseeing carriers' cost allocations have continued to grow, but the staff resources allocated to this function have declined rather than increased. We believe the number of FCC auditors remains inadequate to provide a positive assurance that ratepayers are protected from cross-subsidization.
>
> Increasing the FCC staffing level * * * would cave little or no impact on the federal budget because the government would be reimbursed for its on-site audits and reviews of the CPA audits through fees FCC is authorized to collect from carriers. Given this trend of increasing work load and decreasing resources, the potential exists for ratepayers to be *more* vulnerable, to inappropriate charges resulting from cross-subsidization in the future. (GAO Report at 12, emphasis added).

The separate subsidiary requirement is intended as a regulatory device to assure that the BOCs do not abuse their monopoly position. Heretofore, much of the public debate has focused on whether the local exchange monopolies have abused or would abuse that position. Once the local bottleneck is broken, however, that debate becomes irrelevant. Therefore, at the point there is effective competition, you may wish to sunset the separate subsidiary provision.

Both during the transition and in the long run, the best remedies for redressing unfair competition are the damage and injunctive provisions of the Sherman and Clayton Acts. There is no question that private antitrust cases are not perfect, but they are surely preferable to relying on government enforcement of regulations. I sincerely commend you for not in any way diminishing the applicability of the antitrust laws. You may wish to reinforce your intent by including a strong antitrust savings clause with a narrowly crafted exemption for RBOC activities specifically mandated by the FCC.

S. 1086 addresses another important issue—privacy. By virtue of their monopoly status, the RBOCs have unique access to a -huge amount of information regarding their customers' network usage and calling patterns. This knowledge, if misused, poses a substantial threat to individual privacy as well as offering the RBOCs a significant competitive advantage. West and the EPG believe that customer privacy should be of paramount importance, and commend the authors of S. 1086 for their obvious appreciation of the American public's privacy rights. We also strongly support S. 1086's approach of making sure that the RBOCs and their competitors are subject to the same rules. However, as written, section 234 of S. 1086 would impose ironclad strictures that may prove too inflexible in practice. The Electronic Publishers Group has submitted alternative language to the Subcommittee for its consideration. Our draft CPNI language seeks to allow fair access for competitive providers to customer information that does not raise substantial privacy concerns, and we hope that the Subcommittee will work with us in improving this important section of the legislation.

CONCLUSION

In sum, West and the Electronic Publishers Group believe that effective local competition is far superior to regulation as a national telecommunications and information services policy. However, competition will not be created overnight. Until meaningful, effective competition exists in the local exchange, strong structural safeguards must remain.

Senator INOUYE. I now call on Mr. Kimmelman.

STATEMENT OF GENE KIMMELMAN, LEGISLATIVE DIRECTOR, CONSUMER FEDERATION OF AMERICA

Mr. KIMMELMAN. Thank you, Mr. Chairman.

Mr. Chairman, Senator Danforth, and Senator Packwood, I guess I am representing the customers of monopolists anonymous today, representing consumers. I do not think I can outdo Mr. Opperman.

But I am going to abandon what I have prepared, because I take Senator Danforth's comments to mean that really what you are interested in is a working session, to get amendments for a markup.

It seems to me it would be better to focus on observations on the first panel, because I think it got to the heart of the matter. The committee seemed to raise the questions that were crucial to working out this issue.

It seems to me that it is obvious that everyone believes the answer is somehow local competition. It provides the consumer a choice. It provides all the providers a choice. It allows the local phone companies to get away from this bottleneck problem that has been the basis of litigation, and enables them to get into every market.

The problem really is, as we know, competition is usually evolutionary. Competition does not necessarily become available to everyone universally across the country in each market.

And we have heard from many, as Senator Burns has pointed out, that there are claims that local rates could go up significantly where competition does not develop, or where there may be subsidies involved, and that we need to update our insurance policy, that we have always called universal service, to make sure that we do not have that dislocation.

And we have the problem of measuring competition, as you pointed out, Senator Danforth.

Now, S. 1086, I think, lays out the framework for dealing with all these problems, and it gives us the broad outlines of a new policy. But what we are concerned about, from the consumer's perspective, is getting the refinements that make sure that we really solve these problems.

Everyone this morning spoke about preserving universal service, and if I understood correctly, was committed to doing so. And I would suggest that what the committee needs to do legislatively is, frankly, make them put their money where their mouths are.

Come up with the details of what universal service is legislatively, so we can go forward with that small insurance policy in the areas where competition does not develop.

So, these are our proposals for your legislation. And they are just straightforward amendments that will provide full consumer protection, we believe.

Let us amend the bill, S. 1086, to ensure that we preserve the declining real-cost tradition of local phone service. We need to revamp the high-cost fund for rural areas. We need to modify the life line fund.

We need to make sure that people with disability receive the equipment they need. We need to make sure that average consumers have a new support mechanism, that used to be paid through access charges, in a monopoly environment, with a new mechanism that fits a competitive environment, that does not disadvantage any carriers. This detailed definition of "universal service" would be the first change.

The second change would be to focus on what competition is. We would suggest, as Senator Danforth mentioned earlier, something like from the cable law, or, frankly, it was in your bill, Senator Inouye, in the last Congress, that measures competition, and basically once we have competition with an objective entry test, we allow everyone into every market.

And, third, because there may be difficulties in actually pinpointing actual local competition or maintaining competitive stimuli throughout the market, we need to make sure we have detailed safeguards. Cost allocation rules and competitive safeguards will protect consumers where competition does not develop, will protect competitors, so that we get as much competition in every market that can actually sustain it. Only carefully crafted safeguards will let the market determine where competition can exist.

With these amendments, we would support S. 1086. The framework is there. These are just really putting meat on the bones. And I think the comments from this morning clearly indicate the direction we can go to move this, and have a fully competitive telecommunications environment that preserves reasonably priced phone service for all consumers.

Thank you.

Senator INOUYE. Thank you. I now call on Mr. Ray.

STATEMENT OF WILLIAM J. RAY, SUPERINTENDENT, GLASGOW ELECTRIC PLANT BOARD

Mr. RAY. Thank you, Mr. Chairman.

I am from a small town in Kentucky, and this is pretty oppressive to me, so if I forget to breathe during my statement, I hope you will remind me. [Laughter.]

The electric utility, especially a public power system like Glasgow runs, would appear to be a very small fish in a big sea compared to the people who have been here this morning.

But the reason I felt moved to come here was that I was afraid that there was going to be a King Solomon-type action taken here, and that the future of this technology might be divided among the cable companies, and telephone companies, and others, and some of the real benefactors of this technology might be left out.

To explain that, I am just going to talk about what has happened in Glasgow already. Glasgow, as you might have imagined, is not the technological center of the universe.

It is just a normal town, not unlike 10,000 small towns in the United States. We operate a municipally owned electric utility, a public power system. We buy all our power from TVA.

Over the last decade or so, TVA has sent us price signals, hints, if you will, every month, that a KW demand on peak was something to be avoided. That is not uncommon. Most electric utilities, especially the larger ones, are dealing with this issue of demand-side management and load management.

We decided to build a communications network in order to respond to that need, to position ourselves to offer real-time electric rates, and to automate our electric system, so that we could make the existing facilities last longer, and to mitigate some of the charges we were getting every month from TVA.

We did not know when we built that system that we were building the local portion of an information superhighway. We were trying to build a system, a broadband network, that was interactive, able to move information bidirectionally between the utility and each of our customers.

And the people of the community elected to have us put a competitive cable television service on that same network. It has worked very well for those two purposes.

And we built a system that had enough lanes on it to serve those electric utility needs that we could see now, for competitive cable television service, and other lanes for purposes that were not at that time known.

I can tell you how competition works. With respect to cable television, the way that we have discovered to attract private investment in infrastructure is to create competition. We are served by a cable operator who did not keep the most sophisticated equipment available in Glasgow before competition.

Now that everyone in town has the opportunity to choose between a private cable operator or the municipally owned one, their basic cable service has gone from just over $14 a month for 23-channel service, to a 48-channel service for $5.95 initially and slightly higher rates now. The telephone gets answered now by either company.

In general, the service has ramped up to that which everyone would dream of, and the thing that has caused is competition. It is a model that has proven itself before in our history.

In fact, when I listened today about people speculating on how this telecommunications technology might evolve, it struck me that the best way to look at the future, to predict the future of this technology is to simply look at the past, how electric power became ingrained into the fabric of our daily lives.

The situation was very similar at the beginning of this century. A few very large investor-owned power companies served most of the customers, and the rural communities generally did not get that new technology—electricity—first.

That is what caused public power to develop, was this fierce independence on the part of many small communities to say, if we cannot get you to bring this new technology in, we will do it ourselves. That is what we did with cable television, just copying that same model.

Now, one of those lanes, or a couple of those lanes on our highway, that were not populated in the beginning, are being used to furnish alternate local dial tones. We already have a competitive telephone service in Glasgow.

We use a technology from a company called First Pacific Networks, that allows you to provide telephone service over a broadband system.

GTE is showing a great deal of interest in what we are doing. They have not responded in the same way that the cable operator did, but I think that it is a logical conclusion to draw that once we really roll this out, there will be a similar impact and benefits similar to those resulting from competition in cable television will flow to the people of Glasgow.

We are also doing a 2-megabit-per-second data network on that highway throughout the community. The price we are able to charge for it is rather shocking to the phone company. It is 2 megabits per second. We sell it for $19.95 a month. The phone company classically, for T–1, which is a little slower than that, may charge $1,000 or $1,200 a month.

So, the lesson is vivid there. In this bill, where you ask the question, "How can we get private investments to build this infrastructure?" the answer lies in the history of both public power and now in what hopefully will happen in cable TV. The answer is competition from municipally owned systems.

I would add that the legislation that you pass, hopefully, should be direct and to the point, and not apologetic in nature.

The reason I am making that point is that, with the cable bill that got passed last year, the objective was to foster more competitive systems, like Glasgow's. One of the tools that that bill uses to accomplish that is making sure that the product is available to everyone, that there cannot be exclusive contracts between programmers and cable operator for programming.

I am here to tell you that the final version, after everybody got through with it does not accomplish that goal. There is a large corporation in Atlanta, called Turner Networks, that is still not selling their product as envisioned by the bill.

It is not having the desired effect in promoting competition that it could. I would ask you, or I would urge you to, in this legislation, that if it is truly competition that you want, let us not banter around about it; let us have a forceful statement that all vendors must sell their services in a nondiscriminatory fashion.

We want, for instance, public power to have the capability to act as a catalyst, to satisfy its own needs, and to foster the competition and the private investment that the history has proven will work as it did with electric power.

We would like for that same theme to carry through to cable television and these other telecommunications technologies.

There is a large number of public power systems out there waiting to do that, not speculating on what may happen 4 or 5 years from now, but ready to put wire in the air, and do this tomorrow, if we can just make sure that the same fertile ground that they have had for electricity will exist for these other telecommunications technologies.

Thank you.

Senator INOUYE. Thank you very much, Mr. Ray.

[The prepared statement of Mr. Ray follows:]

PREPARED STATEMENT OF WILLIAM J. RAY

Good morning. My name is William J. Ray. I am the Superintendent of the Glasgow Electric Plant Board in Glasgow, Kentucky, and Chairman of the American Public Power Association Cable Communications Committee. The Glasgow Electric Plant Board is a municipally owned electric utility and member of the American Public Power Association. Our community owned utility is organized just like the other 2000 members of the American Public Power Association. It was set up to provide a vital service, in the manner dictated by the local citizenry, to the people who own the utility. the people of Glasgow. I would like to thank the subcommittee for the opportunity to provide testimony at today's hearing.

SUMMARY

The purpose of my testimony is to inform the members of the subcommittee of the role that public power systems like Glasgow's can play in constructing and operating the "information superhighway" in their communities. In order to illuminate this capability, I will describe the "broadband highway" project that we have already established in Glasgow, Kentucky. Although we did not realize we were building the "information superhighway" when we started our project in 1988, since then it has been widely recognized as a harbinger of things to come in bringing communities into full realization of the "information age". Our project has been recognized this year as one of twenty five finalists in the Innovations in State and Local Government Awards Program of the Ford Foundation and Harvard University. We have also hosted representatives from over 250 other communities that have come to Glasgow to study our project for possible replication in their communities.

GLASGOW'S STORY

In the mid 1980's, the Glasgow Electric Plant Board began to speculate on whether it might be wiser to build communications facilities that would enable people to use less electricity than continuing to build more electric generating capacity. Although little was known about this theory, we decided to embark upon a voyage of discovery. In 1988 we began constructing a community-wide interactive communications-network consisting of 120 miles of coaxial cable which connects to every home and business in the City of Glasgow. Initially we planned to use it for two purposes. The first was to help consumers lower their electric bills by giving them constant information about their rate of energy consumption. The second was to offer a cable television service in direct competition with an incumbent cable operator in order to allow the benefits of competition to flow to the people of Glasgow. Our voyage has been exciting and we have discovered a lot.

From our fairly crude experiments in utilizing the flow of information to offset the need for a larger flow of electricity we have made some astounding discoveries. We have enough experience under our belt to project that we can defer about 3 KW of demand on peak from each home utilizing our information network, We spent about $500 per home passed to install the network. That means we are acquiring electrical capacity at an unheard of rate of $167 per KW! That is likely less than one tenth the cost of constructing new generating capacity! It is results like these that make the electric utility industry one of the most likely purveyors of the "information superhighway". We have the most to gain.

Our experience in competing with a private cable operator also adds credence to this Bill's intention to encourage private investment in the nation's communication networks. Once plans to construct the municipallly-owned broadband network were announced, the private cable operator (TeleScripps Cable Company) suddenly found an interest in reconstructing their outdated plant and drastically lowering their cable television rates. Before construction of the municipal system, the private operator had a system capable of delivering 36 channels of video and sold a basic package consisting of about 24 channels for $14.25 per month. After construction began on our system, overnight they found the capital necessary to upgrade their plant to a state of the art 54 channel system and began offering a basic-package consisting of 45 channels for $5,95 per month. In short, we discovered that competition in the telecommunications marketplace does deliver the effect desired by this Bill.

About two years ago, we were contacted by a company called First Pacific Networks in Sunnyvale, California. They had heard about our broadband network and asked if we were interested in installing their technology on our system. They deported that their technology would allow us to conduct telephone traffic on our broadband network, completely separate and independent of the local telephone company. We were interested, and we have installed this equipment on our network. It is working very well and we are learning a lot about the telephone business. This is where the capacity of public power systems to build the "information superhighway" really begins to take shape. Municipally-owned systems are generally immune from regulation by state public utility commissions, thus, they have the capacity to take bold steps like the ones we have taken without the laborious process of seeking approval from regulatory agencies. Utilizing the First Pacific Networks' technology, we stand on the shore preparing to sail off into the uncharted waters of head-to-head competition for the provision of local dial tone. Meanwhile, GTE (our local telephone company) spends a lot of time gazing at us from the distance, and, we assume formulating their plans for responding to competition. We expect their response to be similar to the cable operators. Suddenly the people of Glasgow will be offered more services at a lower rate than they would have ever thought possible from the phone company. Competition is certainly a magical elixir.

Recently. we have discovered the capacity of our broadband network to carry high-speed data throughout the community. Residents of Glasgow with home computers and access to our broadband network can utilize education software residing on file servers at local schools. access reference material on CD-Roms in the local libraries, query government geographic information data bases and soon do their grocery shopping, banking. and receive health care services all through their home computer hooked to Glasgow's "information superhighway" - These are not pie-in-the-sky predictions or prognostications about what may happen in a few years. These things are already happening today in a small-town in south-central Kentucky that decided to build the "information superhighway" for itself because it was not likely to have access to such a facility anytime during this century otherwise.

ELECTRIC POWER AS A MODEL FOR THE INFORMATION SUPERHIGHWAY

The developments in telecommunications technology in the final years of this century bear a striking similarity to the developments in electric power (the new technology at that time) during the first years of this century. Today it is impossible to read a newspaper or magazine and not come across news of a proposed alliance between a telephone company and a television cable company or a computer manufacturer. These alliances supposedly will bring about unimaginable benefits to our people through mastery of new telecommunications and computer technology.

The early years of the twentieth century were a similarly exciting time for America, From coast to coast word spread about the mind-boggling marvels of electric light and power. National excitement coursed across the country, seemingly as rapidly as electric power today races through modern long distance transmission lines. There is a lesson to be learned in this parallel.

Public power grew out of a concern in many local communities that the benefits of the new technology would not sufficiently serve many smaller communities, so they decided to do it themselves. The electric utility industry heated up in the 1920's as controversy between private power companies and public power systems made headlines and captured the attention of Congress. Controlling by far most of the country's electrical market and wanting more—a handful of powerful utility holding companies waged a campaign to discredit "government in the power business" as a threat to democracy. By 1934 private power holding companies bought out more than 1,500 municipal electric systems to increase their monopolies. Soon 16 holding companies, whose political and economical clout was unrivaled, controlled nearly 85X of the countries entire supply of electricity. Public outrage over this "power trust" prompted Franklin D. Roosevelt to hail the "undeniable right" of a community to establish public ownership of electric service as a "birch rod in the cupboard" to help protect consumers against abuse.

The lessons we learned in democratizing electric power provide a vivid road map that can be utilized in the democratization of today's telecommunications technology. By following this road map, we can avoid the mistakes made in electric powers legacy. This Bill asks the rhetorical question, "Who should be permitted to provide these new services?" We believe that Glasgow's experience indicates that public power is capable of not only constructing the "information superhighway" but indeed has already begun constructing it. We also believe that public power's ability to do so should not be hampered by any federal or state legislation.

Some of the "information superhighway" should be built by a combination of telephone companies and cable television companies and others, but the public should also own and operate a number of these highways following our countries' model for electric power. President Franklin D. Roosevelt helped create this model by pushing for federal hydroelectric developments on the Columbia River, which he said would create "forever a national yardstick to prevent extortion against the public and encourage the wider use of the servant of the people * * * electric power." This "yardstick" has served the people of this country well in the provision of electric power. The same model will work just as well in the twenty-first century for the provision of high technology telecommunications.

Senator INOUYE. Mr. Lasher.

STATEMENT OF GARY E. LASHER, PRESIDENT AND CHIEF EXECUTIVE OFFICER, EASTERN TELELOGICS CORP., AND CHAIRMAN OF THE BOARD OF THE ASSOCIATION FOR LOCAL TELECOMMUNICATIONS SERVICES

Mr. LASHER. My name is Gary Lasher. I am president and CEO of Eastern TeleLogic Corp. I am one of the competitive access providers that Mr. Weiss thinks has taken all of his business from him.

In fact, we are a very small industry of some 30 companies that operate throughout the United States. We do somewhere in the range, in 1993, of $250 to $300 million a year, on a market base of $25 billion for services that we can offer today.

Obviously, our interest is to open up competition in other areas, so that we can offer additional services, and so that we can afford to take them out to some of the more rural areas in the United States.

I might say that many of my fellow companies that are in our organization, the Association for Local Telecommunications Services, operate in many of the constituencies represented by people on this committee; just to name a few, Seattle WA, Portland, OR, Phoenix, Richmond, Omaha, Dallas, Boston, and some other ones.

What we provide is competitive access networks. We build the sophisticated on-and-off ramps to the Nation's telecommunications intercity networks, such as those provided by AT&T, MCI, Sprint, and others.

First, let me say that we support Senate bill 1086, and commend the authors, along with their staffs, which we have had quite a bit of discussion with, for creating a bill that is necessary to ensure that telecommunications support the economic growth that the Nation must make during the nineties and beyond.

An underlying policy and implementation statement, such as Senate bill 1086, is needed now—and I mean now—to set the stage, and provide Federal guidance to the industry, the regulators, both Federal and State, and the country, during a much needed transition from a monopoly to a competitive environment.

If we do not pass this bill, there are forces that will conspire to make the evolution to competition a slow, arduous process, if it ever occurs at all.

I might give you a couple of examples of things that have happened in my area in Pennsylvania. Incidentally, my background is, I come from a monopoly. Originally, I was with CONTEL, which was a telephone company that developed telecommunications in many of the rural areas in the United States.

But let me give you a flavor of the environment that exists out there today. In my own market in Pennsylvania, the existing monopoly continues to game the regulatory system in pricing strategies and procedural filings, designed to hinder development of a competitive market.

My company has been subject to several nuisance complaints by the Bell operating company, complaints filed without any basis, that continually task our limited financial and personnel resources.

The thought of a competitive marketplace is so foreign and so upsetting to some of these people that they will fight it with tooth and nail. In fact, competition will bring more services to more customers, not limit their access.

Competition and universal service are not universally exclusive policies. We, as a company, and we, as an industry, support universal service. And we are prepared to work with restructuring the existing universal service program so that all of these new services can be brought to all areas in the United States.

We do have one issue with that, of course, and that is we do not want to give the money to our competitor, the monopoly carrier, to dole as they deem fit. We think it should go directly to the end user. This is one of the areas we think is important in the bill.

The bill, in order to ensure that the implementation of true measurable local competition is achieved during the follow-on regulatory process, needs further guidance with respect to safeguards, particularly, in the area of, "When does competition exist?"

We are small people out there. If we just open it up to competition in a moment's notice, the monopolistic power could put us out of business in no time at all.

We believe that there must be safeguards, only through the transition period. And once that transition period is completed, then we are prepared to compete in an open market.

We think there are a lot of benefits to Senate bill 1086. If we had to take it as is, we would take it as is, but we believe that with improved safeguards, it will be a much improved bill, and it will be more effective to what the Nation really needs.

Thank you.

Senator INOUYE. Thank you very much, Mr. Lasher.

[The prepared statement of Mr. Lasher follows:]

PREPARED STATEMENT OF GARY E. LASHER

Mr. Chairman and members or the subcommittee: My name is nary Lasher, and I am President and CEO of Eastern Telelogic Corporation, headquartered in the greater Philadelphia area. As Chairman of our industry organization, the Association for Local Telecommunications Services (ALTS), I am honored to testify today in support of Senate Bill 1086, the Telecommunications Infrastructure Act of 1993.

ETC believes that market forces should be used to accelerate private investment in communications facilities and services, without the need for undue government investment incentives. We see the key benefits of the local telecommunications competition fostered by this bill as permitting the development of consumer choice in a critical segment of our economy by speeding the deployment of an advanced information infrastructure. Through the policy prescriptions contained in this bill, we can develop the richest variety of features and functionalities, those vital to the economic health and future growth of our economy and the maintenance of our global leadership in telecommunications.

EASTERN TELELOGIC AND THE COMPETITIVE ACCESS INDUSTRY

Eastern Telelogic was founded in 1986 by entrepreneurs using private venture capital. The company now has over 175 miles of fiber optic loops connecting over 230 commercial office buildings to each other and to interexchange carriers. ETC provides an array of voice, video and data communication services to approximately 400 telecommunications dependent businesses.

My company is part of a new industry, consisting of approximately thirty entities serving over 50 cities, including many with constituents represented by the members of this committee; cities like Seattle, Portland (Oregon), Phoenix, Richmond, Omaha, Dallas and Boston. Like ETC, these companies build, own and operate local digital fiber optic, microwave and PCN networks. We are the creators of the sophisticated on and off ramps to the nation's telecommunications interexchange highways. Our industry is just beginning, having only emerged in the mid-eighties. Our market remains small, relatively a flea on the back of the elephant of the local telecommunications market, but our use of the most advanced applications and emphasis on customer service, particularly for the most telecommunications dependent industries, are what users want. In just a few short years, we have created competitive pressures that have improved overall local telecommunications services. We are only limited in bringing these benefits to as many consumers as possible by the vast array of antiquated regulatory restraints designed for a monopoly environment.

And the truth is that the monopoly local exchange carriers—no matter how well intentioned—simply cannot be all things to all people, everywhere, all the time, perfectly. Consumers want choices; choices of services and choices of suppliers. Thus,

70-334 O - 94 - 8

opportunities have arisen for entrepreneurs and investors to respond to consumer demand by offering new and better services using new and better technologies. Competitive local telecommunications service providers address market needs for service and network diversity that cannot be met by a sole provider. These needs have not been limited to only large users of telecommunications services. Many small businesses, such as florists, pizza parlors, and mail order retailers, are critically dependent upon their phone services and thus are seeking out the type of options supplied by my company as well as other competitive providers.

S. 1086 CONFRONTS CRITICAL POLICY ISSUES FACING THE DEVELOPMENT OF LOCAL COMPETITION AND PROVIDES VITAL AND ESSENTIAL DIRECTION

We commend Senators Inouye and Danforth for their vision. They have recognized that the federal government must help in bringing the benefits of competition to the local telecommunications market, the last bastion of monopoly control. The issues that are under consideration here are critical to the overall development of our country's telecommunications infrastructure and thus our economic development.

The core issues addressed by government policy and regulation in this bill are essentially how to inject competition and prevent monopoly abuse and how to ensure fair and effective opportunities for universal access to vital communications services.

I believe there has developed a consensus that the best way to bring innovation and affordable service to the public and prevent monopoly abuse is to encourage competition. We have definitely seen the public benefits of competition in the communications equipment and long distance markets. However, competition will not come to the local exchange market anytime soon, if at all, by "doing nothing". The current policy structure is stacked against the newcomer. We are often prohibited from entering markets or fully using our facilities. We are left to deal with monopolistic incumbents, who have no incentive to act fairly. And, you must remember, for new players delay is tantamount to death, as a failure to develop new sources and avenues of business results in the lack of a viable market.

Let me give you a few examples of how the current regulatory system is designed to foreclose new entrants to this market. In many states either legislation or public utility regulation specifically preclude the introduction of local or intrastate competition. In many of the states represented on this committee, Arizona, Nevada, South Carolina, South Dakota, Virginia and Hawaii, the benefits of local competition only go to interstate customers. And to the best of my knowledge, even in states that are more pro-competitive, no alternative provider has yet been permitted to complete local calls for end users in competition to the entrenched monopoly provider.

Even the Federal Communications Commission, which has espoused a policy of opening the local interstate access market to competition, has adopted policies that effectively limit the development of that competition. With the Orders mandating collocation, the FCC has also allowed the local companies to initiate unwarranted degrees of pricing flexibility in areas where competition has not yet been truly established, pricing flexibility that actually encourages cross subsidization. udder these policies, only consumers who happen to live in areas where competitors are located benefit from these non-cost based rates offered by the local exchange carriers. And that benefit will be short lived, since competition will not survive.

What government can and must do is ensure that all participants in the marketplace have equal opportunity to serve customers, and none have the opportunity to unfairly preclude competition. Once full and fair local exchange competition is established, it will be appropriate to reduce or eliminate economic regulation of competitive services. However, to deregulate before effective competition is established would be to foreclose the development of competition entirely. The incumbent local exchange company would be able to utilize its monopolistic powers to effectively bar new entrants entirely. As CompTel testified before this subcommittee in July, the market for local competition is still in the most nascent of stages, with less than 1.5 percent of the nation's access to local markets being supplied by entities other than the local exchange provider. We have a long way to go before there can be any claims of the existence of true competition or visible measures of market inroads into the local market by competitors such as ETC.

WHY S. 1086 IS NEEDED AND NEEDED NOW

S. 1086 provides what ETC and the other members of ALTS consider the essential elements for converting an outdated monopoly for local telecommunications into a vibrant competitive marketplace. specifically:

) S. 1086 recognizes the need for the federal government to ensure that competition reaches all potential customers, by allowing all providers to access the essential facilities of the telephone network remaining under monopoly control. All carriers that control essential bottleneck facilities must make those facilities available to other carriers on a nondiscriminatory, unbundled basis, at cost-based rates. Such essential facilities include: switching elements, transport elements, signalling systems, data bases, and rights of way (conduits, poles). By mandating interconnection of competitors with essential local telephone company facilities, barriers to entry are removed.

2) S. 1086 (following on the recently enacted Omnibus Budget Reconciliation Bill) recognizes the need for the federal government to overturn antiquated state and local regulations. As I indicated just a moment ago, progress to date in achieving some form of local competition at the state level has been slow, cumbersome and, in some cases, prohibitively expensive for potential new entrants. Even on an interstate basis, it has taken more than six years just to achieve permission to compete for a small portion of the market. The delays in removing regulatory barriers to entry may force many potential competitors to abandon their efforts. In addition, competition in smaller communities and rural areas will only be possible when companies can economically support the investment, hastened by access to a broader array of services. Without this legislation, it will be well into the next century before the process is complete, if ever.

What does this mean for business in any given location? As businesses seek to locate in friendly, technology-rich areas, it is the locales that offer a variety of telecommunications resources, resources spurred by competition, that will attract new investment dollars. Those communities that lack a diversity of service providers and offerings will see businesses locate elsewhere.

3) S. 1086 recognizes that the nation's telephone numbers are a national resource, which should not be controlled by a single entity or by the historical provider of local service in a particular geographic area. The ability to change service providers and keep one's telephone number places control of the number in the proper hands, that of the consumer. This ability has spurred tremendous competition in the market for 800 services and it will similarly spur competition for local services.

4) S. 1086 properly recognizes that companies do not do business for free. The reciprocal compensation agreements presented in the bill will allow all carriers to be reimbursed for the use of their networks.

5) S. 1086 properly mandates that government must also oversee the setting of interoperability standards so that telecommunications becomes an interconnecting "network of networks". The networks are being built by diverse organizations. Carriers such as the long-distance companies, the Regional Bell Operating Companies, independent telephone companies, cellular companies, and competitive access providers such as ETC, have provided the major portion of the backbone facilities. The ultimate network of networks will also include new technologies from other players such as cable operators, electric utilities and providers of personal communications services. The government must oversee the speedy integration of these diverse networks and technologies by promulgating the rules under which industry groups come together to set the standards necessary for a seamless network of networks.

COMPETITION WILL NOT THREATEN UNIVERSAL SERVICE

Competition and universal service are not mutually exclusive policies. That we already learned during our experience with long distance competition. Yet once again we are hearing, especially from those opposing local competition, that the opening of the local telecommunications market will result in stranding thousands of Americans from access to affordable and technologically advanced telecommunications services. This is simply not true. Actually, competition will enhance their availability. We can easily adapt the mechanisms that we have in place for assisting individuals who cannot reasonably afford the actual cost of their telephone service.

My company and other members of ALTS have long recognized that with the ability to compete in the local market comes the responsibility to support universal service. We are more than ready to contribute our fair share to the costs of providing such support for those individuals who need such assistance. However, we are not willing to contribute to the support of a local telephone company that has not moved to provide service to its users in the most efficient and cost effective manner. Therefore, we believe that the best and fairest way to administer much support is through a neutral third party.

We believe that there will be new companies that will want to compete for the provision of services to customers that local telephone companies complain they are stuck with as "the carrier of last resort". If the subsidies are equally available to all, other carriers may find more efficient and advanced means of providing services to these areas. The current system, whereby the existing monopoly provider determines the extent of the subsidy and receives the subsidy itself, does not encourage the most efficient use of the network. By allowing the consumer to determine the service provider, competition and efficiency will prevail.

CONCLUSION

The history of telecommunications during the last three decades demonstrates conclusively that competition does achieve national objectives better than a regulated monopoly can do. Both political parties have supported the emergence of telecommunications competition, first in customer equipment, and then in long distance service. Now it is time to move swiftly and unequivocally to bring the benefits of full competition to this as yet monopoly controlled local exchange market.

Removing barriers to competition in the local exchange will encourage cost effective investment at no cost to ratepayers or taxpayers, support the rapid deployment of modern high-capacity telecommunications infrastructure at costs appropriate to users' ability to pay, and promote universal service. Removing these barriers can be accomplished solely by regulators but it is unlikely they will do so in a timely fashion without Federal oversight. The goals and policy mandates articulated in S. 1086 ban bring the benefits of local competition to the marketplace. But these objectives will only be achieved if the subcommittee, and ultimately the Congress, have the courage to move the legislation quickly without significantly altering its design. The longer the delay in achieving competitive local exchange telecommunications services, and the larger the number of caveats and obstacles placed in the way of truly opening the market as envisioned by the authors of this bill, the greater the drag on the U.S. economy and the fewer the options for education, health, and public services delivered to ordinary people by telecommunications.

Entrepreneurship is the bridge between technology and its use by all citizens; the policy goals set forth by this Subcommittee and implemented by the Congress through enactment of S. 1086 will encourage this telecommunications entrepreneurship and bring significant impetus to the development of effective competition in the local telecommunications market.

Senator INOUYE. Ms. Easterling.

STATEMENT OF BARBARA J. EASTERLING, SECRETARY-TREASURER, COMMUNICATIONS WORKERS OF AMERICA

Ms. EASTERLING. Thank you.

As secretary-treasurer of the Communications Workers, I represent those workers who are employed at AT&T, the regional Bell operating companies, and many other telephone companies.

While understanding your subcommittee's interest in revising the Nation's telecommunications policy, my organization is concerned about S. 1086 in that the bill, we feel, is perhaps premature, and would not lead to a truly competitive marketplace.

As presently drafted, the bill is an incomplete approach toward establishing a telecommunications policy. All of us associated with the industry—and I think we heard so this morning—are still trying to sort out the full set of implications of the mid-August announcement that AT&T intends to buy McCaw Cellular.

We view the AT&T/McCaw deal as a selective return to local telephone service from which AT&T departed just short of a decade ago.

Ineluctably, this committee and the Congress must evaluate the merits of continued MFJ restrictions on the Bell operating companies. For 20 years the Congress has struggled with the ever developing need for a policy to govern this important industry.

Bills have been introduced in every congress since 1973, with the necessary consensus always unattainable. Currently, several public bodies are involved in addressing telecommunications policy issues, including the two Houses of Congress, the FCC, the NTIA, and the OTA.

And this is in addition to the Department of State, the Department of Justice, the Federal courts, and, of course, the States, which preserve their own regulatory role as well.

We would strongly recommend that before any new legislative efforts are begun, the administration, the Congress, and the industry would establish a broadly based study commission to agree on new policy directions.

Senator Breaux proposed such a study group to make an overall policy recommendation. And we are encouraged that the NTIA has established its own working group to draft the position paper on telecommunications policy issues for presentation to the Congress.

S. 1086, in our view, can be viewed as giving undeserved protection to competitors of local exchange telephone companies. Any legislation should merely protect the process of competition, not aim at guaranteeing market results.

We do not believe that it is the job of Government to promote competition, as some favoring the bill have publicly stated. Promoting competition is the same as creating competition. Government's role must merely be to protect the processes against abuses.

Since the circuit court recently ruled that all carriers must file the tariffs required under present rules, we would view enactment of section 229(h) as decidedly partial to some providers, at the sizable expense of others. Tariff forbearance is a poor idea, as has now in 1993 been decided by the circuit court of appeals.

We believe an unintended side effect of the legislation, and this, of course, is my major reason for being here, would be a sizable reduction in employment in local exchange telephone companies, which would remain the providers of last resort. The new competitors would be free to serve their niche markets, taking only the lucrative businesses.

State regulators, meeting in July at the NARUC summer conference, devoted considerable time to debating the bill, and the need to ensure adequate support within the local rate structure to continue universal service.

One key consideration of NARUC continues to be the viability of the universal service fund, which is unlikely to receive the proper support, unless your legislation is clear and strong in its policy mandates.

As matters have unfolded over the last quarter century, the competition has come in thin wedges, driven in by companies trying to serve only the niche markets.

The overall question of the proper contribution toward the fixed costs of maintaining a first-rate core network, available universally, has yet to be addressed by the Congress over the last two decades.

Section 5 of the bill appears to require all telecommunications carriers to make contributions toward universal service in the present of a competitive environment.

But our analysis does not show that section with enough clarity that each niche competitor would be reached for equitable assessments of funds to help pay the costs of local exchange service.

Where we may have differences with the subcommittee is the belief that this legislation will promote investment in new technology, and at the same time lower rates.

CWA supports legislation that would allow the telephone companies to enter the cable business, since cable companies have recently made the decisive move into selective and lucrative offerings of local exchange services.

The telco-cable provisions of section 8 of the bill are unduly restrictive, in CWA's view. The union instead supports H.R. 1054, legislation offered by Representatives Rick Boucher and Mike Oxley, as a more appropriate means of regulating the enterprise.

We are convinced that the restrictions keeping telephone companies from providing cable services here in the United States have encouraged those companies to invest billions of dollars in foreign enterprise.

The telephone company should not be required to establish separate subsidiaries for offering cable services. Such a requirement would in our view become anticompetitive, in that it would unduly increase the transaction costs of the telephone companies and their competitors, at the same time, would not be similarly burdened.

Finally, we agree with many of the comments that were made this morning. In fact, we found that the questions by the Senators were just as telling as the answers by the panel.

We, in fact, feel that a competitive arena would be a good one, one in which—as I think it was Senator Breaux who said—we open up competition 100 percent to all the comers, and we provide the rules that all will operate under, and it will be a single set of rules.

And to that end, CWA looks forward to working with the subcommittee and the Senate in this undertaking.

Thank you.

Senator INOUYE. Thank you very much.

[The prepared statement of Ms. Easterling follows:]

PREPARED STATEMENT OF BARBARA J. EASTERLING

While understanding your Subcommittee's interest in revising the nation's telecommunications policy, my organization's assessment of S. 1086 is that the bill is premature and will not lead to a truly competitive marketplace. In our view, the Congress and several Administrations have to date attempted to undertake only partial steps, while far more is needed. As presently drafted, we oppose S. 1086, as another incomplete approach toward establishing a telecommunications policy.

One of the major ongoing debates in our country has been to decide and set our telecommunications policy. when the American Telephone and Telegraph Company (AT&T) broke up in 1984, in the largest business transaction in the history of the world, many new and unforeseen problems and issues emerged.

All of us in the industry are trying to sort out the full set of implications of the mid-August announcement that AT&T intends to buy out McCaw Cellular, the nation's leading wireless enterprise. Any move toward legislation now requires an in-depth examination of issues emerging from this multi-billion-dollar transaction. We view the AT&T-McCaw deal as a selective return to local telephone service, from which AT&T departed just short of a decade ago. Ineluctably this Committee and the Congress must evaluate the merits of continued MFJ restrictions on the Bell Operating Companies.

For 20 years the Congress has struggled with the ever-developing need for a policy to govern this important industry. Bills have been introduced in every Congress since 1973, with the necessary consensus always unattainable. The process of legislating a new policy has become extremely complex at a time when the Congress must devote its major energies to other matters presenting themselves as more urgent.

Currently, several public bodies are involved in addressing telecommunications policy issues, including the two Houses of Congress, the Federal Communications Commission, the National Telecommunications and Information Administration in the Department of Commerce, the Congress' Office of Technology Assessment and the Department of State. By enforcing anti-trust law, the Department of Justice and

Federal courts oversee the AT&T and GTE cases, in which telecommunications matters are at the core. Finally, the States preserve their own regulatory role.

In 1967-68, there existed the President's Task Force on Communications Policy, chaired by Eugene V. Rostow, Under Secretary of State for Political Affairs. The Rostow Task Force's final report became a blueprint for new policy directions, including the injection of competition into the telecommunications industry. That Task Force was responding to President Johnson's message keyed to the emerging use of satellites for international and possible domestic use, and developments in common carrier, broadcasting and cable television enterprises.

We would strongly recommend that before any new legislative efforts are begun, the Administration, the Congress and industry establish a broadly based study commission to agree on new policy directions, modeled on President Johnson's task force. Senator John Breaux recently has aptly proposed such a study group to make an overall policy recommendation.

We are encouraged that NTIA has established its own working group to draft a position paper on telecommunications policy issues, for presentation to the Congress. From the agency's announcement, we see a commendable attempt to give a comprehensive coverage of the many complex sub-issues facing government as it regulates the telecommunications industry for the good of all interests in the United States. We strongly support NTIA's efforts to get the policy dialogue moving forward.

S. 1086, in our view, can be viewed as giving undeserved protection to competitors of local exchange telephone companies; any legislation should merely protect the process of competition, not aim at guaranteeing market results.

We do not believe it is the job of government to "promote" competition, as some favoring S. 1086 have publicly stated. Promoting competition is the same as creating competition. Government's role must merely be to protect the processes against abuses. We strongly advocate the Subcommittee strike the bill's proposed subsection 229(h), which deals with "Regulatory Flexibility for Competitive Services." Since the District Court on July 7 ruled that all carriers must file the tariffs required under present law, we would view enactment of 229(h) as decided Iv partial to some providers, at the sizable expense of others. Tariff forbearance is a poor idea, as has now in 1993 been decided by the Circuit Court of Appeals. Since the Supreme Court declined to review that appellate decision, we and many others in the industry contend that tariffing requirements are sound law and policy.

CWA views the August 16 Commission decision in CC Docket 93-36, on the "streamlined" tariffing requirements for "non-dominant" carriers, as yet another anti-competitive action, contravening Section 203 of the Communications Act. The Commission erred in its old "forbearance" policy, begun in 1979, and appears bent on continuing policies which favor the large and powerful rivals of AT&T. The Congress should press the Commission for consistency in tariffing rules.

We believe an unintended side effect of your legislation would be a sizable reduction in employment in local exchange telephone companies, which would remain the "providers of last resort." The new competitors would be free to serve "niche" markets, taking only the lucrative business. In revising telecommunications policy the Congress needs to examine the full range of implications of competition, instead of finding competition a goal in and of itself. In the last decade of "competition," we have seen local telephone rates more than double; we have yet to detect the unqualified success of competitive offerings, especially in the case of ordinary residential telephone users.

State regulators meeting in July at the NARUC summer conference devoted considerable time debating S. 1086 and the need to ensure adequate support within the local rate structures to continue universal service. while the NARUC delegates then were unable to take a unified position on S. 1086, it is clear they perceived the need for extensive work to make the legislation serve its aim of preserving universal service. One key consideration of NARUC continues to be the viability of the Universal Service Fund, which is unlikely to receive proper support unless your legislation is clear and strong in its policy mandates. The competitive access providers have been clear in their opposition to paying their shares of local exchange costs. Any shortfalls of needed local revenues will be made up by residential and small-business users—who will wind up paying billions of dollars to subsidize large business users. This situation will offer almost an exact replay of the network access charge of a decade ago, a means by which billions of dollars of interstate revenues were shifted onto the bills and backs of small users. We all were drowned in rhetoric of "bypass blackmail" in late 1983. Those of us opposed to the access charge at that time correctly noted that the access charge proposal would lead to "bypass," which is in fact happening now. Let the Congress not make matters any worse. Let the Congress listen this time.

In adopting telecommunications policy legislation, our key recommendation is that the Congress squarely face the issue of competition. As matters have unfolded over the last quarter-century, the "competition" has come as thin "wedges" driven in by companies trying to serve only "niche" markets. The overall question of the proper contribution toward the fixed costs of maintaining a first-rate core network, available universally, has yet to be addressed by the Congress over the last two decades.

Section 5 of S. 1086 appears to require all telecommunications carriers to make contributions toward universal service in the presence of a competitive environment; our analysis of Section 5 does not show with enough clarity that each "niche" competitor would be reached for equitable assessments of funds to help pay the costs of local exchange service, which by its very nature is capital- and labor-intensive. Not to address this point clearly in legislation is to invite litigation as to the intent of Congress on subsidies within the rate structure.

If in fact the issue is competition, let the terms of the debate be fully defined and widely understood. who competes with whom and the equity or inequity of terms of competition need a full airing by the Congress and the appropriate Federal and State agencies, to ensure a fair process. We agree with Senator Inouye that "The crucial word is competition," as he noted in the June 9 press conference on the legislation. where we may have differences with the Senator are the beliefs that this legislation will promote investment in new technology and at the same time lower rates.

Whose rates would decrease? Certainly not the rates of the average residential user, whose basic dial-tone charges went up more than 100 percent in only the four-year period starting in early 1982. To date, no entity has shown a credible scintilla of interest in serving the local residential and small-user market for telephone service. We hope the Congress will not lose sight of the present universality of service.

Many aspects of S. 1086 already are within the powers of the Federal Communications Commission. Other aspects are covered by the 1982 Modification of Final Judgment.

CWA supports legislation to allow telephone companies to enter the cable TV business, since cable companies have recently made decisive moves into selective and lucrative offerings of local exchange services. The telco-cable provisions of Section 8 of S. 1086 are unduly restrictive, in CWA's view. The union supports H.R. 1504, legislation offered by Representatives Rick Boucher and Mike Oxley, as a more appropriate means of regulating the enterprise. We are convinced that the restrictions keeping telephone companies from providing cable services here in the United States have encouraged those companies to invest billions of dollars in foreign enterprises. We believe public policies encouraging investment in the United States should be the very first priority.

However, on the cable-telco provisions of both S. 1086 and H.R. 1504, CWA strongly advocates a major amendment. The telephone companies should not be required to establish separate subsidiaries for offering cable services. Such a requirement would, in our view, become anti-competitive in that it would unduly increase the "transaction costs" of telephone companies. Their competitors at the same time would not be similarly burdened with those added duplicative transaction costs, which would become a major disincentive to time competition.

CWA would most strongly suggest the Subcommittee carefully examine the August 24 ruling of the U.S. District Court in Alexandria in the case brought by Bell Atlantic, challenging the 1984 cable communications policy legislation. Section 8 of your bill, S. 1086, would impose numerous conditions on the Bell and other local exchange carriers. Some conditions to be imposed by Section 8 may not be consistent with the District Court's decision. In the event Bell Atlantic begins to move forward with its plans to enter the cable TV business, the Congress may be a placing itself in the awkward position of attempting to legislate a company out of a legitimate enterprise. We believe the Congress instead should concentrate on a single and consistent policy of competition. All providers should be under the same rules, to include tariff-filings and justifications.

In conclusion, the position of CWA is that the Congress should make numerous extensive amendments to S. 1086 to open the competitive arena to all comers, provided all operate under a single set of rules. CWA looks forward to working with your Subcommittee and the Senate in this undertaking.

Senator INOUYE. Ms. Preston.

STATEMENT OF PAULA SMITH PRESTON, VICE PRESIDENT OF ADMINISTRATION, HELLO, INC.

Ms. PRESTON. Thank you, Mr. Chairman.

I am Paula Preston of Hello, Inc., a telephone answering service, located in Richmond, VA. I am appearing on behalf of my company and my fellow members of the Association of Telemessaging Services International, known as ATSI.

ATSI is the national trade association representing the voice mail and live answering industry. The members are virtually all small businesses.

The majority are woman-owned. In fact, my great-grandmother, Margaret Redmond Smith, was one of the pioneers in the industry, having founded our company as a doctor's answering service in 1923.

Today, because of the regulatory bias created by telephone companies' anticompetitive access to and use of customer proprietary network information, or CPNI, as well as joint marketing and cross-subsidization, the future does not look so bright for our industry, or for the American consumer of telemessaging services.

These three advantages, CPNI, cross-subsidization, and joint billing, which are not available to me as an independent enhanced service provider, severely hamper our industry's ability to provide affordable voice messaging services to the American public.

How does the telephone companies' use of CPNI create an unfairness in the marketplace, and cause the American consumer to be the real loser?

Telephone companies can use CPNI to identify potential customers for voice mail or other unregulated services, such as subscribers moving into the area, or adding service locations, before we, as competitors, are aware of them.

Telephone companies can use CPNI to prepare highly effective marketing presentations, based on usage information, such as numbers of missed or blocked calls. This information is unavailable to me. Is it verifiable to the consumer?

Perhaps most frustrating, our supplier is, first and foremost, our competitor. Because of that, telephone companies can identify the customers using competing vendor services by matching up the CPNI of vendors and customers.

The telephone companies want you to believe that these examples are merely abstract "what-ifs." We have found that this is not the case.

ATSI has compiled volumes of incidents where proprietary information has been used in an anticompetitive fashion to discriminate against our industry, and thwart the American public's freedom of choice.

One of the many examples of such use and abuse of CPNI occurred when my father, Edward L. Smith, Sr., called C&P Telephone Co. to have delay, call forwarding service on his home phone.

He was asked what it was to be used for, and he replied that he wanted to have his calls answered by Hello, Inc. He was told that he could not use it with Hello, Inc., specifically, and that it could only be used for C&P's voice mailboxes.

It is impossible to calculate how many times C&P Telephone may have used this marketing strategy. It is equally impossible to know how much business my company has lost to C&P Telephone, after we have marketed our services, and customers who are ready to sign on are told by the telephone company that it cannot be done, except their way.

I do know this: They win. We and the American consumers lose every time.

How do the telephone companies take advantage of their structure? By not providing unregulated services such as voice mail to a separate subsidiary. The telephone companies can add their charges for voice mail onto the basic service mail for virtually no cost.

In many States, the charge for unregulated services, such as voice mail, combined with the basic telephone subscription charge, the consumer is charged one lump sum called monthly service.

The customer is conveniently never again reminded of how much the enhanced service costs, except in some areas where they have an annual itemization.

In fact, the telephone company may increase the charge for the unregulated services as they deem necessary. Because the charge is not itemized, the American consumer is not provided with the information necessary to properly evaluate the cost or worth of the service, nor is the consumer given any incentive to privately price comparable services, and decide what is best for his given situation.

The telephone companies are also not averse to underpricing their voice mail services in an attempt to price us out as their competition.

In fact, in May 1992, the Georgia Public Service Commission found evidence that Southern Bell could not possibly offer its voice mail service, which is known as Memory Call, for the price that it charged. It was offered at far below cost. We cannot compete with these rampant abuses.

Such lack of fair competition, as I have outlined, will return telecommunications in America back to the dark ages, before the breakup of the AT&T monopoly. Ultimately, the American consumer will once again be the loser.

Thank you for the opportunity to testify. ATSI welcomes this opportunity to work with the committee in securing enactments of the safeguards contained in S. 1086.

On behalf of my fellow members in the telemessaging industry, I urge that the joint marketing and separate subsidiary rules be expanded to apply to the telephone answering voice mail industry.

Thank you.

Senator INOUYE. Thank you very much, Ms. Preston.

[The prepared statement of Ms. Preston follows:]

PREPARED STATEMENT OF PAULA SMITH PRESTON

My name is Paula Smith Preston, of Hello, Inc., a telephone answering service, located in Richmond, Virginia. I am appearing on behalf of my company and my fellow members of the Association of Telemessaging Services International known as ATSI.

ATSI is the national trade association representing the voice mail and live answering industry. The members are virtually all small businesses. The majority are woman-owned. In fact my great-grandmother, Margaret Redmond Smith, was one of the first In the telemessaging industry, having found our company as a doctors' answering service in 1923.

Today, because of the regulatory bias created by telephone companies' anti-competitive access to, and use of, customer proprietary network information

(CPNI)—as well as joint marketing and cross-subsidizationthe future does not look so bright for our industry or for the American consumer of telemessaging services. These 3 advantages CPNI—crosssubsidization, and joint billing—which are not available to me as an independent enhanced service provider, severely hamper our industry's ability to provide affordable voice messaging services to the American public.

How does the telephone companies' use of CPNI create unfairness in the market place and cause the American consumer to be the real loser in receiving advanced customized telemessaging services?

- Telephone companies can use CPNI to identify potential customers for voice mail or other unregulated services, such as subscribers moving into the area or adding service locations, before we, as competitors, are aware of them;
- Telephone companies can use CPNI to prepare highly effective marketing presentations based on usage information (such as number of missed or blocked calls). This information is unavailable to me. Is it verifiable to the consumer?
- And perhaps most frustrating, our supplier is, first and foremost, our competitor. Because of that, telephone companies can identify the customers using competing vendor services by matching up the CPNI of vendors and customers.

The telephone companies want you to believe that these examples are merely abstract "what ifs." That is not the case. ATSI has compiled volumes of incidents where proprietary information has been used in an anticompetitive fashion to discriminate against our industry and thwart the American public's freedom of choice.

One of the many examples of such use and abuse of CPNI occurred when my father, Edward L. Smith, Sr., called C&P Telephone Company to have Delay, Call Forwarding service on his home phone. He was asked what it was to be used for, and he replied that he wanted to have his calls answered by Hello, Inc. He was told that he couldn't use it with Hello, Inc., specifically, and that it could only be used for C&P's voice mailboxes!

It is impossible to calculate how many times C&P Telephone may have used this marketing strategy. It is equally impossible to know how much business Hello, Inc. has lost to C&P Telephone after Hello has marketed its services, and customers who are ready to sign-on are told by the telephone company that it can't be done—except their way!

I do know this: They win. We and the American consumer loses. Every time.

How do the telephone companies use joint marketing to create unfair advantage in the marketplace and cause the American consumer to be the real loser in receiving advanced telemessaging services? Through Computer III, the FCC has specifically allowed the telephone company to use its regulated service sales personnel to sell competitive services. With few limitations, a new phone subscriber can be sold voice mail at the same time he is requesting basic phone service. This is before he as a consumer even knows there are competitors offering the same or better service to meet his needs Furthermore, these same phone company personnel can sell voice mail when subscribers call the telephone company for any reason— new service, repairs, or most unfairly, when they order the services link—call forwarding—which they need to receive voice mail service from independent providers.

How do the telephone companies take advantage of their structure? By not providing unregulated services such as voice mail through a separate subsidy. The telephone companies can add their charges for voice mail on the basic service bill for virtually no cost. In many states, the charge for unregulated services such as voice mail is combined with the basic telephone subscription charge. The consumer is simply charged one lump sum and it is called "monthly service". The customer is conveniently never again reminded of how much an enhanced service costs except in some areas where there is an annual itemization. In fact, the telephone company may increase the charge for the unregulated service as they deem necessary. Because the charge is not itemized, the American consumer is not provided with the information necessary to properly evaluate the cost—or worth—of the service. Nor is the consumer given any incentive to privately price comparable services and decide which is best for his given situation.

The telephone companies are not averse to underpricing their voice mail services in an attempt to "price us out" as their competition. In fact, in May 1992 the Georgia Public Service Commission found evidence that Southern Bell could not possibly offer its voice mail, Memory Call, for the price it charged. It was offered at far below cost. We cannot compete with these rampant abuses. This is not a level playing field.

Such lack of fair competition as I have outlined in my testimony, will return telecommunications in America back to the dark days before the breakup of the AT&T monopoly. Ultimately, the American consumer will once again be the loser.

Thank you for the opportunity to testify. ATSI welcomes this opportunity to work with the committee in securing enactments of safeguards contained in S. 1086. On behalf of my fellow members in the telemessaging industry, I urge that the joint marketing and separate subsidy rules be expanded to apply to the telephone answering voice mail industry.

Thank you.

Senator INOUYE. The word "competition" has been used by every witness many, many times. And I will take the suggestion of the author, Senator Danforth, very seriously.

I think it has become obvious that the crux of the whole measure may lie in the definition of the word "competition." Mr. Lasher has suggested that the sudden onslaught of competition may be devastating, that there must be some time to prepare oneself for competition. I believe Ms. Easterling has said that the definition in this bill is unclear.

So, all of you have testified on the word "competition." I would suggest very strongly that you come forth with your definitions of what competition should mean, because, otherwise, we will hear from RBOC's at AT&T, and they may set the standard and definition.

And I listened—Mr. Ray, I was very much impressed, because up until now we have been told that private industry can do the job better than Government, and apparently it is not so in your part of Kentucky. Congratulations.

Mr. RAY. Thank you.

Senator INOUYE. I did not quite get the number. You said that in one business you can do it for less than $50, when private industry would charge over $1,000; is that right?

Mr. RAY. Yes. On the communitywide local area network that we have, we have established a 2-megabit-per-second network that anyone with a home computer, or other businesses that want to tie their networks to create virtual networks, and once the highway is constructed, the addition of these services, the incremental cost is fairly insignificant.

I do not pretend to know why the phone company charges for these services the way they do. I have my suspicions. But I know that, a T–1—there is a tariff for T–1, which is 1½ megabytes per second in Kentucky, and it is in the $1,200 range.

And we charge $19.95 for it, and make probably more margin on that service than we do on cable TV or electricity, for that matter.

Senator INOUYE. Mr. Opperman, we have been advised that the RBOC's has been discussing your support of their entry into long distance, and with your support they will support your safeguards for electronic publishing. What is your interest in long distance?

Mr. OPPERMAN. Well, it is really historic, in this sense, that when the monopoly was broken up, as we all well know, long-distance charges dropped by approximately 40 percent in a very short period of time. That pretty much unleashed our product, WESTLAW, and it unleashed much of our competitors, the electronic product.

We have an interest in seeing competition at all levels. And our interest in long lines issue is merely that as we move to a competitive model, we see the need for some transition rules, but not transition rules forever.

If I can explain that, it just seems to us that the same great experience we have had so far in the last 12 years will be replicated many times over if we can follow that same process as we break up the local bottleneck monopoly.

Now, obviously, our interest in breaking up the local bottleneck monopoly is the same as the one we had in breaking up the long-distance monopoly. We are a customer. We cannot compete with these people. We are a customer.

And we get better service and cheaper prices when the long-distance monopoly broke up, and we are pretty darn sure that we are going to get lower prices and better service when the local bottleneck gets broken up. That is really our interest.

Senator INOUYE. Thank you.

Mr. Kimmelman, you touched upon universal service. This is another area where definition may be very important. What is your definition of "universal service"?

Mr. KIMMELMAN. Mr. Chairman, I do not think there is any one definition. I think that it evolved over time. But I would treat it from this perspective: I think the American people have come to have an expectation that local telephone service is a necessity for everyday life, dealing with family, friends, business contacts.

And it has been one of the best bargains that we have had in our economy. The real price of local phone service has actually fallen 60 percent since the 1930's.

As I look at the desire to move to competition, I believe the American public will say, "What is in it for me? What will I get?"

Well, there are a lot of promises about new services and where we are going. But I think the one thing that will seem anomalous, if not totally contradictory, is that the price of what they view as a necessity actually goes up.

So, my suggestion, on a practical level, based on how the public seems to respond to their needs, is to put in place a directive from Congress to the regulatory bodies to preserve that pricing tradition.

We need a new mechanism, when there is not one monopoly carrier, when there are multiple carriers. They will need to all contribute, as your legislation calls for.

I would just urge you to put more detail in, to ensure that if competition does not exist everywhere, or does not actually drive down prices for local phone service, that we have a backdrop, that we have guaranteed that we do not have the unintended effect of actual price increases.

Subsumed within that is the traditional problem, raised earlier by Senator Stevens and Senator Burns, of rural communities having higher costs, the need to spread costs across the entire country, low-income households that cannot even afford a phone, persons with a disability, who may need special equipment to functionally "have" basic phone service.

I think all that can be subsumed under a new universal service mechanism, that is fair to all providers, is not expensive, but provides that insurance policy, that local rates will remain at about their historical level.

Senator INOUYE. Thank you.

Mr. Lasher, you expressed some fear that if competition suddenly came upon you, it might hurt your business.

Mr. LASHER. It may be the ambiance of this room, but I guess I did not express that the way that I wanted to express it.

My concern would be that if we just open it up to go-fight-it-out competition, that a monopoly would use their monopolistic powers to take companies like mine, that are startup companies, and companies getting into the business, and use their monopolistic power to put us out of business.

So, what we are interested in is the safeguards during a transition period, to the point that once competition comes to a point where it is real true effective competition, then take off the safeguards and no-holds-barred at that time. But until that takes place, there has to be restrictions on a monopoly during the transition period.

Senator INOUYE. I will be looking forward to receiving your paper on competition. From the top of your head, how long should this transition period take?

Mr. LASHER. I think it might take place at different times in different places, depending on how fast competition comes into the area.

And there should probably be some type of test that would establish when competition exists in that particular service area or geographical area.

Senator INOUYE. Thank you.

Ms. Easterling, you have indicated that we are not quite ready to be addressing this type of measure, that we should have a commission. Listening to the testimony here, do you not think we have enough information to proceed?

Ms. EASTERLING. My concern has always been—and we have promoted the commission concept. I think we have done that because of the length of time that we have been talking about doing something, changing the law that was created in 1934, and not seeing anything being done.

But I think what really drives us is the fact that in this day and age we are talking about fiber, and copper, and microwave, and cellular, and all of those things, and no one has sat down to say, "Maybe we ought to devise a telecommunications policy for the country," just as we had a policy for designing the highways in this country.

Everybody just did not go out and build a highway wherever they wanted one. To that extent, we think that there is a lot of waste on the building of a lot of the fiber that is being laid, that we are going to get to the point where we could not possibly use all the fiber that is out there, which is a tremendous cost.

We think we ought to be looking at the future, and that somebody ought to be looking down the road somewhat to say, "What are the highways of the future?" and to help construct that in a better fashion, where some heads of different organizations, companies, and so forth, could sit down and do it; and not so much in an emotional state, where we have fibers, 50 companies coming forth to tell you what their needs are in a bill, because this is what happened every time.

And certainly you have seen me up there many times, and we all come up here, and we all share with you what our thoughts are and what we think should be done to better the telecommunications in this country, and nothing ever happens.

I think sometimes it is because we have created such a lead balloon that we cannot get it off the ground. So, that is the reason that we have been promoting, that the Communications Workers have been promoting a telecommunications study, so that we would be looking at the overall picture, and also at the future.

We are concerned about it. We are concerned about the employees, No. 1. That is whom we represent. We do not sell any telephones, or make any money off any of that. Our commodity is the employees who are there. And we are concerned about them.

We are very concerned about the business that is going overseas, because the companies are not permitted to perform their work in the United States. And we think we ought to be creating those jobs in the United States, instead of sending them overseas.

Those are a lot of our concerns. For that reason, I guess we are not ready just to move ahead on a bill that we find some problems with, but we are willing to work with the committee.

I do not want you to think that we are not. We are. And we are willing to work with your staff, as we have in the past, to try to perhaps devise some amendments that would be acceptable to the committee and acceptable to us as well.

Senator INOUYE. Thank you very much.

Ms. Preston, you are quite fearful of what the telephone companies can do to you. Are the safeguards in this measure sufficient to protect your small enterprises?

Ms. PRESTON. Well, they do not specifically address voice mail. I think electronic publishing is addressed. But electronic voice mail is not specifically mentioned in the legislation. And that was what, really, the thrust of my testimony was.

We have been in the voice mail business longer than the telephone company in our area, for some years. They went into it after we were in it, and as well as other competitors we have in the area.

After the telephone company came in, the market changed drastically. That is why we felt that the voice mail industry specifically should be mentioned.

Senator INOUYE. Thank you.

I wish to thank all of you for your participation this morning. I should tell you that I do have one concern. I am impressed by the dynamic and everchanging nature of this industry.

That being the case, somewhere in the back of my mind there is the fear of putting everything in concrete at this stage. If you can help me and try to make it sufficiently flexible, so that this measure will accommodate, or take into consideration some new technology that may come about 30 days from now, because every time we have hearings of this nature, something new comes up.

I remember when I first joined this committee, fiber optics was not even mentioned. Now, it is commonplace. So, I hope you can do me that service, of taking a careful look at the measure.

That is my concern, that we might ask for a lot of trouble if we put everything in concrete. I hate to come back every 6 months with amendments to the measure.

Senator Danforth.

Senator DANFORTH. Mr. Chairman, thank you.

Thank you to all members of the panel for your very useful testimony.

I just have one question, and it is to Mr. Kimmelman. It concerns the recent case, the Bell Atlantic case relating to telephone entry into the cable business.

This legislation deals with that same subject, and this is a subject that has been near and dear to the hearts of various Members of Congress. Senator Burns, for example, has been very interested in the so-called telco entry.

However, we have endeavored to provide safeguards, and specifically what we have done in this legislation is to say that we encourage the telephone companies to get into the cable business, provided that it is competitive.

I mean, it does not—our view was that it does not create competition, for a telephone company to buy the cable company.

Now, it is my understanding that this recent court case, which allows telephone company entry into cable, has no limitation whatever on the telephone company simply buying out the cable company, and, therefore, expanding a single monopoly into a dual monopoly. Do you share my view?

Mr. KIMMELMAN. Absolutely, Senator Danforth. In light of that court decision, there is no doubt that there is a danger now, without legislation, that you could take two individual monopolies, a cable monopoly and a telephone monopoly, and they could merge and create one larger monopoly, whether it is the phone company that buys out the cable company, or vice versa.

In an area in which we could have competition, that would be very unfortunate. So, your legislation clearly would prohibit that, and that is a major step forward.

There is one addition I would suggest to the legislation. Again, it is really a clarification. The question has always been: When you are investing to expand a telephone system into a video capable system, who pays?

And we have always been concerned that the telephone ratepayer, who just buys local phone service, should not have to pay for this video investment. Further clarification of that, I think would both protect ratepayers and ensure fair competition.

So, there is no doubt that right now, with that court decision, this legislation is even that much more important and necessary to promote competition.

Senator INOUYE. Thank you very much. Senator Robb.

OPENING STATEMENT OF SENATOR ROBB

Senator ROBB. Thank you, Mr. Chairman. I will be very brief.

Unfortunately, I have had four different important events all occurring at precisely the same timeframe, and have missed parts of the testimony, some of which I had very much looked forward to, and I will review it in the record.

The statement that the chairman just made a minute ago about the concern about obsolescence and what you do reminds me of an occasion earlier this morning, that I happened to be with the President and the Vice President, when they were talking about the leadtime that the Government takes to order computers.

I do not know all the specifics, but it seems to me that within the private sector it takes a year or so, but within the Government I think the timeframe that they referred to was 49 months to get a computer.

And the suggestion was made that by the time the computer arrived, it was two generations obsolete, in terms of its ability to fulfill whatever needs the ordering entity or Government agency had.

That clearly is a concern here, in terms of what the Government role ought to be. I was interested, when I was here with the earlier panel, when our colleague, Senator Breaux, asked the question, which I guess goes to the same point that has been raised by almost everyone, and that is "What would happen if Government stepped back altogether?

Had I had an opportunity to ask a question at that point, I would have asked for some explanation, because I think that is the only way that those of us who are not technical can address this particular question.

I would ask, not so much in the form of a question for today, but for some issues for the record, if you would address the question of what kinds of safeguards ought the Government to be most concerned about, particularly with regard to the timeframe of any transitional period.

Some have argued for the absence of any restrictions to competition, however it is defined; others would define "competition" in various different ways.

But it seems to me that if we could have a listing, to the extent it is not already explicitly covered in your testimony, some idea of the kinds of safeguards, and the kinds of timeframes that would be absolutely critical, in your judgment, to the industry or the sector of the whole question that you represent, and any discussion of how that is going to impact on other players in the industry.

Again, we had quite an exchange between two folks on the earlier panel, in terms of having an issue apparently joined, and yet not so clearly joined, that it was easy to draw a line, and determine exactly what the Government policy should be.

If you could assist us in that regard, and attempt to, at the same time, answer the chairman's question about obsolescence, and whether or not anything that we do now is going to inhibit the development of the industry, and the expansion of the competition for which we are looking. Something along those lines would be very much appreciated.

Mr. Chairman, this panel has been very patient. They have been here for I guess——

Senator INOUYE. Five hours.

Senator ROBB [continuing]. About 5 hours already. I do not think that trying to go through a detailed litany of some of the things that I am asking right now would serve any purpose, or generate a great deal of goodwill with the panel, or others who are witnessing this particular hearing.

But I would make that request, and I would conclude by thanking you and the ranking member for not only putting this piece of legislation into force, but for calling this and the previous hearing, and offering to hold additional hearings, because I think that after the Congress disposes of the whole question of health care reform, that the questions that are explicit and implicit in this legislation, and the related legislation that Senator Danforth referred to a minute ago, I believe, are going to be the most important questions that this Congress has to wrestle with, and ones which we are going to need the most assistance from experts, and people who have technical expertise, as well as those who can advise us with regard to policy.

So, I would say that I think we are on to what is a very important topic, and will be for a long period of time. And I commend both the chairman and the ranking member for pursuing it, and I look forward to pursuing it with you, and will simply make that request.

And with that, Mr. Chairman, I will yield.

Senator INOUYE. All right. Thank you very much.

Before recessing this hearing I would like to announce that we will have another hearing, at least one more. We will have to accommodate the last panel, as I promised.

In addition to that, we had at least 12 requests to appear from very competent and very reliable witnesses, but we could not accommodate them today. We will most certainly accommodate those witnesses.

I would like to point out to the administration that I would hope that the administration will participate in the next hearing. I am certain that the administration noted, or should have noted, that there is a lot of interest on this committee.

I can assure you that the authors of this measure have no intention to just have hearings. We will report on this measure. We will have a markup. And that is why your participation and your assistance is very important.

The administration should know that. And that being the case, I would hope that they will participate in the process and the deliberations.

With that, I would like to, on behalf of the committee, thank all of you for your patience and your help, and we look forward to working with you.

[Whereupon, at 2 p.m., the hearing was adjourned.]

APPENDIX

Prepared Statement of Dean J. Miller, Commissioner, Idaho Public Utilities Commission

Good morning and thank you for giving me the opportunity to comment on the Telecommunications Infrastructure Act of 1993, S. 1086. My name is Dean J. (Joe) Miller. I am a member of the Idaho Public Utilities Commission and Chair of the National Association of Regulatory Utility Commissioners (NARUC) Committee on Communications, on whose behalf I appear today.

As you may know, the NARUC is a quasi-governmental, non-profit organization founded in 1889. Within our membership are the governmental agencies of the fifty States, the District of Columbia, Puerto Rico and the Virgin Islands which are engaged in the regulation of utilities and carriers. Our chief objective is to serve the consumer interest by seeking to improve the quality and effectiveness of government regulation in America. More specifically, the NARUC is composed of inter alia, State and territorial officials charged with the duty of regulating the telecommunications common carriers within their respective borders. As such, they have the obligation to assure the establishment of such services and facilities as may be required by the public convenience and necessity, and the finishing of service at rates that are just and reasonable.

The NARUC congratulates this committee for beginning a dialogue on issues which are becoming increasingly important in the telecommunications industry and regulatory field—infrastructure development and competition. The NARUC supports your efforts to encourage the development of an advanced telecommunications infrastructure and to promote competition, consistent with the preservation of universal service, reasonable rates, and maintenance of high quality service.

At our summer meetings this year, the members of NARUC spent a great deal of time discussing the proposed legislation, S. 1086. We passed a resolution which supports Congressional attention on this important topic but expresses our concerns with the bill as drafted. While some state regulators may disagree with some proposals contained in S. 1086, we all agree that for a successful transition to an advanced and competitive telecommunications network, careful monitoring of the industry and cooperation between federal and state regulators is necessary. The NARUC resolution has been attached as Appendix A for your review.

There is much within the bill's stated purposes and findings with which we can agree. NARUC generally endorses most of the purposes as stated in Section 3. We too, want a wider range of choices in the provision of telecommunications [Sec.3(2)], to preserve universal availability of telephone service [Sec.3(5)], and to protect the privacy interests of telecommunications users [Sec.3(7)]. State commissions also want to encourage investment in the Nation's telecommunications infrastructure through the development and deployment of advanced and reliable capabilities and services in telecommunications networks [Sec.3 (1) and (6)], and the interconnection and interoperability among telecommunications carriers [Sec.3(4)].

The knowledge gained from our experience in the day-to-day business of regulating the telecommunications industry is consistent with may of the bill's findings. We agree that additional development is needed to bring advanced services to certain populations [Sec.2(4)], that telecommunications infrastructure development is a means of enhancing the quality of life for many areas of the country [Sec.2(1)] and that all telecommunications providers should contribute to universal service [Sec.2 (10)]. Most importantly, we agree that competition must proceed under rules that are fair to all telecommunications carriers and protect consumers [Sec.2(9)].

While we recognize that in the transition to a regulatory format based more and more on market reliance there may need to be some adjustments in federallstate responsibilities, we are greatly concerned about the bill's proposal to accomplish this by pre-empting the authority of the states [Sec. 3(3)] to deal with barriers to competition and otherwise [Sec.2(12)] curtailing the ability of the States to respond to local conditions. Congress should recognize that there is some tension between the twin goals of universal development and universal competition. As a result, affordable—universal availability should remain a primary concern for policy makers.

I will discuss these concerns further in a moment. First, thought, let me touch on activities that have already occurred in the states which both promote infrastructure development and encourage market-based solutions.

As state regulators, the members of the NARUC have been working on many of the issues contained in S. 1086 on various levels in our respective states. Like you, we seek to improve the telecommunications technology available to consumers while keeping telephone rates affordable for all. State commissions have been using innovative regulatory approaches to improve infrastructure and move toward a competitive environment for some time. Many states have facilitated competition in certain segments of the telecommunications marketplace and have developed incentive ratemaking approaches to spur efficiency and improve service. States have used primarily three methods to achieve the goals of improving the infrastructure: alternative regulation (which relaxes pricing restrictions on new services and encourages competition); setting specific timelines for the deployment of advanced technologies; or a hybrid policy.

According to the draft of NARUC Report on the Status of Competition in Intrastate Telecommunications (1993 Update), all multiple LATA states allow intrastate interLATA competition, 45 states allow at least partial competition in the intraLATA toll market, and 22 states allow at least partial competition in local exchange service.

Furthermore, 39 state commissions have enacted alternatives to ratef-return regulation, 19 state commissions have acted to allow some form of collocation, and 36 states have ONA (requirement to unbundling network functions) tariffs approved or pending.

Now being implemented as a result of efforts by the state legislature and Board of Regulatory Commissioners is "Opportunity New Jersey." This plan requires broadband service availability everywhere in the state within 18 years. The phone company is required to accelerate deployment of four technologies: advanced intelligent network; narrowband digital services; wideband digital services; and broadband digital services. Protection of universal service is aided through a cap of 25 cent per month price increases on residential services in any one year.

Tennessee has developed a 10 year master plan, "FYI Tennessee", in an effort to modernize the state's telecommunications network. Regulatory reform proposals include incentive sharing and the flexible pricing of services. Small LECs, those under 70,000 access lines, have their choice of regulation plans. The Master Plan for Technology Deployment accelerates deployment of 557 technology, ISDN, and broadband capabilities. Specific dates for deployment are established for urban, suburban, and rural areas.

Pennsylvania is another state which recently signed into law legislation allowing deregulation of competitive services while retaining price-based regulation of noncompetitive services. The PUC is prohibited from rate regulating competitive services, but retains the flexibility to shift services between competitive and non-competitive classifications as appropriate, according to established criteria. In exchange for alternative regulation, telephone companies must commit to deploying broadband capabilities to all customers by the end of 2015.

Long considered a leader in telecommunications regulation, the New York PSC was the first regulatory body, federal or state, to authorize competition in the provision of special access and switched access transport services. In addition, the PSC recently permitted competition for end-user access services, requiring New York Telephone to unbundle its residential and business loops. In taking these actions the PSC has continued to ensure universal access to basic services, reasonable rates, and maintenance of quality service through a comprehensive lifeline program and careful monitoring during the transition to a more competitive industry.

Most recently, the Oregon PUC was authorized to allow local exchange competition by service or by geographic area if competition can be found to serve the public interest. The law allows the PUC to approve competition in all or part of a local exchange and to establish rules and standards for the local exchange competitors. Small telephone companies, those under 15,000 access lines, are exempt from competition until 1998. In addition, the Oregon PUC currently completed a docket on unbundling and interconnection issues. The results allow a competitor to purchase individual components of the network and allow customers of a LBC to interconnect to the local network in a variety of ways.

Because the removal of regulatory barriers does not guarantee meaningful competition, states have been transitioning to competition with care given to the potential impacts on universal service and service quality issues. One trend among the states is to reduce regulation on services as they become increasingly competitive.

Another is to delay competition mandates in rural areas. Rural areas, in most state plans, are expected to achieve the same level of technology as urban areas but are given an extended time period. This type of consideration allows regulators to learn from urban markets and make modifications for rural areas. States have found their actions make the transition to competition more effective while giving consumers better protection.

Let me now address our concerns regarding this legislation. They fall into four categories.

PREEMPTIVE REMOVAL OF ENTRY BARRIERS

The NARUC strongly questions the bill's findings that "removal of all state and local barriers to entry * * * are essential to the development of a national interstate telecommunications infrastructure" [Sec. 2(12)] along with the addition of a new section to the Communications Act prohibiting any limitation on the "ability to of any entity to provide interstate or intrastate telecommunications services" [Sec .229(a)].

There are, no doubt, many markets in which competitive entry would have the beneficial effect of lowering prices, expanding choices for consumers, and increasing efficiency on the part of the incumbent. We are fearful, though, that rural or less-dense markets may not be capable of sustaining robust competition and that competitive entry would result only in selective by-pass by a few large customers while remaining customers are saddled with the lost revenue requirement.

Rather than the virtually immediate and blanket prohibition against entry barriers contained in the bill you might consider one of two alternative approaches. The first would be to limit the application of Sec. 299(a) to statutorily defined markets in which competition is likely to be effective. This, we recognize, would present a challenge in drafting statutory language that would correctly define those markets. A second and preferable approach would be to delegate to the state commissions the responsibility to make case-by-case findings regarding the advisability of removing entry barriers for specific markets.

This sort of delegation is not without precedent. In both the Public Utility Regulatory Policy Act (PURPA) and the Energy Policy Act of 1992, (EPACT) Congress established broad federal policy preferences but delegated to the states responsibility for implementing those preferences, taking into account local conditions and circumstances. PURPA is widely viewed as a success, and as my summary of activities already occurring in the states with respect to telecommunications indicates, states can be expected to act responsibly in the telecommunications area as well.

We also suggest amending this section to make it clear that conditions of entry, such as revenue requirements for telephone relay service (TRS) or telephone service cut off rules, are not considered barriers to entry or limitations on the ability to provide service. These conditions ensure a basic level of service for all consumers and ensure that consumer rights are not overlooked. Such conditions of entry would be applied equally to all providers of telecommunications service.

RE-DEFINING THE FEDERAL/STATE RELATIONSHIP

While NARUC can agree that national policy is needed in telecommunications [sec. 2(11)] and that this will carry with it some re-definition of Federal and State responsibilities [sec.2(13)], we have several concerns in this area.

S. 1086 provides a unique opportunity to clarify the role of state and FCC officials during this transition to competition and to recognize that the states have the expertise in oversight of the local telecommunications services. The bill offers the opportunity to explore ideas for strengthening federallstate coordination, and establishing a new framework that preserves state policy prerogatives while recognizing the need for greater consistency as new technologies and markets evolve.

We believe, however, that in its current form the bill tends simply to shift responsibilities away from the states and to concentrate them at the FCC rather than encouraging coordinated federal/state actions. Examples of shifted responsibilities include: the federalization of entry regulations [Sec. 229(a)]; preemption of interconnection rules that are inconsistent with federal rules [Sec. 229(c)]; federalization of local number portability [Sec. 229(f)]; preemption of state rural safeguard programs the FCC finds to be inadequate [Sec. 230]; federal authority to define services subject to price flexibility [Sec. 229 (h)(2)]; and the elimination of state authority over information service regulation [Sec .223].

This concentration of authorities at the FCC raises two types of concerns. The first is administrative. As implementation of the Cable Act of 1992 illustrates, budget and staff limitations place a very real limit on the ability of the Commission to implement even well-intentioned legislation. There is no reason to think it will be any different with S. 1086.

The second has to do with institutional capabilities. Even if the FCC doubled in size, its ability to respond to local concerns would be less than currently exists in the states. State commissions are the ones on the front lines, the ones who receive calls about poor or unavailable service or high cost of service. Because we hear from the consumers, because state commissions will see firsthand the emergence of competition, and because we are sensitive to the infrastructure needs of the state, state commissions must retain authority over intrastate telecommunications. Even as competition emerges, the state commissions need to retain authority to resolve subscriber complaints, intercompany disputes and other administrative matters. Most importantly, States need to retain their authority to re-regulate if quality, availability, or terms and conditions of deregulated service deteriorate or if competition does not develop. State regulatory authority will enhance, not impede the sound and smooth transition to competition.

Finally, the NARUC supports enhanced processes to strengthen federallstate coordination. The 1934 Communications Act recognized that there may be disputes between these policy makers and provided for the Federal/State Joint Board process, a statutorily-sanctioned forum of representatives from the FCC and selected States, all dedicated to resolving territorial concerns. We believe this process, with a panel of experts working toward solutions, is far superior to granting one jurisdiction preemptory power over the other.

PRESERVING UNIVERSAL SERVICE

NARUC endorses the bill's stated goal of ensuring the universal availability of telephone service [Sec.3(5)], although we note with some dismay the bill's failure to make a specific finding that preservation of universal service is in the public interest, giving rise to the fear that this may be a lower priority than some of the bill's other goals.

Our concern here is heightened by the complicated and ambiguous interplay between Section 229(d)—Universal Service, Section 229(h)—Regulatory Flexibility, Section 230(a)—Rural Markets and Noncompetitive Markets, Section 229(a)—Removal of Barriers to Entry and Section 229(c)(1) Obligations of Telecommunications Carriers.

One hand, Section 229(d) requires that States "in combination with the Commission, shall ensure the preservation and advancement of universal service." This bold directive, however, is unaccompanied by any guidance as to its attainment.

Notwithstanding this general directive, other provisions of the legislation take away the tools traditionally used to achieve this goal. For example, it can be argued that a necessary corollary of freedom of entry is a freedom of exit thus raising doubts about the ability of states to enforce carrier of last resort obligations. And, while we recognize that Section 229(h), dealing with regulatory flexibility is permissive, not mandatory, there is some tension between price flexibility and preservation of averaged rates in rural and noncompetitive areas. Section 229(h) acknowledges this tension but provides no help in resolving it.

The concept of universal service traditionally carries with it the idea that customers in rural and noncompetitive areas should have equality of access to telecommunications services and networks. This thought appears to be embodied in Section 230(a) wherein the States are directed to undertake actions to achieve that goal, and the FCC is authorized to take remedial action in the event of failure by a state to achieve the goal. The section, though, requires that prices for rural facilities be based on "reasonably identifiable costs of providing such services" [Sec. 230(a) (2)], thus suggesting that urban/rural de-averaging is required. This conflicts, at least in concept, with the mandate of Section 229(h) that States shall ensure that rates for services that are not competitive remain "just and reasonable and that universal service is preserved and enhanced." Before state entry regulation is preempted, we must ensure that all customers will be served by a telephone company or telecommunications provider.

The expansive requirement for interconnection in the interstate jurisdiction envisioned by Section 229(c)(1) carries with it the possibility of cost shifts to the intrastate jurisdiction if significant volumes of traffic leave the incumbent's network. This when combined with price flexibility for competitive areas suggest these costs may be born by rural and non-competitive areas, again threatening the goal of universal service.

Finally, although some large telecommunications users may have need for network redundancy [Sec.2(8)] and multiple sources of supply, steps should be taken to ensure that basic residential customers do not pay for redundancy they do not use and that does not benefit them.

LACK OF SUITABLE TRANSITION MECHANISMS, MONITORING AND ABILITY TO RE-REGULATE IN THE EVENT OF COMPETITIVE FAILURE

S. 1086 needs to give some thought to the transition between the existing regulatory structure and a competitive market. State commissions will play a large role in the transition because we will see the trends and problems long before they become egregious enough to attract the FCC's attention. Consumer education will be very important as well. Studies continue to show that consumer understanding of telecommunication lags behind the technology. For example, many consumers still believe that AT&T is their telephone company. State commissions will be instrumental in providing the assistance and information that consumers will need to make the best choices in selecting their telecommunications services. Consumers will also be counting on us to help them solve the problems they encounter with telecommunications services and providers.

The NARUC supports the inclusion of stronger monitoring provisions in the bill. Such information will help us determine the progress that is being made toward improving the infrastructure or increasing competition. Should modifications to the legislation be necessary in the future, such data could be very helpful in determining the best course of action. Such data would also be extremely helpful in determining the level of competition faced by services or companies, which could be used for a variety of purposes including pricing flexibility. Lastly, monitoring information will be helpful in the enforcement of rules, is a key aspect to making competition a reality.

Finally, we think there should be some recognition that de-regulation might not work. Everyone hopes the legislation's result will be effectively competitive markets. Equally possible, however, is the emergence of unregulated monopolies or near-monopolies. We suggest there be conscious consideration of whether existing anti-trust law and enforcement is adequate to guard against this possibility or whether additional safeguards should be included in this legislation.

The success of any national plan to achieve a competitive, advanced telecommunications network will need the assistance of autonomous state regulatory bodies to monitor the transition and enforce regulations as necessary. I, and all the other members of NARUC, look forward to working with the members of this committee in clarifying these issues.

Thank you again for the opportunity to testify on this matter.

APPENDIX A—RESOLUTION REGARDING S. 1086, "THE TELECOMMUNICATIONS INFRASTRUCTURE ACT OF 1993"

Whereas, On June 9, 1993, Senators Danforth (R-MO) and Inouye (D-HI) introduced S. 1086, the "Telecommunications Infrastructure Act of 1993"; and

Whereas, The bill's stated objectives are to: encourage the continued development and deployment of advanced and reliable capabilities and services in telecommunications networks; ensure the availability of the widest possible range of competitive choices in the provision of telecommunications and cable television services and encourage the universal availability of telephone service; and

Whereas, Numerous states are at the forefront in upgrading their telecommunications infrastructure, encouraging new services, and promoting competition; and

Whereas, The characteristics of the marketplace and the costs of providing telephone service vary geographically and affect the competitive and regulatory environments among the states; and

Whereas, Cooperation between state and federal regulators is the best means to ensure that the benefits of competition and infrastructure development will be realized, consistent with the preservation of universal service, reasonable rates and maintenance of high quality service; and

Whereas, Until such time as competition is achieved, states must have the authority to ensure against the emergence of unregulated monopolies; and

Whereas, The bill as currently drafted contains numerous provisions of concern to state regulators including, but not limited to, the following examples:

- requiring that within one year, no State statute or regulation, or other state legal requirement shall prohibit or limit the ability of any entity to provide telecommunications services;
- directing that interconnection to a carrier's facilities at any technically feasible point for provision of telecommunications services must be permitted;

• authorizing the FCC to preempt state regulatory measures if determined by the FCC that the measures prevent the realization of the goal of ensuring that customers in rural and noncompetitive markets have access to high quality telecommunications network facilities and capabilities, regardless of the states evaluation of the propriety of such measures;

• providing the FCC with the authority to define intrastate competitive services for purposes of pricing flexibility;

• identifying universal availability, rather than universal service, as one of the seven stated purposes of the bill;

• requiring state and federal coordination in preserving universal service, without clarifying whether the FCC could establish universal service policies which the state must implement;

• providing no clearly defined role for states: (1) in assuring that appropriate intrastate compensation arrangements between telecommunications carriers are implemented; (2) developing intraLATA equal access and presubscription requirements for cellular and two-way wireless service providers;

• preempting state regulations relating to the provision of intrastate information services;

• allowing local exchange carriers to provide video services without addressing how the cost of network facilities used jointly to provide basic telephone services and video services will be allocated;

• promoting network redundancy without addressing potential impacts on the rate making process;

• encouraging pricing flexibility for competitive services without addressing the effects of deaveraging on rates in noncompetitive and high cost areas; now, therefore be it,

Resolved, That the Executive Committee of the National Association of Regulatory Utility Commissioners (NARUC), convened in its 1993 Summer Meetings in San Francisco, California, supports the efforts of Congress to encourage the development of a modern, efficient and high quality public telecommunications infrastructure and to promote competition, consistent with the preservation of universal service, reasonable rates and maintenance of high quality service; and be it further

Resolved, That S. 1086 be amended to address the bill's lack of clear definition regarding the states' authority to prescribe intrastate telecommunications policies and other concerns expressed in this resolution; and be it further

Resolved, That S. 1086 preserve the states' authority to implement policies that reflect market conditions within each state, consistent with the public interest objectives of encouraging infrastructure development and promoting competition, the preservation of universal service, reasonable rates and maintenance of high quality service; and be it further

Resolved, That NARUC encourages cooperation between state and federal regulators and welcomes the opportunity to work with Congress in order to achieve these public interest objectives; and be it further

Resolved, that the NARUC Washington office be directed to convey these views on S. 1086 to Congress.

Prepared Statement of William F. Squadron, Department of Telecommunications and Energy, the City of New York

Mr. Chairman, Members of the Subcommittee, on behalf of the National Association of Telecommunications Officers and Advisors ("NATOA"), I thank you for the opportunity to testify today concerning S. 1086, "The Telecommunications Infrastructure Act of 1993," introduced by Senators Inouye and Danforth. NATOA represents local franchising authorities in more than 4,000 jurisdictions, which collectively regulate cable television systems that serve an estimated 40 million cable subscribers. NATOA is affiliated with the National League of Cities, for which it serves as a technical advisor on cable television and other telecommunications matters.

My testimony will focus on those provisions of S. 1086 relating to the provision of video programming services by telephone companies ("telcos"). This testimony is based on NATOA's initial review of S. 1086. We look forward to sharing with the Subcommittee in the near future any additional comments we might have on the bill.

NATOA was a strong advocate of the Cable Television Consumer Protection and Competition Act of 1992 ("1992 Cable Act"). From Jefferson City, Missouri and Gillette, Wyoming to New York City and Miami, consumers told us loud and clear that they wanted protection from skyrocketing rate hikes, poor customer service, and the other monopoly practices of many cable operators. We believe that the 1992 Cable Act will provide consumers most of that protection.

NATOA now urges Congress to take the next step necessary to protect consumers. That step is to enact legislation that will further promote competition in the provision of cable service—which we believe will also reduce the need for regulation. NATOA has consistently argued that meaningful competition would increase the chances that subscribers would receive quality, yet affordable, cable service. We believe the only significant way that meaningful competition to a cable system can occur is if there is at least one other multichannel video programming distributor that offers a full array of services and programs comparable to those that traditional cable operators provide.

At this time, telcos have the capability of being one of the few viable alternatives to traditional cable operators. In addition to promoting competition, however, a goal of telco legislation also should be to ensure that the public receives the benefits and protections they expect from commercial users of public property. In order to ensure that the public receives such commitments, it is critical that telco legislation recognize the legitimate and historical role of franchising authorities in the dual federal-local structure for the regulation of multichannel video programming service providers. In short, local governments serve as trustees for their communities to guarantee that their rights-of-way and other property are used properly and in the community's interest.

When it enacted the Cable Communications Policy Act of 1984 and the 1992 Cable Act (the "Cable Act"), Congress determined that local governments, by imposing franchise requirements, could best ensure that safeguards necessary to protect against consumer abuse are in place, and that the public receives the benefits to which they are entitled for the use of their property. Congress recognized that "city officials have the beet understanding of local communications needs and can require cable operators to tailor the cable system to meet those needs." H.R. Rep. No. 934, 98th Cong., 2d Sees. 24 (1984). NATOA believes that any telco legislation enacted by Congress should continue to recognize the critical role of local governments in ensuring that "local communications needs" are met.

The best way to ensure that telcos providing video programing services meet local communications needs is to subject them to the same franchising structure and public interest regulations that have operated successfully with respect to cable operators—namely, the Cable Act. By subjecting telcos to the Cable Act, Congress would ensure, among other things, that telcos: (1) provide community programming and local information services, such as, but not limited to, public, educational and governmental access ("PEG") channels, facilities and equipment; (2) provide compensation to local governments for the for-profit use of local public property and rights-of-way; and (3) are subject to local customer service standards and other consumer protection measures.

For the reasons stated above, NATOA has advocated the provision of video programming services by telcos in order to promote competition and technological innovation, so long as appropriate regulatory conditions and safeguards are established and enforced to ensure that the public receives the benefits and protections that they have a right to expect from commercial users of public rights-of-way. NATOA believes S. 1086 is a substantial step in the right direction toward making the benefits of increased competition and technological innovation, which telephone companies claim they can make, available to the public. Such competition and innovation must be done in a manner consistent with the public interest.

NATOA is encouraged that S. 1086 recognizes that a telco providing video programming services within its service area is a "cable operator" under the Cable Act. As cable operators, such telcos should be subject to local cable franchising requirements. By identifying such telcos as "cable operators," the bill apparently intends that such telcos would be subject to local cable franchising requirements. We believe that S. 1086 should be amended to make that intent crystal clear. Franchising requirements that S. 1086 should make clear apply to telcos providing video programming include, but should not be limited to: (1) requirements for PEG facilities, (2) franchise fees, and (3) customer service standards. Moreover, both the video ink non-video programming services provided as part of a telco's cable service should be subject to franchising requirements.

We believe it is essential that local governments be permitted to require franchises and impose franchise-related requirements. They are in the best position to understand and respond to the cable television needs and interests of consumers within their jurisdiction.

NATOA has urged that if telcos begin to provide cable television service, they should do so through a separate video programming subsidiary. We are pleased that S. 1086 includes a requirement that video programming be provided through a separate subsidiary. Such structural separation facilitates the audit process and provides some assurance that cross-subsidization is avoided. Telephone service customers should not be forced to finance a telco's entry into the cable business. S. 1086 also properly prohibits telcos from engaging in practices prohibited by the Federal Communications Commission ("Commission") or a state with regard to cross-subsidization.

NATOA is concerned that S. 1086 may not subject to local franchising requirements telcos providing transport services for so-called "video dial tone" service. As indicated above, NATOA believes that telcos providing video programming services, whether through a subsidiary or on a common carrier transport basis as part of "video dialtone" service, should be subject to appropriate franchising requirements under the Cable Act. Local governments have a responsibility toward the property that they manage on behalf of their communities. Only through the franchising process can a community be sure to receive appropriate benefits for the commercial use of its property.

Moreover, as a potential competitor of a franchised cable operator, it is only appropriate that "video dialtone" services be franchised. Among other things, local governments should be able to ensure through franchise requirements that a telco providing "video dialtone" transport service: (1) makes transmission facilities available to public, educational and governmental users; (2) provides service to subscribers throughout a franchise area on a nondiscriminatory basis; (3) complies with consumer protection requirements concerning such matters as service, installation and billing; and (4) constructs its facilities using public streets in a safe and reasonable manner.[1]

S.1086 sets forth other safeguards to telco entry that NATOA supports or believes are steps in the right direction. The bill severely limits the right of a telco to purchase, or enter into a joint venture or partnership with, any existing cable system within its local telephone exchange service area. NATOA believes that this provision would be improved if telcos were totally prohibited from having any ownership interest in an existing cable system. The five percent ownership interest permitted under S. 1086 may result in reduced incentives for full and fair competition between a telco and the cable system in which it has an ownership interest. We believe that a total ownership ban on the purchase of existing systems is a critical restriction if telco entry into the cable market is to achieve the desired goal of increasing competition.[2]

There are several provisions in S. 1086 that NATOA asks that the Subcommittee consider amending as it reviews this legislation.

First, we do not believe that telcos providing cable service in rural areas should be exempted from the provisions in the bill. Cable consumers in rural areas should not be treated as second class citizens and should be entitled to the same protections granted under the bill to cable subscribers served by telcos in other areas.

Second, the bill would preempt state or local regulation of intrastate telecommunications services that are inconsistent with federal regulations or S. 1086. NATOA hopes that Congress will continue to respect the traditional, long-standing right of state and local governments to regulate intrastate communications in the best interests of their localities.

Third, NATOA is concerned with the provision in the bill that would preempt state and local regulations or statutes regulating interstate telecommunications services that are "inconsistent" with federal regulations and S. 1086. NATOA does not suggest that state and local governments should be allowed to regulate in a manner that undermines the goals of S.1086. NATOA only suggests that Congress grant state and local governments some flexibility in regulating interstate services.

---

[1] In this regard, NATOA particularly is concerned with the Commission's decision to allow telcos to provide video dialtone service without a franchise pursuant to the Cable Act. The FCC, in an interpretive ruling, concluded that a "video dialtone" service provider is not a "cable operator" required to have a franchise under Section 621(b) of the Cable Act. This ruling is being challenged in the U.S. Court of Appeals for the District of Columbia circuit by NATOA, the city of New York, other local governments, and cable industry representatives.

NATOA believes that the FCC's interpretive ruling is contrary to the express language of the Cable Act and Congressional intent. We believe that Congress intended, when enacting the Cable Act, to subject all multichannel video programming distributors that use public rights-of-way to the Cable Act.

NATOA hopes that Congress, by passage of S. 1086 or some other legislation, will clarify that such video dialtone service providers are cable operators subject to franchising requirements.

[2] However, we do believe that in exceptional circumstances local governments should be allowed to approve the purchase of an existing cable system by a telco. Such circumstances may exist where, for example, a telco is the only entity capable of assuming the operation of a poorly performing or abandoned cable system.

We recommend that the bill be amended to preempt only those state or local statutes or regulations which are "irreconcilable" with federal law or S. 1086.

NATOA believes that S. 1086 takes a positive and significant step towards introducing competition into a monopoly cable industry, broadening the diversity and quality of programming available to consumers, and ensuring that the provision of video programming services by telcos is consistent with the public's right to a dividend from commercial users of public rights-of-way. NATOA applauds Senators Inouye and Danforth for their efforts to advance the country's communications infrastructure and promote competition in the cable industry. We look forward to working closely with the Subcommittee on these issues to make consumers the true beneficiaries of competitive multichannel video programming technologies.

Prepared Statement of Gene Kimmelman, Legislative Director, Consumer Federation of America

The Consumer Federation of America (CFA) agrees with the goals articulated by Senators Danforth and Inouye in S. 1086, the "Telecommunications Infrastructure Act of 1993." We share the belief that a pro-competitive telecommunications policy framework that protects universally available and affordable basic phone service, provides thorough protection against cross-subsidization, preserves principles of common carriage and nondiscrimination and dismantles all inappropriate impediments to vibrant competition, is in the public interest.

CFA believes Congress has a unique opportunity to offer the American people the possibility of robust telecommunications competition in all markets without abandoning the social policy gains achieved under the Communications Act of 1934. Marketplace developments, technological advances, the AT&T break-up, and passage of the 1992 Cable Act make it possible to offer consumers:

1. a continuation of the decline in real prices for local telephone service (since passage of the 1934 Communications Act, local telephone rates have fallen about 60 percent in inflation adjusted dollars);
2. an updated definition of "universal service" that guarantees affordable prices for an evolving array of services (e.g., lifeline, local exchange, video, digital) and equipment (e.g., telephone, television, computer, and equipment for special needs groups) that ensure multimedia connectivity for all citizens and vital institutions (e.g., the poor, rural communities, persons with disabilities, schools, health care facilities);
3. increased choice in all communications services and equipment to enhance consumer sovereignty over how, when and what to communicate;
4. lower prices for local exchange, video, and information services that have not been subject to either strong competition or effective regulation; and
5. numerous new services and combinations of services, with appropriate customer privacy protections, that a competitive environment should foster.

While CFA is confident that the sponsors of S. 1086 share this vision, their legislation will require significant alteration to meet these goals. Unless S. 1086 is amended to include a precise, updated definition of universal service, consumer-friendly cost allocation principles (as proposed in § 201 of Chairman Inouye's S. 2112, the "Information Services Diversity Act of 1991"), and more thorough safeguards against anti-competitive practices (like in § 201 of S. 2112), the legislation cannot accomplish its stated goals.

I. THE INGREDIENTS OF A PRO-COMPETITIVE, SOCIALLY RESPONSIBLE TELECOMMUNICATIONS POLICY

The fundamental public policy dilemma in attempting to promote free-wheeling telecommunications/multimedia competition is how to get from "here" to "there." As we found with the Cable Act of 1984, reduced regulation in the name of promoting competition can lead to the opposite: entrenched, expanded monopolization.[1] Also, history demonstrates that, contrary to superficial logic, structural prohibitions on diversification (e.g., the Modified Final Judgment line of business restrictions, the telephone company-cable cross-ownership restriction) help foster the growth of competition in communications markets.

To meet the goals of S. 1086 it is therefore critical that Congress base its policy directives on actual marketplace experience, rather than superficially attractive rhetoric. As S. 1086 generally acknowledges, local telecommunications competition can only succeed if local telephone companies are required to open their networks

---

[1] See Conference Report to accompany S. 12. Cable Television Consumer Protection and Competition Act of 1992, Report 102-862, Sept., 14, 1992.

to competition on nondiscriminatory terms and conditions. The bill also appropriately calls for breaking local bottleneck control over telecommunication transmission before allowing the local exchange carriers to expand into an adjacent market like cable television. Unfortunately, the bill does not provide adequate safeguards to ensure the development of local competition, or the preservation of affordable local phone service, before allowing local exchange carrier expansion into the video market.

II. UNIVERSAL SERVICE: PRESERVING A SOCIAL POLICY TRADITION

It is of course no accident that local telephone service is one of the best "buys" in today's marketplace. The various elements of "universal telephone service"—e.g., "just and reasonable" prices for basic phone service, "lifeline service" for low-income households and a "high cost fund" to moderate prices in rural America—emanating from the Communications Act of 1934, state regulatory statutes, corporate strategy and regulatory policy, reflect America's historical commitment to making prices for basic communications services a bargain for all citizens. The ongoing success of our universal service ethic is reflected in simple statistics; since the 1930's, the price of local phone service has declined, in constant dollars, by about 60 percent, making it possible for about 93 percent of households to have a telephone today.[2]

Unfortunately, many telephone industry officials claim this universal service ethic cannot be preserved in an increasingly competitive telecommunications market. They in essence relegate our nation's most important and successful tool that promotes ubiquitous egalitarian dialogue, discourse and increasingly essential communications options, to the position of an anachronism suitable only to a monopoly environment. According to this logic, a competitive market cannot coincide with the low priced local phone service that our universal service policy currently ensures.[3]

If the industry view of the telecommunications market is correct, then consumers have more to lose than gain from a simple pro-competition policy. Under this scenario, the most essential local phone services could face steep price increases, with competition driving down prices for large volume users of data, long distance and other sophisticated high-end communications applications. The vast majority of consumers that principally rely on the phone to reach family, friends and local business contacts would pay more than they currently do under our universal service ethic.

On the other hand, if Congress preserves our universal service tradition as it opens the door to full-scale telecommunications competition, consumers would be insulated from local rate hikes and could receive significant benefits from competitive pressures to develop more useful information age services. This is the environment that CFA believes, with the appropriate universal service and safeguard amendments, S. 1086 could promote.

Since local telephone companies believe they should raise local rates significantly to respond to competitive pressures,[4] it is unlikely that S. 1086's vague universal service language will provide consumers any protection against rate increases unless the bill is amended to specifically cap local rates as part of its universal service mandate. In the transition to a competitive environment, this can be accomplished by continuing to require all telecommunications service providers that interconnect in any fashion with the local exchange carrier's networks, to pay an access charge that supports universal service with contributions to joint and common network costs.

As new service providers begin replacing the local exchange telephone company as consumers' local carrier, this access charge can evolve into a public network transmission surcharge (based on the provider's net public switched transmission revenue) that would replace today's high cost fund, lifeline fund and preserve the tradition of declining local telephone rates. Because this mechanism would treat all telecommunications providers alike, it would in no way reduce or handicap the development of competitive local telephone service. Most importantly, as competition grows—most likely in an uneven and unpredictable fashion—this universal service mechanism would continue to guarantee that low-income, rural households and consumers with disabilities have access to the same or functional equivalent of local phone services that all other consumers receive for a reasonable price.

---

[2] See Kimmelman and Cooper, "Divestiture Plus Five," Dec. 1988. We highlight the fact that, despite the success of our universal service policy, it remains a "work in progress" since about 20 percent of households below the poverty-line remain phoneless, and half the states do not provide subsidized "Lifeline" programs for low-income families.

[3] See Statement of Gary McBee, USTA, on S. 1086 before the Senate Commerce Committee, July 14, 1993 at 6.

[4] Id.

If the Bell telephone companies are allowed to maintain or expand their information services business, and all local exchange carriers are permitted to enter the video business in their telephone service territory, as proposed in S. 1086, consumers are likely to face dramatic overcharge for local phone service (until effective local competition develops). As we have pointed out in great detail to this Committee, consumers have been increasingly abused through overcharges by local telephone monopolies that use ratepayer funds to expand into new markets.[5] As we have also pointed out, the greater the incentive to shift costs onto monopoly ratepayers, the less the possibility for state and federal regulators to uncover the abuse and protect consumers.[6]

In the case of video service, the risks to telephone ratepayers are enormous. Even if the local exchange carriers take five to ten years to expand their telephone networks to provide video services, using the cheapest means available, experts estimate that these companies will need an additional $10-20/month from each telephone subscriber to cover their costs.[7] Even Bell Atlantic's recent statement, that its video subsidiary will pay for $500/line in network upgrades, could leave telephone ratepayers paying and additional $1,000/line to make the phone system capable of transmitting video signals:

> Bruce L. Egan, a research fellow at Columbia University's Institute for Tele-Information, recently estimated the amount of money needed by the cable and telephone industries to upgrade their services.
>
> To reach the modest goal of supplying both conventional cable television and ordinary phone service over the same line, telephone companies would have to spend about $1,500 * * *[8]

Surprisingly, the Federal Communications Commission (FCC) has failed to develop rules to deal with the hundreds of billions of dollars involved in local exchange carriers' network alterations to provide video services. Rather than develop clear guidelines to govern critical consumer protection issues, the FCC decided to address "video dialtone" issues involving federal/state jurisdiction, cost allocation, pricing and consumers safeguards on an ad hoc, application by application basis.[9] So far, local exchange companies have proposed charging their telephone ratepayers most of the costs associated with putting fiber optic "trunk" lines in place to provide video services.[10] And so far, the FCC has not objected to this flagrant rip-off of telephone ratepayers.

Unfortunately, S. 1086 does not explicitly insulate telephone ratepayers from paying for infrastructure investments designed to provide information or video services (See § 8's and § 11's general reference to subsidies involving "the improper assignment of costs"). To protect telephone ratepayers from inflated local rates, S. 1086 must be amended to include S. 2112's cost allocation rules (§ 201 amending § 227 (f) and (g) of the Communications Act of 1934) which directly prohibits the loading of "joint and common" costs related to information services onto local telephone rates. CFA believes similar specific statutory language is needed to prevent cost misallocations related to video services.

III. PROMOTING COMPETITION

So long as Congress ensures that universal service (as defined in section II, supra.) is preserved, CFA strongly endorses S. 1086's goals of promoting local telephone competition and devising fair competitive safeguards for all telecommunications providers and users. While we believe S. 2112's market entry test would provide the greatest assurance of robust competition, S. 1086's competitive safeguards can be fine-tuned to protect information services and video competition.

By eliminating barriers to entry into local telecommunications markets, and preventing local telephone monopolies from buying local cable monopolies (or vice versa), S. 1086 provides an essential, fair starting point for pro-competitive tele-

---

[5] See Statement of Gene Kimmelman before the Senate Communications Subcommittee on S. 1981, May 9, 1990.

[6] Id.

[7] Edmund L. Andrews, "A Baby Bell Primed for the Big Fight," New York Times, Feb 21, 1993, citing Bruce Egan at Columbia University's Institute for Tele-Information. See also Dr. Mark Cooper, Developing the Information Age in the 1990's, CFA, June 8, 1992; and William Page Montgomery, Accelerated Broadband Networks, ETI, Feb., 1992.

[8] Andrews, op.cit.

[9] Telephone Company/Cable Television Cross-Ownership Rules, Second Report and Order, 7 FCC Rcd. 5781 (1992).

[10] In the Matter of the Application of the New Jersey Bell Atlantic Co. (Florshan System), File No. W-P-C-6858, Nov 16, 1992 at 5, (Dover System), File No. W-P-C-6840, Dec 15, 1992 at 5.

communications policy.[11] Although it is unclear how competitive the local telephone and video markets will become in the future, S. 1086 appropriately prevents either local telephone companies or cable companies from collaborating to cut off competition before it has a chance to get started.

Given the Bell telephone companies' history of anticompetitive practices,[12] S. 1086 does not do enough to maintain or enhance competition in the information or video services that the legislation would allow local phone companies to offer. S. 1086's competitive safeguards have no meat on their bones. CFA therefore supports more detailed and through separate subsidiary, auditing and privacy requirements for all information and video services offered by the Bell companies. By augmenting S. 1086 with the safeguards from § 201 of S. 2112 and similar legislative proposals under consideration in the House of Representatives,[13] Congress can maintain and expand competitive market forces in all aspects of telecommunications as the Bells move into new markets.

CONCLUSION

While CFA concurs with the goals of S.1086, we cannot support this legislation unless it is amended with a specific comprehensive definition of universal telephone service, cost allocation rules that preserve the declining-price tradition of local phone service, and more specific competitive safeguards. Unless opening the telecommunications market to more competition, including local phone company provision of cable television service, involves continued real-price reductions for local phone service along with broader choices of information age services, consumers may have more to lose than gain from a restructuring of the telephone industry. 3y augmenting S. 1086's broad policy tlirectives with a precise definition of universal service and S. 2112's more specific cost allocation and competitive safeguard provisions, Congress can usher in a new era of falling prices and maximum competition in all communications markets.

LETTER FROM GARY W. MCBEE, VICE CHAIRMAN, BOARD OF DIRECTORS, UNITED STATES TELEPHONE ASSOCIATION

JULY 28, 1993.

The Honorable DANIEL K. INOUYE,
U.S. Senate,
Washington, DC 20510

DEAR CHAIRMAN INOUYE: Thank you for the opportunity to testify on S. 1086. I hope that the testimony will be helpful in your deliberations on the bill.

I would like to clarify two issues we discussed during the hearing. It is my understanding that the hearing record will remain open until July 28th and that these additional comments will be part of the hearing record.

I. USTA POSITION ON RBOC INTER-LATA RELIEF

Discussions are ongoing at USTA on the issue of supporting inter-LATA relief for the regional Bell operating companies. While no consensus has yet been reached on full inter-LATA relief, the membership of USTA does support the specific inter-LATA relief provisions contained in S. 1086.

II. RECOVERY OF COSTS AND THE EFFECT OF COMPETITION ON UNIVERSAL SERVICE

There also appeared to be confusion over the recovery of a telephone company's costs in providing residential and rural telephone service.

To assist the Committee in understanding the significant level of contribution to universal service that is provided by revenues generated by other local exchange carrier services, I have attached a series of studies that were released recently by USTA. Among these materials are the Strategic Policy Research and BellCore studies I referenced in my testimony.

---

[11] Preventing the combination of telephone and cable monopolies in any market is particularly significant given the recent Federal District Court decision that the cable-telephone company cross-ownership restriction is unconstitutional. *C&P Telephone Co. of Va.* v. *U.S.*, Cir. No. 92-1751-A, E. Dist. Va., Aug 24, 1993. While we believe this ruling will be overturned on appeal, it opens the door to multimedia monopolies in the interim.

[12] See Dr. Mark N. Cooper, Divestiture Plus Eight, CFA, Dec., 1992.

[13] H.R. 3515 (102nd Congress, 1st Session), H.R. 5096 (102nd Congress, 2nd Session) and Rep. Markey's Discussion Drafts, Aug. 1, 1991, et al.

These studies illustrate that basic local exchange service is priced below actual cost and is highly subsidized. The level of subsidy differs among various regulatory jurisdictions.

While each of these studies is instructive, the conclusions of the Strategic Policy Research [SPR] study are particularly compelling.

The SPR study explains that regulatory policy has promoted universal service by maintaining artificially low rates for rural areas and for residential customers. In a competitive marketplace, however, prices for services will move towards cost. The study states such changes could have a "seismic intensity" on the price of local telephone service in that as much as $21.8 billion of internal subsidy generated by access charges and intra-LATA toll would be jeopardized.

Rather than view this issue in a theoretical or national model, the telephone industry would offer the island of Hawaii as an example of the gap between what a local telco charges for local residential services and the cost of providing that service. In a recent filing with the Hawaiian Commission, the Hawaiian Telephone Company estimated the Long run incremental costs of providing an individual line to be $60.37 per month. The rate the Hawaiian Telephone Company is permitted under tariff to charge a residential customer for that $60.37 a month line is $12.25.

The subsidies to fund this gap have been maintained by policy makers in order to ensure that telephone service is universally available at reasonable rates. The emergence of new providers of local service threatens the viability of this subsidy system. If policy makers are going to encourage open competition in the local exchange market, they must first confront this matter. If policy makers decide that subsidization is still good public policy, then all firms that provide telecommunications services or connect with the public network must share the burden.

S. 1086 promotes local competition, but in USTA's view, fails to adequately protect universal service.

GARY W. MCBEE,
Vice Chairman, Board of Directors.

[The material referred to above may be found in the committee files.]

LETTER FROM JOHN F. STURM, SENIOR VICE PRESIDENT, GOVERNMENT, LEGAL AND PUBLIC POLICY, NEWSPAPER ASSOCIATION OF AMERICA

JULY 28, 1993.

The Honorable DANIEL K. INOUYE,
U.S. Senate,
Washington, DC 20510

DEAR CHAIRMAN INOUYE: On behalf of the Newspaper Association of America (NAA) I would like to thank you once again for considering our views on S. 1086, the Telecommunications Infrastructure Act of 1993, as set forth in the testimony of Uzal H. Martz, Jr., President and Publisher of the Pottsville (PA) Republican and a member of the NAA Board of Governors.

As noted at the hearing, NAA strongly encourages the development of local loop competition but also recognizes the dangers that will be inherent in the RBOCs' telephone exchange monopolies until competition becomes a reality. In particular, NAA believes that if the RBOCs are to engage in electronic publishing, much stronger safeguards are needed than those now in effect under FCC and state regulation. In broad outline, we support the framework for safeguards spelled out in S. 1086. As indicated in our July 14 testimony, we are eager to work with you and your staff to develop the concepts outlined in S. 1086 into comprehensive yet workable safeguards that will permit electronic publishing to continue to flourish.

In that spirit, we have a number of suggestions where we believe the provisions of S. 1086 can be made more effective, especially by including more detail and specificity, as was done in S. 2112 in the last Congress.

Entry test. NAA believes that the goal of encouraging development of a competitive electronic publishing industry is not well served by broad RBOC participation when market conditions are still susceptible to abuse. Thus, we would urge you to include an entry test, such as that contained in section 201 of S. 2112. We would welcome an opportunity to work with you and your staff to refine the S. 2112 entry test to more precisely match entry with the development of competition in various submarkets, so that the RBOCs will be neither handicapped nor advantaged.

Sec. 233(b)(1).[1] *Separate subsidiary.* S. 1086 requires the Bell Telephone Companies and their affiliates to provide electronic publishing only through a separate subsidiary. We support this safeguard as an essential improvement over current FCC policy. However, and particularly in light of the FCC's reluctance to require separation,[2] we urge you to spell out in some detail the organizational and business areas that must be separated. In particular, and as was provided in S. 2112, the separate subsidiary should have separate directors, officers, and employees, separate property, and financing, especially of debt. In addition the subsidiary should have a separate corporate identity, such as trademarks and servicemarks.

Sec. 233(b)(2). Transaction requirements. The bill prohibits transactions between the telephone operating companies or their affiliates and the separate subsidiary from being based on discrimination or preference arising out of their affiliation. This requirement will be difficult to administer and enforce. We strongly suggest that an objective requirement be added that all such transactions be at "arms length" and financially sound, so that the telephone companies must deal with their subsidiary in the same manner as unaffiliated parties deal with one another. Also, in order that any violations of the requirements may be discovered and remedied, we urge you to include the requirements from S. 2112 that transactions be pursuant to publicly available contracts or tariffs and be fully auditable.

Sec. 233(b)(3). Separate operation and property. In addition to the restriction against joint ventures and partnerships, as indicated above, there should be requirements for separate employees, financial structure and property.

Sec. 233(b)(4). Separate commercial activities. We strongly support the bill's provision requiring separate marketing and sales. To preclude indirect cross-subsidies, other activities should also be separate, namely, accounting, hiring and training of personnel, purchasing, installation, and maintenance, and research and development. Several of these items were included in S. 2112.

Sec. 233(b)(5). Books, records and accounts. The requirement for separate books, records, and accounts is essential, but it should be supplemented by public reporting requirements to act as a further check on abuses and to provide an additional means of detecting violations, e.q., unconsolidated financial statements and SEC-type reports (both of which were required by S. 2112).

Sec. 233(b)(6). Provision of services and information. The requirement that services or information that a Bell Telephone Company provides to its subsidiary also be provided on the same terms and conditions to other publishers should not be evaded by indirect transactions through other affiliates via the parent holding company (RBOC).

Sec. 233(c). Prevention of cross subsidies. The bill requires a cost allocation system that is intended to prevent cross subsidies, but gives the FCC complete discretion to implement it. As was provided in S. 2112, the legislation should spell out how transactions between the operating company and its subsidiary will be valued to prevent cross subsidies.

Enforcement. Due to the size and complexity of the RBOCs, it may be difficult or impossible to detect violations of the safeguards quickly. Therefore, as was provided in S. 2112, annual compliance audits should be required. In addition, because com-

---

[1] References are to sections of the Communications Act as it would be amended by S. 1086.

[2] The FCC's Computer I decision required "maximum separation" between a telephone company's regulated services and data processing services. Regulatory and Policy Problem Presented by the Interdependence of Computer and Communications Services and Facilities, 28 FCC 2d 291(1970) (Tentative Decision), 28 FCC 2d 267 (1971) (Final Decision), affd in part and rcvd in part, *GTE Service Corp.* v. *FCC,* 474 F.2d 724 (2d Cir. 1973). In Computer II, the FCC refined its policy to require separation between basic communications services and "enhanced services," which correspond to information services and include electronic publishing. Amendment of Section 64.702 of the Commission's Rules, 77 FCC 2d 384 (Final Decision), recon., 84 FCC 2d 50 (1980), further recon., 88 FCC 2d 512 (1981) (Further Reconsideration), aff'd, *Computer & Communications Industry Ass'n* v. *FCC,* 693 F.2d 198 (D.C. Cir. 1982), cert. denied, 461 U.S. 938 (1983).

The FCC reversed its long-standing policy requiring separate subsidiaries in Computer Ill. Amendment of Sections 64.702 of the Commission's Rules and Regulations (Third Computer Inquiry) (CC Docket No. 85-229), Phase 1, Report and Order, 104 FCC 2d 958 (1986), recon., 2 FCC Rcd 3035 (1987), further recon., 3 FCC Rcd 1135 (1988), second further recon., 4 FCC Rcd 5927 (1989), Phase 11 Report and Order, 2 FCC Rcd 3072 (1987), recon., 3 FCC Rcd 1150 (1988), further recon., 4 FCC Rcd 5927 (1989). This decision was vacated by the court of appeals due to the FCC's faulty decision to remove structural separations. *California* v. *FCC,* 905 F.2d 1217 (9th Cir. 1990).

On remand from the court of appeals, the FCC reaffirmed its decision to eliminate structural separations. Computer III Remand Proceedings: Bell Operating Company Safeguards and Tier 1 Local Exchange Company Safeguards (CC Docket No, 90-623), 6 FCC Rcd 7571 (1991). NAA and other have appealed this decision. *California* v. *FCC,* No. 92-70083 (9th Cir.).

petitors may be injured by violations, the bill should provide for a private right of action that would include injunctive relief and damages, again as was done in S. 2112.

Once again, we look forward to meeting with you and your able staff on this issue. If you have any questions, please direct them to the undersigned.

JOHN F. STURM.

LETTER FROM TONDA F. RUSH, PRESIDENT AND CEO, NATIONAL NEWSPAPER ASSOCIATION

JULY 28, 1993.

Honorable DANIEL K. INOUYE,
U.S. Senate,
Washington, DC 20510

DEAR MR. CHAIRMAN: On behalf of the National Newspaper Association, I am writing to ask you to include the text of the Burns amendment for local information providers as an addendum to the record of the July 14 hearing on S. 1086. We would appreciate your including this letter and the attached amendment language in the hearing record.

As our witness, Mr. Dalton Wright, testified at the hearing, the language embodied in the Burns amendment is of paramount importance to small, community newspapers because of its guarantees of access and equal rates for local information providers. We feel that it is essential that the members of the subcommittee have the language of the Burns amendment available to them in the record so that they can consider the importance of including statutory protection for local information providers in legislation designed to create national telecommunications policy.

Thank you for the courtesy extended to the National Newspaper Association at the July 14 hearing and throughout the legislative process.

Sincerely,

TONDA F. RUSH,
President and CEO.

PROPOSED CHANGES TO S. 1086, TELECOMMUNICATIONS INFRASTRUCTURE ACT OF 1993

On page 6, line 15, insert the following:

"(26) local information services are essential to maintaining local communities."

On page 9, line 12, insert the following:

"(oo) 'Local provider' means any publisher or broadcaster of an electronic service generated within and intended primarily for distribution within a local government jurisdiction."

On page 11, line 11, insert the following:

"(6) nondiscriminatory access to any of the carrier's telecommunications facilities and information necessary to the transmission and routing of any telecommunications service or information service and the inoperability of both carriers' networks; and guaranteed nondiscriminatory access to local information providers in any situation where the facilities or services are limited by physical or technological barriers."

On page 14, line 17, after "providing such services," insert the following:

"provided that a local information provider may be required to pay a nondiscriminatory rate no higher than any rate offered to any other information provider in the market."

PREPARED STATEMENT OF THE PUBLIC SERVICE COMMISSION OF THE DISTRICT OF COLUMBIA

INTRODUCTION

The Public Service Commission of the District of Columbia (D.C. PSC) hereby submits these comments on S. 1086, the Universal Service and Telecommunications Infrastructure Development Act of 1993.

The D.C. PSC is concerned that S. 1086 would significantly reduce state commission jurisdiction over telecommunications. The D.C. PSC submits that state commissions play an important role in supporting universal service. In the D.C. PSC's view, the provision of Section 229(d)(2) which provides that states "shall ensure the pres-

70-334 O - 94 - 9

ervation and advancement of universal service," rings hollow if states are not permitted to take action to preclude impacts on local customers. In particular, the D.C. PSC is concerned that states are precluded from determining whether entry of competitive providers should be permitted, under what conditions telecommunications carriers should have pricing flexibility, what advanced services should be provided to subscribers, and at what rates, terms and conditions information services should be offered. The D.C. PSC also submits that where separate subsidiaries are required for Bell Telephone Company information services, states should have the ability to audit the books of the subsidiary in order to preclude cross-subsidization. If these changes are not made, there is little that the D.C. PSC can do to protect universal service or prevent cross subsidization. The District of Columbia has had a significant decrease in the telephone penetration rate since 1984, resulting in a reduction in universal service, and since there is no long-distance service in the District of Columbia, the D.C. PSC cannot set long distance rates to cover costs of low-income subscribers to basic exchange service. Therefore, the D.C. PSC contends that state commissions should be able to determine (1) whether entry of new competitors and provision of new services will increase costs which will result in reduction of telephone usage by low income residents, (2) whether a local market is or is not so competitive as to warrant flexible pricing, and (3) whether Bell Telephone Company provision of information services will result in cross-subsidization. The D.C. PSC therefore proposes amendments to the bill as set forth below.

In addition, the D.C. PSC points out that the proposed amendment Sec. 613(b) would require tariffs to be filed with states whose laws do not provide for the filing of tariffs for video programming. Instead, the legislation should contain a provision requiring the convening of a Joint Board to separate the costs of video programming, as is contained in H.R. 1504.

PROPOSED CHANGES

Sec. 229(a) Removal of Barriers To Entry

The D.C. PSC urges that this section be eliminated, since it precludes state commissions from considering the impact that new entry would have on universal service. At the very least, it should provide that state commissions may only prohibit entry upon a showing that new entry would preclude the preservation and advancement of universal service.

Sec. 229(b)(2) Pricing Flexibility

This section should be amended to add the words "or state commissions" after the words "the Commission" in line three. As presently set forth, the language permits only the Federal Communications Commission (FCC) and not state commissions to determine whether services are competitive and warrant flexible pricing. Competition for local exchange service is likely to be different in different localities, and state commissions should be permitted to determine its existence based on their familiarity with local conditions.

S77. 230(B) Full Effectuation

This section permits the FCC to determine what high quality services are required and to preempt any state action precluding the offering of such services. Citizens in particular localities should not be required to pay for advanced services which they do not need and state commissions should have a role in determining what services will be available. In that connection, Section 230(c), State Regulatory Incentives, is a better way of achieving advanced services than FCC preemption. The D.C. PSC hereby urges the elimination of Section 230(c)

Sec. 613(b)(2)(B) Restrictions on Ownership and Control of Cable Television Systems by Telephone Companies

The D.C. PSC opposes Section 613(b)(2)(B). This section requires that video programming by common carriers shall be preceded by a tariff filing which has been approved by the state in which the common carrier provides service. The District of Columbia has no law which requires the filing of tariffs for video programming service, and therefore this provision would require State action which is not consistent with State law. Instead, the D.C. PSC urges that the bill adopt the provision in H.R. 1504 which contains a requirement that a Joint Board be convened to determine the jurisdictional separation of video programming. Such a requirement would enable state commissions to prevent cross subsidization of video programming by telephone exchange service. The language is set forth below:

Joint Board—The Commission shall, within 30 days after the date of the enactment of this part, convene a Federal-State Joint Board under the provisions of section 410(c) for the purpose of establishing the practices, classifications, and regulations as may be necessary to ensure proper jurisdictional separation and allocations of the costs of providing video programming.

Sec. 233(B) Separate Subsidiary

This section is unclear, since it applies only to electronic publishing service, while section 233(c), which states that it relies in part on "the subsidiary requirements of subsection (b)," relates to all information services. Further, the D.C. PSC urges that the following language be added:

(7) States may determine the amount of electronic publishing revenues that will be used to contribute to universal service.

As stated above, states have a limited amount of revenue to support universal service, and should be permitted to direct the Bell Telephone Companies to use all or part of their electronic publishing revenues to make a contribution, as required by Section 2(10).

Sec. 233(c) Prevention of Cross Subsidies

While this section prohibits subsidies from exchange and exchange access service to Bell Telephone Company information services, it does not provide an enforcement mechanism, other than the provision in Section 230(b)(5) for the FCC to prescribe accounting rules for electronic publishing services. The D.C. PSC submits that states, which regulate exchange and exchange access services, can prevent cross subsidies if they are given the tools to do so. Accordingly, at the end of Section 230(c), the following sentence should be inserted:

States may audit the books of any such affiliate to determine whether its services have been subsidized by revenues from telephone exchange service or telephone exchange access service.

Sec. 13 Jurisdiction

This section should be eliminated because it preempts state regulation of information services. The D.C. PSC submits that states should be able to regulate information services, and may need to do so to preclude cross subsidization of these services by exchange and exchange access services.

PREPARED STATEMENT OF JAMES E. LEVIN, VICE PRESIDENT, GOVERNMENT AFFAIRS, SPRINT COMMUNICATIONS CO.

Pursuant to discussions with the Subcommittee staff, on behalf of Sprint Communications Co. ("Sprint"), I hereby submit our views on S. 1086 for the hearing record. Sprint is supportive of the need for legislation in this area and of this bill in particular. We would, however, suggest a number of changes to several sections of the bill, which we believe are needed to foster a healthy, dynamic and competitive telecommunications marketplace in the future. If these changes or similar modifications to S. 1086 are made, Sprint can enthusiastically support the legislation. Our suggestions are outlined as follows:

I. DEFINITIONS (SECTION 4 OF THE BILL)

We believe that the definition of "telecommunications service" should be expanded to include entities that provide access or access-like services but don't directly interconnect with a LEC. Suggested language is as follows:

(ii) "Telecommunications service" means the offering of—

(1) telecommunications facilities that (A) are owned or controlled by a provider of telephone exchange service or (B) interconnect with the network of a provider of telephone exchange service, or (C) interconnect with other facilities that interconnect with the network of a provider of telephone exchange service and provide access services (or the functional equivalent thereof), as defined by the Commission; or

II. TELECOMMUNICATIONS COMPETITION (SECTION 5 OF THE BILL)

Section 229(c) should be modified to permit economic considerations when contemplating interconnection to local telephone networks. Suggested language is as follows:

(1) interconnection to the carrier's telecommunications facilities at any technically and economically feasible point within the carrier's network;

(3) nondiscriminatory access where technically and economically feasible to poles, ducts, conduits, and rights of way owned or controlled by the carrier;

Section 229(d) should authorize the FCC to determine the impact of competition on universal service. Suggested language is as follows:

(2) The Commission shall, within ninety (90) days after enactment of this Act, institute a notice of proposed rulemaking that convenes, under Section 410, a Joint Board to determine the impact of competition on universal service and to develop rules to preserve and advance universal service. Notwithstanding any other provision of this Act, such Joint Board shall complete its proceeding and issue its report within eighteen (18) months of its creation and the Commission shall complete its rulemaking within twelve (12) months after receipt of the Joint Board's report.

(3) In administering this subsection, the Joint Board and Commission shall ensure that universal service funding methods provide support for: (A) individuals that cannot afford the cost of their telecommunications service or equipment, and (B) local exchange carriers the Commission determines serve high-cost areas. In determining which carriers serve high-cost areas, the Commission shall consider, among other relevant facts, each local exchange carrier's local loop costs compared to the national average local loop costs.

Section 229(h) should definitively authorize local pricing flexibility. Suggested language is as follows:

(2) The Commission and the States shall permit telecommunications carriers to have the ability to price flexibly, but not discriminatorily, for services the Commission finds are competitive. In recognition of the increasing competition in the provision of local exchange services, the Commission shall institute a rulemaking to determine the appropriate form and level of future regulation, if any, of local exchange carriers and local exchange service competitors.

III. INFRASTRUCTURE INVESTMENT (SECTION 6 OF THE BILL)

Section 230(a) should emphasize economic considerations in the deployment of technology. Suggested language is as follows:

(a) The Commission and state regulatory authorities shall ensure, and the Commission shall promulgated regulations that ensure telecommunications carriers provide consumers in rural markets and noncompetitive markets with access to high quality, technologically advanced and economically sustainable network facilities and capabilities which—

Section 230(c) should authorize the FCC to develop regulatory incentives to promote infrastructure development. Suggested language is as follows:

(c) REGULATORY INCENTIVES.—The Commission and the States are encouraged to implement incentives, including incentives providing for accelerated capital recovery, to promote the development of high quality telecommunications network facilities and capabilities. If regulatory incentives fail to result in the deployment of high quality telecommunications network facilities and capabilities in rural markets and noncompetitive markets, the States may adopt other methods to ensure that the goal of subsection (a) is achieved.

Section 230(d) should be changed to authorize local telcos to share network facilities. Suggested language is as follows:

(d) INFRASTRUCTURE SHARING.—To the extent it finds it reasonable, necessary to achieve the goal of subsection (a), and otherwise in the public interest, the Commission may promulgate regulations authorizing local exchange carriers to share or jointly provide, on terms the Commission finds to be reasonable, network facilities and capabilities in rural and noncompetitive markets.

IV. IMPLEMENTING REGULATIONS (SECTION 7 OF THE BILL)

This provision should be revised to accommodate FCC and Joint Board rule-making activities. Suggested language is as follows:

(a) Except as provided in Section 229(d)(2), the Commission shall issue regulations to implement this Act within 12 months after the date of enactment of this Act. Such regulations shall take effect within 6 months after their issuance, except that the Commission may extend such effective date for up to 24 additional months for any small carrier providing telephone exchange service in rural areas, upon a showing by the carrier that compliance would not be technically or economically feasible without additional time.

V. CABLE TELEVISION SYSTEMS (SECTION 8 OF THE BILL)

Section 613(b)(1)(B) should be revised to permit joint ventures between telephone and cable TV companies, but only to the extent that they do not amount to a circumvention of the ownership restriction. Suggested language is as follows:

enter into a joint venture or partnership with such cable system if the purpose, result or effect of such joint venture or partnership is to enable either party to avoid the ownership restriction contained in subsection (b)(1)(A).

Section 613(b)(1)(A) should be amended to apply the cable TV ownership restriction in areas with greater than 10,000 population. Suggested language is as follows:

Except as to proposed service areas containing any incorporated or unincorporated place of 10,000 inhabitants or more, or any part thereof, no carrier shall:

Section 613(b)(2)(B) should be deleted to allow local telephone companies to provide programming services to subscribers in the same way as cable TV companies are now able to provide such services.

VI. INTEREXCHANGE SERVICES (SECTION 9 OF THE BILL)

Section 231 should codify the AT&T Consent Decree's restriction against the provision of interLATA services by the Bell Companies, if exceptions to that restriction are to be permitted. Suggested language is as follows:

(i) Except as provided in these sections, no Bell Telephone Company or affiliate thereof shall directly or indirectly provide any interexchange telecommunications or interexchange information service.

(ii) No earlier than 10 years after the enactment of this section, a Bell Telephone Company or its affiliate may apply to the Commission for authorization to provide interexchange telecommunications or interexchange information services.

(iii) The Commission shall grant authorization to a Bell Telephone Company or its affiliate that has applied under subsection (b) only to the extent that such company proves that there is no substantial possibility that such company could use monopoly power to impede competition in any relevant market for the activity to which the application relates.

(iv) The determination by the Commission regarding an application made under subsection (c) shall be final unless, within 60 days after such determination, any person injured by the determination commences a civil action against the Commission in any district court of the United States for a de novo determination regarding the application.

(v) The court shall enter a judgment granting the authorization for which the Bell Telephone Company or its affiliate applied to the Commission only to the extent that such company proves that there is no substantial possibility that such company could use monopoly power to impede competition in any relevant market for the activity to which the application relates.

Section 231(b)(2) should be revised to clarify that the Bell Companies are being authorized to provide interLATA cable service only to the extent that the MFJ Court has already authorized them to do so. Suggested language is as follows:

may use inter-LATA distribution facilities only insofar as the federal District Court in D.C. has granted waivers of the inter-LATA restriction contained in the Modified Final Judgment in the case of the *United States* v. *Western Electric Co. and AT&T* (Civil Action No. 82-0192), as of the date of enactment of this Act.

VII. CELLULAR MOBILE RADIO SERVICES (SECTION 10 OF THE BILL)

Section 232(b) should be deleted or revised. Considering that the FCC is already deliberating whether to mandate conversion to cellular equal access by all carriers, Congressional action at this time is probably not needed. But, if Congress is to act, cellular equal access (where not required by Consent Decree) should be contingent upon market demand, and the FCC should be authorized to reconsider the need for equal access for emerging wireless technologies as they develop and are implemented. Suggested language is as follows:

(1) When adopting rules for equal access to wireless services, the Commission shall consider differences in market demand in different areas, costs of conversion and other economic and technical factors, and may provide that cellular or other wireless carriers in those areas are required to convert to equal access technology within a certain period of time only after bona fide requests for the service are received from a material percentage of customers.

(2) The Commission shall also evaluate the need to impose equal access requirements on new forms of wireless communications as they may be authorized to provide commercial service, and may exempt such new services from equal access requirements if it can be shown that they are competitive and do not provide the carrier with a means to impede competition in the relevant market.

VIII. CUSTOMER PROPRIETARY NETWORK INFORMATION (SECTION 12 OF THE BILL)

Section 234(b) should be modified to clarify that a customer's opportunity to limit disclosure of subscriber lists does not apply to the routine publication of telephone directories. Suggested language is as follows:

> Nothwithstanding subsection (a), a LEC shall provide subscriber list information under nondiscriminatory, and reasonable rates, terms and conditions to any person upon reasonable request. A LEC may publish directory listings, but if it proposes to use subscriber list information for any other commercial purpose the LEC shall provide each of its subscribers with the opportunity to prohibit or limit disclosure of his or her subscriber list information.

LETTER FROM STEVEN J. METALITZ, VICE PRESIDENT AND GENERAL COUNSEL, INFORMATION INDUSTRY ASSOCIATION

JULY 26, 1993.

The Honorable DANIEL K. INOUYE,
U.S. Senate,
Washington, DC 20510

DEAR MR. CHAIRMAN: On behalf of the Information Industry Association (IIA), thank you once again for your invitation to testify before the subcommittee on July 14 on S. 1086, the Telecommunications Infrastructure Act of 1993. I am pleased to supplement IIA's testimony with the following information.

Our testimony on S. 1086 focused on four critical issues: (1) the scope and implication of the definitions contained in S. 1086; (2) the appropriateness of strong antitrust protections; (3) the need for strong direction to ensure cost-based pricing of basic services offered by local exchange carriers and avoidance of enhanced service provider access charges; and (4) the importance of a clear and unequivocal statement of Federal preemption of any State utility and entry/exit regulation of enhanced services.

On July 23, IIA and other groups submitted to the subcommittee staff specific suggested amendments to S. 1086. These address most of the issues highlighted in our testimony. We believe these suggestions will clarify and strengthen the bill, and we urge you to give them serious consideration.

The suggested amendments retain the terminology of "telecommunications" and "information service" which were employed in S. 1086 as introduced. However, the amendments would revise the definitions of these terms to track more closely the existing distinction between "basic" and "enhanced" services. As IIA noted in its testimony, this distinction, which the FCC has employed for several years, has worked well. Incorporating its basic thrust into S. 1086 will provide the greatest benefit to users and providers alike.

Our testimony of July 14 stressed the importance of more specific safeguards to assure market fairness and prevent discrimination, cross-subsidization, or other anticompetitive behavior. The suggested language changes submitted to the staff recommend suitable steps to prevent such abuses, including more detailed structural separation and open network architecture requirements. Such clear prescriptions, coupled with fair enforcement, will improve the industry for all consumers and providers.

IIA's testimony also noted the importance of customer proprietary network information (CPNI). The suggested language addresses both the competitive and privacy concerns inherent in this complex issue. It will offer broad privacy protections to consumers and ensure the commercial viability of CPNI to business users. At the same time, it provides the Commission with needed flexibility to determine how best to balance the interests of privacy and free flow of information, by authorizing it to adopt an "opt-out" approach to the dissemination of appropriate elements of CPNI.

Finally, as noted by many of the witnesses at the subcommittee hearing, we support all that this bill can do to set the stage for improved and widespread availability of digital transmission facilities for all Americans. The adoption of standards to ensure the integrity and quality of the network, as proposed by the suggested amendments, are a key element of the effort to make the maximum benefits of advanced telecommunications infrastructure available to rural, inner city, and all other markets.

IIA commends the leadership that you and Senator Danforth have shown on these critical issues. We look forward to working with you as this measure moves forward.

Sincerely yours,

STEVEN J. METALITZ,
Vice President and General Counsel.

LETTER FROM GERALD J. KOVACH, SENIOR VICE PRESIDENT, EXTERNAL AFFAIRS, MCI COMMUNICATIONS CORP.

SEPTEMBER 17, 1993.

The Honorable DANIEL K. INOUYE,
U.S. Senate,
Washington DC 20510

DEAR MR. CHAIRMAN: On July 14, 1993, Mr. James G. Cullen, President of Bell Atlantic, testified before your Subcommittee's hearing on S. 1086, the Telecommunications Infrastructure Act of 1993. Certain representations were offered as part and parcel of a concerted effort by the Bell Operating Companies (BOCs) to secure legislative relief from the line of business restrictions contained in the Modification of Final Judgment (MFJ). Mr. Cullen and others would have the Subcommittee and the public believe that the local telephone markets are subject to competition, while the long distance business is not. Neither of these allegations is true.

Among the assertions made by Mr. Cullen was that interexchange carriers (IXCs) do not pass through to consumers the benefits of access charge reductions they receive from local exchange carriers (LECs). To support his allegation, Mr. Cullen pointed to a single switched access charge filing made by LECs in April 1993. While that one filing did in fact result in a decrease in access charges, it does not support his conclusion. Mr. Cullen did not reveal several switched access charge increases made in the last year that offset that one reduction. Indeed, interstate access charges to the interexchange industry went up in the relevant period. Thus, the implication that IXCs had increased rates while their costs had actually declined is untrue.

Since divestiture, competition has dramatically reduced the price of long distance service providing positive consumer benefits. In the last seven years, the largest IXCs as a group have decreased prices significantly more than LECs have decreased access charges. Between 1985-1992, MCI estimates that overall domestic switched long distance rates for the three largest IXCs decreased by 43 percent more than reductions in related access charges. In nominal terms, long distance rates fell by nearly 19 cents per minute, while the LEC's access charges only dropped by 13 cents per minute. After adjusting for inflation, interexchange rates fell by 53 percent more than comparable access expenses. Clearly competition in the long distance market has driven actual rates down faster than access, and suggestions to the contrary are false.

Erroneous information regarding the long distance industry should not distract the Subcommittee from focusing on the current monopoly that the LECs, including the Regional Bell Operating Companies (RBOCs), maintain in the access market. Of the access charges paid by MCI in the last year, an overwhelming majority, 99.4 percent, went to LECs. An infinitesimal percentage, .06 percent, went to their competitors. This fact alone places the tale of vibrant local access competition in the proper light.

Not only are the RBOCs monopolists in their serving areas, but they have earned substantial sums from their interstate access business. In 1992, they achieved net income of over $3 billion from interstate access, with a return on investment of nearly 12.5 percent. This latter figure is a full 125 basis points greater than the interstate authorized rate of return which the FCC has determined is fair. Despite RBOC claims of being faced with competition, the facts are: they control over 99 percent of the access market; they price access very near the allowable maximum rates as determined by price caps; and they continue to earn excessive rates of return.

In addition to producing dramatic pricing declines, competition in the long distance marketplace has also been responsible for numerous service enhancements and new service offerings. Residential customers have access to new services, such as personal 800 service, and discount plans, such as MCI's Friends and Family plan. Business customers have been afforded specialized billing arrangements, network reconfiguration, and virtual private networks to satisfy increasingly sophisticated telecommunications needs.

The public has been well served by the establishment and maturing of competition in the long distance marketplace. To promote the public interest, it is of utmost importance that policymakers introduce competition into the local exchange marketplace. MCI looks forward to working with you in these endeavors.

Sincerely,

GERALD J.KOVACH.

LETTER FROM JAMES G. CULLEN, PRESIDENT, BELL ATLANTIC

SEPTEMBER 2, 1993.

The Honorable DANIEL K. INOUYE,
U.S. Senate,
Washington, DC 20510

DEAR MR. CHAIRMAN: At the July 14, 1993 hearing before this subcommittee on S. 1086, I stated that the interstate long distance companies had not reduced prices to customers by the same amount that the local exchange companies have lowered interstate access prices. In support of that claim I attached a Merrill Lynch analyst's report that stated:

"In another sign of oligopolistic tendencies in the long distance (LD) business, AT&T has informed the FCC that it does not plan to reduce its long distance prices on July 1 despite the fact that its largest cost, local telco access charges, are scheduled to fall by at least an annualized $250M on that date."

AT&T, through Thomas Norris, its Executive Vice President, challenged that claim in an August 2, 1993 letter to this subcommittee. This letter responds to the misstatements contained in the AT&T letter.

In his letter,, Mr. Norris claims that the "access charges in recent LEC filings have increased, not decreased, AT&T's access expenses." One look at AT&T's most recent quarterly report, however, confirms that the Merrill Lynch report was accurate. AT&T's own report shows that its access costs for the most recent six months, when compared to the same period a year ago, have decreased by $298 million. Over the same period, AT&T's revenues from telecommunications services increased by $76 million. See Attachment 1. I have also attached an analysis of the most recent access tariff filings by the local exchange companies (Attachment 2). That analysis also confirms that AT&T's rates have been reduced this year by $250 million.

These access charge reductions are not an isolated occurrence. As Figure 1 shows, local exchange company access prices to long distance companies have dropped substantially and consistently since divestiture. Figure 2 shows that, even when adjusted for lost market share and increased demand, AT&T's access expenses have decreased significantly since divestiture. Specifically, AT&T's annual access charges have been lowered by $10.03 billion since divestiture. Surprisingly, however, AT&T has only lowered its rates by $8.37 billion over the same period. When one considers that $310 million of AT&T's rate reduction represented FCC-mandated rate reductions not related to access costs, AT&T has failed to pass on nearly $2 billion in access price reductions.

In a truly competitive environment, changes in prices would tend to reflect changes in costs and not allow providers to pocket a staggering $2 billion. As I testified before this subcommittee, the Bell companies can add real competition to the long distance marketplace. Competition from the Bell companies does not necessarily mean duplicating the long distance networks of the other carriers. Permitting the Bell companies to resell the services of other carriers will benefit consumers by giving them the information they need to choose the lowest priced carrier on every call instead of being stuck with one carrier for all calls.

Thank you for this opportunity to set the record straight and I stand ready to assist this subcommittee in any further deliberations on this matter.

Yours truly,

JAMES G. CULLEN.

[The attachments and figures referred to may be found in the committee files.]

LETTER FROM THOMAS H. NORRIS, VICE PRESIDENT, FEDERAL GOVERNMENT AFFAIRS, AT&T

AUGUST 2, 1993.

The Honorable DANIEL K. INOUYE,
U.S. Senate,
Washington, DC 20510

DEAR MR. CHAIRMAN: During recent hearings before your Subcommittee on S. 1086, the Telecommunications Infrastructure Act of 1993, James G. Cullen, President of Bell Atlantic, made several highly inaccurate and misleading statements about AT&T's pricing. This attempt to shift the Subcommittee's attention away from local exchange competition and the monopoly that his company and the other Regional Bell Operating Companies (RBOCs) enjoy in local telecommunications must be corrected.

Mr. Cullen confused the issues raised in S. 1086 by selecting an issue that is inordinately complex, even to those in the industry. Specifically, he testified that the local exchange carriers (LECs) reduced access charges to long distance carriers, or IXCs, by approximately $250 million and that AT&T did not reduce its long distance prices by a comparable amount. In fact, over the recent period reflected in AT&T's price changes, LECs increased their switched access charges to IXCs by $20 million.[1] Mr. Cullen's statement is therefore not only a gross overstatement, it is also completely contrary to the actual direction of LEC access pricing.

Mr. Cullen's first mistake was to use a single access reduction to disguise multiple recent IXC access charge increases and then blaming long distance carriers for failing to pass through reductions we never received. He further misled Members of the Subcommittee by blaming AT&T for failing to pass through 100 percent of the fictional IXC access reduction. In fact, of course, AT&T pays only about 60 percent of IXC access charges and could hardly be expected to account for the entire reduction even if therehad been one. The basic fact remains that the access charges in recent LEC filings have increased, not decreased, AT&T's access expense.

Mr. Cullen's testimony also included a chart from the February 1993 issue of Business Communications Review (BCR) which he claimed showed the IXC industry is engaged in oligopolistic pricing. He selectively used a chart but failed to include the entire article, which would have disproved the precise point he sought to make.

Had he done so, the Members of the Subcommittee would have seen that "basic long distance charges" as defined by the author, are the undiscounted per minute rates for AT&T Megacom, MCI Prism 1 or Sprint Ultra WATS for a "typical" customer with 1,000 hours of daytime traffic per month and include the cost of T1 access 24 voice grade lines) at one end.

The very point of the BCR article was to inform business telecommunications managers that there is substantial competition for such services and that they can and should negotiate discounts off these "basic" rates. The author wrote, "Tariffed or not, today's basic long distance service prices are just a *starting point* (my emphasis) in determining what your particular company *might pay*." This point is constantly reiterated throughout the article with further statements such as "base rates are just the 'sticker price'; pricing plans and special deals make the difference."

Nor are customers for the services whose base rates were charted representative of customers in general. But, in any case, every residential and business customer has a plethora of discount plans available, from AT&T's *i* plan to Sprint's The Most to MCI's Friends and Family to Tariff 12 and its clones. Customers have unique requirements, and Mr. Cullen's implication that long distance pricing can be described accurately in one chart shows either a deliberate attempt to confuse or a gross misunderstanding of how long distance competition works.

If Members of the Subcommittee and other policy makers are distracted by specious arguments such as Mr. Cullen's, they will fail to focus on the real problem and public policy need—the absence of competition in the local information infrastructure. It is noteworthy that nowhere in his testimony or rebuttal did Mr. Cullen refute the fact that the entire IXC industry is totally dependent on the LECs for access to our customers. Well over 99 percent of AT&T's access expense in 1992 went to the LECs; only $19 million out of $14 billion, or 0.14 percent, went to their competitors while 99.86 percent stayed with the LECs.

The debate in Congress is, properly, how to ensure that competition flourishes in all segments of the telecommunications and information industry. Competitive conditions already exist in the interexchange industry and need to be established, if

---

[1] A $250 million reduction in switched access charges to the IXC industry was made as Part of the LECs' annual tariff filing effective July 1, 1993. But the $250 million number reduction in carrier switched access was more than offset by $270 million increases in access expenses.

possible, in non-competitive segments—notably that of local exchange telephone companies.

AT&T looks forward to contributing to the fact-finding and deliberations of the Subcommittee and full Committee. You may rely on us to furnish facts as clearly as we know how.

Sincerely,

TOM NORRIS.

ATTACHMENT A

Below are LEC access expense changes that have an impact on the entire LEC access customer base. AT&T's Price Cap Regulated Basket 1 and 2 Services are a subset of this total. LEC access charge changes affecting AT&T's Price Cap Regulated services are referred to as "Delta Y", and serve as input to AT&T's Price Cap Indices (PCT). AT&T's PCI is typically updated twice per year to include accumulated LEC access changes, inflation (GNP-PI), and FCC sanctioned exogenous cost adjustments. Delta Y is computed using 1992 Base Period Price Cap volumes and is relegated to AT&T's Price Cap Baskets using an "As Allocated" methodology consistent with FCC Docket 87-313 rules.

SWITCHED ACCESS CHARGE CHANGES	EFFECTIVE DATE	ESTIMATED IXC INDUSTRY IMPACT ($M)
LEC INFORMATION DATABASE	FULLY DEPLOYED MID, 1992	$100
GTE CCL INCREASE (TRANS. #778)	2-Apr-93	$33
LEC 800 NUMBER PORTABILITY	1-May-93	$74
LEC ANNUAL ACCESS FILING	1-Jul-93	($250)
GENERAL SUPPORT COSTS REALLOCATION & UNIVERSAL SERVICE FUND ADJUSTMENTS	1-Jul-93	$63
*SUM OF ACCESS CHARGE CHANGES		$20

LECs reduced total access expenses by $184 million in their annual filing. The net reductions to industry switched access services were $250 million, apportioned to switched access categories as follows: carrier common line charges reduced by $229.1 million, switched traffic sensitive charges reduced by $21.1 million. (Note: LEC Information Database charge changes were absent from AT&T's prices during 1992) Including the additional access charge changes, industry switched access expenses increased by $20M. AT&T has included approximately $9M worth of LEC switched access charge changes in its recent PCI filing.

PREPARED STATEMENT OF ASSOCIATION OF TELEMESSAGING SERVICES, INTERNATIONAL

On behalf of the Association of Telemessaging Services, International ("ATSI"), we welcome and appreciate the opportunity to submit this statement regarding S. 1086, the Telecommunications Infrastructure Act of 199 3, as it relates to Customer Proprietary Network Information ("CPNI"). As owner of C&J Telecommunications, Inc., which provides live answering services in Honolulu, Hawaii, and President of Available Communications, Inc., which provides voice and live answering services in St.

Louis Missouri, respectively, we are keenly aware of the importance of this matter to our businesses, and we support restrictions on the use of CPNI.[1]

ATSI is the only national trade association representing the voice mail and live answering industry. As the Washington eyes, ears, and voice of the telemessaging industry for over 50 years, the association's rocts are in live telephone answering services. With industry growth and progress, so too has the association grown to include businesses that also offer a wide variety of telemessaging services such as voice messaging, paging, and fax services.

Before presenting our association's views on CPNI, we would like the Subcommittee to consider the following scenario: Imagine, if you will, that you are competing for re-election with only one opponent in your District. Now imagine that one person has but a single advantage: He or she is also the registrar of voters. Only he or she knows the make-up of each household—how many qualified voters, how they voted, if they vote independently or the party line, who they voted for, those important socio-economic demographics of age, education, habits—maybe even zip code—that help you better understand the needs of your constituents. Now take this unfair advantage just two steps further: The voters must personally ask for their ballots from your opponent! And you don't have this information!

That's not unlike the unfair advantage that the telephone companies enjoy through the unlimited and unregulated use of CPNI.

CPNI is more than an abstract telecommunications acronym to us and our industry. Under the FCC's current regulations, the Bell Operating Companies ("BOCs") enjoy virtually unlimited and unregulated use of CPNI, an important tool the BOCs use to gain an unfair advantage over other information providers in identifying, targeting, marketing, and selling their own enhanced services to potential customers. Defined by the FCC to include "all information about a customer's network services and the use of those services that a Bell Operating Company (BOC) possesses by virtue of its provision of network services," CPNI includes information such as: what types of regulated services a customer uses; the number and identity of the customer's service locations; when and where a customer calls and how long he speaks, the amount of the customer's monthly telephone service bills; and, how much a customer uses the different regulated services to which he subscribes.

It is also important to note what CPNI does not include, because the BOCs often have used this information without giving customers any opportunity to object, and without making it available to competing information providers. Under the FCC's rules, the BOCs are free to exclude from CPNI a customer's name, address, telephone number, credit information, and information regarding a customer's use of non-regulated services.

The FCC's CPNI rules allow BOC information services personnel to obtain this valuable information at will—unless the customer first objects. However, unaffiliated vendors, such as our colleagues among the telemessaging providers, cannot obtain this same information unless they first secure the customer's affirmative permission. In other words, affirmative action must be taken by the consumer before a BOC competitor may obtain information that the BOC may obtain at will.

And perhaps the most insidious portion of these so-called "safeguards": The BOCs don't even have to notify residential and single-line business customers of their right to prevent disclosure of their proprietary network information. The FCC's rules merely require the BOCs to notify business customers with more than 20 lines of their CPNI rights.

CPNI is both highly confidential and competitively significant, and the FCC's rules gives the BOCs a substantial marketing advantage. CPNI and related information can be used to:

- identify new potential customers, such as subscribers moving into the area or adding service locations, before competitors become aware of them;
- prepare highly effective marketing presentations based on usage information, such as the number of missed or blocked calls, that is unavailable to competitors;
- engage in more timely and cost-effective marketing because they need not secure customer authorizations;
- readily identify the appropriate customer contact, while competing vendors often experience substantial delay in doing so;
- identify customers who currently use competing vendors by matching up the CPNI of vendors and customers;

[1] Our support of the bill's proposed restrictions on the use of CPNI is based on our understanding that the bill's prohibition on disclosure by a "telecommunications carrier" of "any customer proprietary network information to any person" without customer consent, would preclude a telephone operating company from disclosing CPNI to its own voice messaging operation, whether operating within the same corporate entity or through an affiliate.

• enjoy virtually unrestricted access to the CPNI of residential and small business customers; and

• spare the expense of obtaining data—such as credit information—from external sources, while their competitors must do so.

These rules have been used in an anticompetitive fashion to discriminate against our industry, and in many ways, the American public. For example:

• In Albuquerque, New Mexico the local BOC, using CPNI, contacted businesses using answering services and urged them to switch their business from the answering service to the BOC. How were they able to identify these customers? Through CPNI!

• In Denver, Colorado, Alert Telephone Answering Service, Inc., keeps detailed records of customers lost to "poaching" by U.S. West solicitations. For example, customers of competitors have been told that U.S. West would waive installation charges on its voice messaging service. Also, messaging customers have been solicited for U.S. West's voice messaging service when they call to request the removal of call waiting, a relocation of service to another address, or the addition of phone or fax lines at existing addresses. The manager of one telephone answering company who called U.S. West to report loss of service on her telephone was offered a free trial of U.S. West's voice messaging in lieu of credit on her telephone bill. How did the telephone company know that there was a credit on this person's bill? Through CPNI!

• In Wilmington, Delaware, a customer who asked Diamond State Telephone Company to have call forwarding put on his phone line so that his calls would be answered by Phonepower, Inc., learned upon checking his new service that Diamond State had hooked him up to their own "Answercall." How did they know he was a good credit risk? How did they know he was the appropriate contact? Through CPNI!

• In St. Louis, Missouri, one of our members received a direct mail piece mailed to his home promoting Southwestern Bell's Messaging Service. He pays an additional amount for a non-published number, and has the expectation that this number will not be given to the public or used for sales solicitation. In this case, Southwestern Bell utilized the non-published CPNI information for solicitation by their voice mail subsidiary. They even state in their direct mail that "we'll call you in a few days." Utilizing the non-published data base to get the telephone numbers for the telephone calls means that they used CPNI!

• And in Boston, a customer to whom a member sold an answering service, contacted New England Telephone (part of NYNEX) to have call forwarding put on her phone so that she could forward her calls to our service. She later called the member back to say that she wouldn't be taking Ansaphone's service after all. She was going to wait a few months until New England Bell installed their answering service in her area. Clearly this was a case of New England Telephone pre-selling a service that didn't currently exist in that area. How did they know that she would be a good customer? Through CPNI!

We strongly support the Subcommittee's efforts to develop legislation to correct these inequities. The Subcommittee's draft legislation would place the BOCs and their information provider competitors on equal footing when competing for new business opportunities.

In considering S. 1086, we urge the Subcommittee to focus on the following questions:

1. Can the government achieve a competitive, vibrant telecommunications market if the government affords one player in the enhanced services market the kind of marketing advantage obtained from preferential access to customer proprietary network information?

2. Should a single competitor in the information services market be subsidized by information obtained from, and paid for by, rate payers?

3. What kind of government telecommunications policy will invite small businesses to participate in the information infrastructure?

4. And what would happen to America's freedom of choice and voting rights as we know them today if the same rules that apply to CPNI were applied to elections in your District?

On behalf of the entire telemessaging industry we urge Congress to restrict the use of CPNI for marketing purposes, so that a telephone operating company could disclose CPNI to affiliated or unaffiliated providers of unregulated services only if required by law, or if the customer requests disclosure of the information for itself or for a service provider designated by the customer.

Thank you.

LETTER FROM CHARLES H. TOWER, COORDINATOR, ELECTRONIC PUBLISHING GROUP

JULY 28, 1993.

Senator DANIEL INOUYE,
U.S. Senate,
Washington, DC 20510

DEAR CHAIRMAN INOUYE: The Electronic Publishing Group, a coalition of those involved in electronic information content creation and distribution, strongly supports the overall thrust of S. 1086.

We regard the bill as a major step forward in the development of a widely available telecommunications infrastructure based primarily on marketplace competition. This, in turn, will encourage the expanding competitive diversity of information products' and services which should be the hallmark of the information age.

While we have a substantial general interest in all parts of your bill, our particular concern relates to Section 12. That section deals with the privacy of customer proprietary network information, often referred to as CPNI.

I am enclosing with this letter a suggested redraft of Section 12 of S. 1086. We believe that our proposed changes will clarify and strengthen this section of your bill. They are similar to those recommended by two of the larger electronic information industry associations, ITAA and IIA.

The bi-partisan leadership which you and Senator Danforth have provided in the development of this important legislative proposal and the dedicated effort of both majority and minority staffs working together deserve the recognition and support of your colleagues. We shall do all we can to bring about that result.

We request that this letter and the enclosure be made part of the record now being developed in your Subcommittee's consideration of S. 1086.

Respectfully yours,

CHARLES H. TOWER.

AMENDMENT TO THE COMMUNICATIONS ACT OF 1934

[Bold is deleted material and italic is added material]

SEC. 234. PRIVACY OF CUSTOMER PROPRIETARY NETWORK INFORMATION.

(a) PRIVACY REQUIREMENTS FOR **Telecommunications** *Local Exchange* CARRIERS.—A **telecommunications** *local exchange* carrier—

(1) shall not disclose any customer proprietary network information to any person *that is engaged in the provision of other than telecommunications services,* except—

(A) as required by law; **or**

(B) upon affirmative written request by the customer to which it relates; *or*

(C) pursuant to rules established by the Commission, if—

(i) such information is of a type that the Commission has determined may be generally disclosed without raising substantial customer privacy concerns; and

(ii) the local exchange carrier has provided notice to its customers of the general availability of such information and of the opportunity to object to such disclosures;

(2) shall disclose such information, upon affirmative written request by the customer, to a service provider designated by the customer;

(3) shall, whenever such **telecommunications** *local exchange* carrier provides any aggregate information based on customer proprietary network information or any data base or other compilation of customer proprietary information to any person, notify the Commission of the availability of such aggregate or compiled information on **the same** *reasonable* terms **and,** conditions, *and price* to any other service provider upon reasonable request therefor; **and**

(4) shall not discriminate between affiliated and unaffiliated service providers in *soliciting consents regarding,* providing access to, *formatting, categorizing, collecting, compiling, aggregating or disaggregating, indexing,* or in the use and disclosure of, individual and aggregate or compiled information made available consistent with this subsection*; and*

(5) shall ensure that its own or any affiliate's information service personnel no longer have access to, or use of, directly or indirectly, any customer proprietary network information that was disclosed to such personnel without the customer's affirmative written request before the effective date of this section.

(b) PROVISION OF SUBSCRIBER LIST INFORMATION.—Notwithstanding subsection (a), a local exchange carrier shall provide subscriber list information under nondiscriminatory and reasonable rates, terms, and conditions to any person upon reasonable request. A local exchange carrier shall provide each of its subscribers with the opportunity to prohibit **or limit** disclosure of his or her subscriber list information.

(c) DEFINITIONS.—As used in this section:

(1) The term "customer proprietary network information" means—

(A) information which (i) relates to the quantity, technical configuration, type, destination, and amount of use of telecommunications service subscribed to by any customer of a **telecommunications carrier** *local exchange* carrier, and (ii) is available to the **telecommunications carrier** *local exchange carrier* by virtue of the **telecommunications carrier** *local exchange carrier*-customer relationship;

(B) information contained in the bills for telecommunications service received by a customer of a **telecommunications carrier** *local exchange carrier*; and

(C) such other information concerning the customer as is (i) available to the **telecommunications carrier** *local exchange carrier* by virtue of the customer's use of the carrier's services, and (ii) specified as within the definition of such term by such rules as the Commission shall prescribe consistent with the public interest.

(2) The term "aggregate information" means collective data that relates to a group or category of services or customers, from which individual customer identities or characteristics have been removed.

(3) The term "subscriber list information" means information identifying a local exchange carrier subscriber's name, telephone number, address, billing name and address, or primary directory advertising listing, or any combination thereof.

(d) Rulemaking Schedule.—The Commission shall, within 90 days of the enactment of this section propose, and within 270 days of enactment adopt rules that:

(1) implement the requirements of this section, including:

(A) specifying the types of customer proprietary network information that may be generally disclosed without raising substantial customer privacy concerns; and

(B) specifying the types of customer proprietary network information that may not be disclosed absent a requirement of law or affirmative written request by the customer to which it relates; and

(2) ensure that any customer proprietary network information or aggregate or compiled information or data base made available to an unaffiliated service provider shall be in such form as the provider shall reasonably request.

LETTER FROM WILLIAM L. WEISS, CHAIRMAN AND CEO, AMERITECH

OCTOBER 20, 1993.

The Honorable JOHN C. DANFORTH,
U.S. Senate,
Washington, DC 20510

DEAR SENATOR DANFORTH: On behalf of the seven regional companies, I am responding to the request you put to me and Bob Allen at the hearing on September 8 to develop a definition of local exchange competition. I hope the following information is helpful to you and the Subcommittee as you continue your deliberations on S. 1086.

In developing a definition of local exchange competition, we must first determine the purposes for which such a standard might be used. Tests for competition have been developed, as in the cable reregulation legislation, to determine the appropriate level of regulation, of the prices of a telecommunications provider. For example, Illinois law provides the following standard of effective competition which, when met, relieves a provider of a service of certain regulatory pricing burdens:

> "Competitive Telecommunications Service" means a telecommunications service, its functional equivalent or a substitute service, which, for some identifiable class or group of customers in an exchange, group of exchanges, or some other clearly defined geographical area, is reasonably available from more than one provider. 220 ILCS 5/13-209.

This type of test is not appropriate as a test for entry into a market. An entry test, based on antitrust principles, must focus on conditions in the market that one is seeking to enter. The Modified Final Judgment (MFJ) provides just such a test. Recognizing that excluding a competitor from a market harms consumers, the MFJ provides that the line of business restrictions, including the long distance prohibition, shall be removed when there is "no substantial possibility that a (regional company) could use its monopoly power to impede competition in the market it seeks to enter." This standard does not require the elimination of the local exchange monopoly. Indeed, it assumes the continuation of substantial market power, if not a de jure monopoly. Instead, relief is mandated if there is no substantial possibility that any existing monopoly power in the local exchange will impede competition in the market the local exchange company seeks to enter. The Court of Appeals has interpreted "impeding competition" to mean the ability to increase price or restrict output. This means, for example, that there must be a significant threat that the regional companies will dominate the long distance market.

The regional companies believe that existing conditions in today's telecommunications marketplace satisfy the test for entry set out in the MFJ. Regardless whether the Subcommittee agrees with that proposition, there can be no doubt that the unbundling requirements of S. 1086 justify the elimination of the long distance ban. It is Ameritech's position that the unbundling requirements of S. 1086 reduce barriers to entry and eliminate any remaining argument that the regional companies could act anti-competitively in the long distance business.

Predatory pricing by the regional companies in the long distance market is not feasible due to the scale economics of long distance carriers such as AT&T, MCI, and Sprint, and the fact that the regional companies would start with zero market share. Such a pricing strategy would fail because the long distance carriers could withstand any losses from matching below cost prices and could not be driven from the market. Access discrimination would be impossible to implement due to the current equal access regulations in place and the unbundling provisions of S. 1086 bill which would make any attempted discrimination much easier to detect. Accordingly, the seven regional companies urge the Subcommittee to mandate long distance relief in S. 1086.

For this Subcommittee to establish a new test for long distance entry based on market metrics raises several concerns. First, local exchange competition will occur at different times for different groups of customers in different geographic areas. This has already been the experience in the development of long distance competition. To permit entry by the regional companies only to those customer groups and geographic areas for which competition exists—whether defined as the existence of a substitute service or some specified level of market share—will result in piecemeal entry that will not be in the best interests of consumers. This approach will increase customer confusion as to what carriers provide such services a at given point in time, and could cost consumers millions of dollars in foregone savings that would result from full regional company entry. Even worse, delaying entry until some overall metrics is satisfied will delay entry in the most contestable arena far beyond any reasonable time. Ironically, a metrics test has the effect of placing the public policy decision of competitive entry into the hands of the incumbent providers who can control the entry of competitors into their own businesses by their decision as to whether or not, and on what scale, they choose to enter the local exchange business. These effects conflict with the main objective S. 1086—to facilitate the development of universal access to an advanced telecommunications infrastructure.

In addition, continued or piecemeal exclusion of the regional companies from the long distance market would have a serious impact on the types and quality of services offered to consumers. For example, Ameritech has developed the Wisconsin Health Information Network (WHIN), linking doctors, hospitals, and insurance carriers in a network that reduces the cost of health care services while increasing the responsiveness of the industry to the health care needs of Wisconsin. As the current debate over health care attests, this type of service is critical to our nation's ability to provide quality health care services to all Americans. Other of the regional companies are offering or are planning to offer similar services. Due to the long distance restriction, the regions are unable to serve smaller, less populated areas because of the high cost of replicating a network for each area. As a consequence, the benefits derived from WHIN will be provided only to people in the larger cities of Wisconsin, such as Madison and Milwaukee, while people in less populated areas—those in the greatest need of improved health care services—are excluded. Removal of the long distance restriction would allow the regions to serve less populated areas using facilities based in larger cities, thus extending the full benefits of the network to all consumers and reducing the costs to everyone.

In conclusion, we urge the Subcommittee to recognize that opening all markets to all competitors offers the best hope of developing the nation's telecommunications infrastructure for the benefit of all citizens and therefore, to amend S. 1086 to eliminate the long distance ban of the divestiture decree.

Sincerely,

W.L. WEISS.

LETTER FROM ROBERT E. ALLEN, CHAIRMAN OF THE BOARD, AT&T

OCTOBER 5, 1993.

Senator DANIEL K. INOUYE,
Senator JOHN C. DANFORTH,
U.S. Senate,
Washington, DC 20510

DEAR SENATORS INOUYE AND DANFORTH: As promised at your September 8th hearing on S. 1086, I am pleased to provide you with legislative language, which we believe appropriately defines "effective competition" in local telephone services. While we have discussed the language and sought input from several industry participants, the proposal is AT&T's alone. Other parties may wish to comment or submit their own language.

I have asked my Senior Vice President and General Counsel, John Zeglis, and Mike Brown of our Washington Office, to work with your staff in developing language for the bill. I appreciate your interest and personal involvement in these important issues.

Very truly yours,

R.E. ALLEN.

INTEREXCHANGE SERVICES (SECTIONS 9 & 10)

Add a new provision (to section 231 or 232), as follows:

"(a) Except as provided in these sections, no Bell Telephone Company or affiliate of a Bell Telephone Company shall engage in the provision of interexchange telecommunications services.

"(b) No earlier than seven (7) years after the enactment of this section, a Bell Telephone Company may petition to the Commission for authorization to provide interexchange telecommunications services.

"(c) The Commission may grant authorization to a petitioning Bell Telephone Company to provide interexchange telecommunications services upon a showing by the Bell Telephone Company (i) that there is no substantial possibility that the petitioning Bell Telephone Company or its affiliates could use monopoly power to impede competition in the provision of any interexchange telecommunications services and (ii) that effective competition in telephone exchange and exchange access services exists in the region in which the Bell Telephone Company provides telephone exchange and exchange access services. Effective competition can be found to exist only if:

"(1) all the regulations required by section 229 of this Act have been adopted by the Commission and the States and the requirements of subsection 229(c) and number portability have been fully implemented by the Bell Telephone Company and its affiliates in the exchange areas of the Bell Telephone Company;

"(2) 30 percent or more of the telephone subscribers in the exchange areas of the Bell Telephone Company obtain telephone exchange and exchange access services exclusively from an alternative provider; and

"(3) 75 percent or more of the telephone subscribers in the exchange areas of the Bell Telephone Company may obtain, from two or more alternative providers, telephone exchange and exchange access services that are like the services of the Bell Telephone Company and comparable in quality, coverage, price and capability.

"(d) The determination by the Commission regarding an application made under subsection (c) shall be final unless, within 60 days after such determination, any person injured by the determination commences a civil action against the Commission in the district court of the United States for the District of Columbia for a de novo determination regarding the authorization. The Court shall enter judgment granting authorization only to the extent it finds the Bell Telephone Company has made the showing required by subsection (c). A judgment entered under this paragraph shall be stayed until any appeals from the judgment have been exhausted or the time for filing appeals has expired.

"(e) DEFINITIONS.—AS USED IN THIS SECTION:

"(1) The term "alternative provider" means a provider of telephone exchange and exchange access services

"(a) that is not affiliated with the Bell Telephone Company providing such services in that area continuously since January 1, 1984, and

"(b) that provides exchange and exchange access services

"(i) that are like those provided by that Bell Telephone Company without making use of the switching, transmission or other facilities of that Bell Telephone Company, and

"(ii) that permit the alternative provider's subscribers to place and receive calls to and from any of its other subscribers without making use 6f the switching, transmission or other facilities of that Bell Telephone Company, and

"(iii) that provide its subscribers with exchange access to interexchange carrier networks without making use of the switching, transmission or other facilities of that Bell Telephone Company.

"An alternative provider may interconnect with facilities of the Bell Telephone Company solely for the purpose of allowing calling between telephone subscribers of the alternative provider and those of the Bell Telephone Company.

"(2) The terms "exchange access", "exchange area", "interexchange carrier", and "interexchange telecommunications services" all have the meaning given those terms in the Modification of Final Judgment and judicial decisions interpreting that judgment."

DESCRIPTION OF EFFECTIVE COMPETITION AMENDMENT

(a) No interexchange services would be permitted without passing the test, except for the current cellular and video authority in the bill.

(b) A seven (7) year waiting period is required to permit time for local competition to develop. During this time the RBOCs' incentives to open their local networks would not be diminished by anticipation that interexchange relief is imminent. This time frame will allow for the adoption of state and federal regulations to open local networks to competition, the development of a new universal service structure, and implementation and network re-engineering by the telephone companies. Seven years is a reasonable time for competition to develop from essentially a standing start and was, in fact, proposed last year in the House (H.R. 5096).

(c) Any FCC approval requires passing an effective competition test and the current decree's "impeding competition" requirement. Both are important. The former would have the FCC find that effective competition has replaced monopoly in local services. The latter would guarantee the Commission an opportunity to examine whether, notwithstanding compliance with the effective competition metrics, other factors indicate that the RBOCs remain a threat to the continuation of vigorous interexchange competition.

The effective competition provision requires:

1. Full implementation by the local Bell Company of the open network and number portability provisions in Section 229 of the bill;

2. 30 percent of subscribers actually relying upon alternatives to the Bell Company in its region; and

3. Two or more competitors offering services like those of the Bell Company to 75 percent of subscribers in the region.

(d) De novo court review permits the court and case law to play its proper role.

(e) An "alternative provider" is defined as a full facilities-based competitor for the Bell Company, using only its own connections to the customer, switching and local transport. Arrangements would be permitted for calling between the systems. If competitors continue to be dependent on the local telephone company to offer services to customers, then true competition will not have developed.

Four other terms ("exchange access", "exchange area", "interexchange carriers" and "interexchange telecommunications services") are defined with reference to the MFJ as is now done with "LATA," again utilizing existing case law.

QUESTIONS ASKED BY SENATOR PRESSLER AND ANSWERS THERETO BY AT&T

QUESTIONS FOR PANEL 1

I am concerned not only with access to affordable phone service. I care about the quality and diversity of services available in small states like South Dakota. If access to the information superhighway becomes an essential means of fully participating in society, we may need to redefine universal service.

Question. What services do you think should be universally available?

Answer. The universal service obligations of the local exchange carriers have typically included provision of voice grade dialtone, repair service, and annual white pages directories. Telecommunications services in the future will be increasingly more multimedia in nature, mixing voice, data and video to the home. It is difficult to predict what new services may eventually be considered part of "basic service" in the future, but it is reasonable to expect that at least some local exchange services will depend on increased bandwidth in the local loop, central office digital switches, and interoperable networks.

Question. What level of competition can we expect to see in smaller states?

Answer. If broad-based local exchange competition were to become feasible, it would be likely to come first to the more densely populated metropolitan areas of the country. However, rural areas would be apt to be affected as well. The level of competition would be dependent in part on the pace at which legal and regulatory barriers to competition are removed. These barriers include—but are not limited to—franchise and right-of-way restrictions that advantage the incumbent local exchange carrier, the inability of customers to change local carriers while retaining their current local telephone number, lack of reasonable arrangements for alternative carriers to interconnect with the incumbent carrier, and regulated pricing structures that subsidize local rates. The pace at which these barriers are removed in rural areas would have to reflect the continuing need to preserve universal service objectives.

Question. What is the best way to ensure that we do not become a nation of information haves and have-nots?

Answer. The goals of infrastructure development and nationwide access to new services in both metropolitan and rural areas can best be achieved by a regulatory process that encourages competition while protecting targeted high cost rural customers. Universal service requirements ensuring that local exchange carriers provide access to the networks and services of long distance carriers and alternative providers, through interoperability and open interconnection requirements, will ensure that all customers have access to information-rich services without regard to their location.

Already there have been significant efforts to modernize rural exchanges. The National Exchange Carriers Association (NECA) has found that current trends in rural infrastructure investment indicate that by 1994, 97 percent of the central offices of the 1,000 small companies surveyed will have converted to digital switches. Significant gains are also expected in the availability of fiber optic interoffice facilities and signaling system 7 deployment.

Customer targeted support programs such as the federal and state Lifeline and Link-Up programs, as well as the Universal Service Fund (USF) program for high cost local exchange carriers, may continue to be a necessary supplement in delivering services to economically disadvantaged and hard-to-serve areas. In general, these targeted programs should not be jeopardized by the potential of increased competition as long as regulators adopt funding mechanisms that are equitable and broad-based across all industry participants.

QUESTIONS FOR ROBERT E. ALLEN—AT&T

Local telephone companies currently have universal service obligations. They argue that opening the local loop will allow competitors to "cream skim" or "cherry pick"—that is, serve the most lucrative customers, while ignoring the others. S. 1086 attempts to address this problem by requiring all carriers to contribute to the universal availability of affordable phone service. You suggest that any subsidies should generally be provided directly to the end user.

Question. Would subsidies to end-users ensure that new services are brought to high cost service areas—for example, sparsely populated areas of South Dakota?

Answer. In those cases where subsidies are needed, flowing subsidies to end-users is good public policy because it gives the LECs the incentive to set service rates at more cost-based levels. Cost-based rates will provide the proper economic conditions upon which potential competitors can determine whether to enter the market with competing services.

In those areas where competition may be slow to emerge, e.g., low density rural/high-cost areas, providing subsidies directly to end-users may also help competition to develop, since it establishes incentives for cost-based rates. The incumbent local carrier will need to reduce costs and provide new services to maintain customer satisfaction and loyalty in the face of potential competitive entry.

Question. A United States Telephone Association study estimated that local telephone service is subsidized in the amount of $20 billion annually. Are you familiar with this study (USTA)? Do you agree with its findings? If not, what do you estimate the amount of subsidy to be?

Answer. AT&T agrees with USTA that there are significant subsidies built into the current local pricing structure. While the overall subsidy may not reach the

$20B USTA estimate, it is nevertheless high. Regulators need to begin now to adopt a regulatory structure based on cost-based pricing, in order to provide potential competitors an appropriate basis for successful entry—supplemented by targeted support programs to protect rural and low income customers. A competitive marketplace should be no bar to the continuation of appropriate subsidies designed to promote universal service, provided they are imposed in an even-handed way, are based on financial need, and are administered effectively.

We disagree with the report's suggestion that the LEC would have to choose between maintaining the current subsidy system and losing significant profits. It is likely to be many years before all the necessary conditions for local exchange competition are implemented, and before competition makes any measurable inroads—if it emerges at all—into the local exchange market. The LECs would therefore have a reasonable period within which to make investments in the infrastructure that will improve productivity, to eliminate other inefficiencies, and to rebalance rates to more accurately reflect the costs of providing individual services. Moreover, as the prices of toll and access are reduced, demand will increase, partially offsetting any lost contribution.

Question. 6. S. 1086 is designed to open the local telephone loop to competition. You have stated that competitive forces are the best way to spur delivery of advanced telecommunications services. For example, when AT&T faced stiff competition in the long distance market, it accelerated investment in digital fiber technology and wrote off billions in older, analog—but still serviceable—plant. Why would local telephone competition encourage similar investment?

Answer. For the same reason it has stimulated investment in the interexchange industry: to permit the carrier to remain competitive with services offered by competing carriers based on current state-of-the-art technology. Failure to provide new services made possible by the latest technology (e.g., digital switching, out-of-band signaling, and fiber optic transmission) would be a severe competitive handicap.

The National Telecommunications and Information Administration (NTIA) concluded in its October 1991 study, The NTIA Infrastructure Report: Telecommunications in the Age of Information, that local exchange competition "should have beneficial effects on local prices, the variety and quality of available services, *and infrastructure development*" (NTIA Report page 266, emphasis added). The report went on to note that Metropolitan Fiber Systems estimates that a $10 million investment by an alternative carrier stimulates a $100 million investment by a LEC.

Question. Who should pay for the investment—rate payer or shareholders?

Answer. Prices for competitive services will be dictated by the market, and competition will ensure that investments are justified by expected demand.

With respect to non-competitive services, regulators have recognized that what constitutes "a reasonable investment" for ratepayers to fund is subject to wide interpretation. This remains a difficult issue that generally must be considered on a case-by-case basis. Nevertheless, regulators have at least partially addressed the problem for the large LECs by establishing price cap regulation, which caps LEC rates irrespective of their underlying costs, thereby limiting to some degree the LECs' ability to recover the costs of uneconomic investments while allowing them to retain the earnings generated from more efficient investments.

Question. The Bell telephone companies argue they need additional incentives—such as the opportunity to provide video programming and long distance services—to justify significant investment in their networks. A federal district court recently struck down the ban on telephone companies provision of cable television. The Bell companies are asking Congress to codify this decision. What is your view?

Answer. The RBOCs' claim is not correct. The local exchange companies already provide originating and terminating access for essentially all long distance calls in their areas. Their monopoly over local access gives them a revenue stream from long distance carriers of about $26 billion each year ($14.2 billion from AT&T in 1992), in addition to over $50 billion in revenues from local service. This fact, and the desire to provide customers with the services they want, should give them adequate incentive to upgrade their networks. Nevertheless, until the local exchanges experience real competition, they will not have the full marketplace incentive they need to invest more wisely and to be more responsive to their customers' needs for a fully capable local infrastructure.

Question. You state that the long distance marketplace is robustly competitive. Today, there are four major nationwide digital fiber networks, eight large regional fiber networks, and almost 500 long distance carriers. The Bell companies request permission to provide long distance service. Would Bell company entry into the long distance market threaten this robust competition?

Answer. Premature MFJ relief would strike at the very heart of competition in the long distance market. As long as the Bell companies retain monopoly control over the local exchange bottleneck, they will have the incentive and ability to leverage that monopoly to discriminate in the provision of local exchange access to unaffiliated long distance carriers, and to cross-subsidize their own long distance services with monopoly revenues. Such anticompetitive conduct would have a chilling effect on competition in the long distance market by inhibiting the ability of long distance carriers to effectively compete, and by discouraging new entrants. Because local exchanges are so complex and technologically dynamic, and characterized by enormous joint and common costs, no foreseeable regulatory scheme can effectively prevent conduct resulting in such discrimination and cross-subsidization.

As Judge Robert Bork recently noted: "If the BOCs were allowed into long-distance service, they could, and almost undoubtedly would, injure competition and ratepayers in both that market and in the local service markets. Any contrary view must rest upon the assumption either that the BOCS would not understand their opportunities or that they would refrain from exploiting those opportunities out of devotion to the public interest. Neither assumption is warranted". (See the attached letter from Judge Bork at page 5)

Question. You suggest that the MFJ line of business restrictions could be lifted if the local exchange becomes truly competitive. You propose there should be agreed upon standards at the federal level defining effective competition in the local market. How would you define effective competition?

Answer. The test for effective competition for MFJ relief must meet certain minimum criteria: (1) that all legal and regulatory impediments to local competition are eliminated; (2) that customers have a real choice of carriers for comparable local exchange services; (3) that customers in fact exercise this choice, so that a substantial percentage of the local exchange market is served by providers other than the local exchange carrier in that area; and (4) that competitors have the capacity to absorb a substantial amount of the local exchange carrier's remaining traffic if it raises its prices above competitive levels. These last two criteria are especially important, because it is the customers' behavior in the marketplace that will confirm whether or not effective competition truly exists to serve their local communications service needs.

At the September 8, 1993 hearing, AT&T and the RBOCs' representative, Ameritech, were asked to meet with the sponsors of S. 1086 to help develop a legislative test for effective competition. We understand that other industry representatives will also be involved in this effort. We are in the process of preparing our specific proposal. We look forward to working with the sponsors and the rest of the Subcommittee in this regard.

[The Bork letter may be found in the committee files.]

Questions Asked by Senator Pressler and Answers Thereto by Mr. Opperman

LOCAL COMPETITION/UNIVERSAL SERVICE

Question. I am concerned not only with access to affordable phone service. I care about the quality and diversity of services available in small states like South Dakota. If access to the information superhighway becomes an essential means of fully participating in society, we may need to redefine universal service. What services do you think should be universally available?

Answer. Clearly, the concept of universal service—as the telephone system itself—exists within a dynamic, ever-changing universe. The services that should be universally available are not the same as those that were available twenty years ago and are not the same as those that should be universally available even in the short term future. I think we can all agree that the vast majority of Americans should (within reason) have access to the harvest of benefits that will come from the technological revolution now in progress. In my view, the only way to ensure this result over time as technology evolves is through effective, sensible state and federal regulation.

Question. What level of competition can we expect to see in smaller states?

Answer. There is already a tremendous amount of competition in the information services and long distance markets, including smaller states. Therefore, I interpret your question to ask what level of competition can we expect to see in the local exchange. I think the real question was not with the size of the state but with the dispersion of its population. Clearly, the answer to this question depends in part on technologies that continue to evolve such as wireless communications. My own prediction is that there will be competition even in states with dominantly rural, dispersed populations. In fact, it is not unreasonable to expect that the communications revolution will affect population density. A recent cover story from Time (9/6/93) discussing the boom in some Rocky Mountain States attributes that boom to the telecommunications revolution.

Question. What is the best way to ensure that we do not become a nation of information have and have-nots?

Answer. In my judgment, we must make sure first that we do not become a nation of information "have nots". Competition is the best way of making sure that all Americans can participate in the communications revolution. Competition has provided Americans with extremely diverse choices among manufacturers, long distance providers and information services providers. However, in addition, appropriate regulation is essential to ensure universal service to any residual market that competition does not adequately serve.

Question. S. 1086 would allow telephone companies pricing flexibility to respond to their new competitors. It is always a difficult regulatory problem to introduce competition into what was previously a monopoly market. We all know the FCC's resources are severely strained. State regulators also have limited resources. How can we ensure fair competition?

Answer. The nation's flexible antitrust laws are the best protection against anticompetitive conduct. They do not require any federal regulatory body to enforce. In addition, S. 1086's approach of breaking down the legal barriers to competition in the local exchange is essential. Once there is effective competition in the local exchange, there will be no need for federal or state regulatory agencies to ensure fair competition and the antitrust laws should be all that is necessary. Until there is effective competition, however, strong structural safeguards are necessary.

Question. Is there a danger that telephone companies could cross-subsidize competitive services with ratepayer revenues?

Answer. As long as local exchange companies continue to have a monopoly, that danger exists. Given the history of the ineffectiveness of regulation in the telephone industry anti competitive cross-subsidization is inevitable. However, once there is effective competition in the local loop, telephone companies providing local exchange service will no longer have a monopoly position and will have to price local telephone calls competitively, making cross subsidization economically infeasible. Therefore, once there is effective competition in the local loop, the danger of cross subsidization diminishes considerably.

PROTECTION FOR RURAL AREAS

Question. S. 1086 requires regulators to ensure that rural customers have access to high-quality telecommunications services. The bill specifically directs the FCC to ensure that local telephone companies in the same geographic area engage in joint coordinated planning and design of the telephone network. However, the bill does not provide much further direction. In your opinion, what steps can state and federal regulators take to ensure that new services reach rural areas?

Answer. I am not an expert in this area. In my view, there will be substantial competition for rural customers. Just as numerous major national retailers have undertaken to serve rural customers, so will major phone companies.

MFJ ISSUES: LONG DISTANCE

Question. S. 1086 would amend the AT&T consent decree's (MFJ's) so-called "interLATA" (long distance) restriction to allow the Bell Companies to provide certain long distance services. Specifically, it would permit them to provide some cellular and cable television services across LATA boundaries. Essentially, the bill would codify limited waivers currently pending before the Department of Justice. I understand that the Bell Companies would like the waiver completely lifted. Their competitors may agree to these limited provisions, but want to make sure Long distance relief goes no further. How much competition is there today in the long distance market?

Answer. There are numerous major national carriers of long distance. West Publishing Company had no problems In receiving quality service at competitive rates in long distance.

Question. How would Bell Company entry affect that competition?

Answer. As long as the Bell Companies have a local exchange monopoly, their entry into long distance would likely diminish, not increase, competition in long distance.

MFJ: INFORMATION SERVICES

Question. The Courts have lifted the information services restriction. S. 1086 would not reimpose the restriction, but it would impose certain safeguards designed to prevent the Bell Companies from cross-subsidizing their "information services", and from discriminating against their competitors. It would also require the Bell Companies to use separate subsidiaries to offer electronic publishing. Are these safeguards necessary?

Answer. Yes. We believe that they should be significantly strengthened.

Question. How effectively will the safeguards prevent cross-subsidization and discrimination?

Answer. Strong structural safeguards combined with vigilant enforcement of the antitrust laws will provide competitors the greatest protection.

Question. The Bell Companies have been able to offer information services in the absence of these safeguards for almost two years. Do we have any evidence of the Bell Companies crosssubsidizing or discriminating in the provision of information services during this period?

Answer. Yes. There are numerous allegations that the Unity Coalition has compiled. Enclosed please find a copy of their report.

Question. If not, do you think the safeguards are needed? Why or why not?

Answer. I think there has been evidence of abuse. I believe safeguards are necessary to prevent further abuses.

QUESTIONS ASKED BY SENATOR PRESSLER AND ANSWERS THERETO BY CORNING, INC.

Question. What services do you think should be universally available?

Answer. The precise definition of universal service should not and, in fact, cannot be static. Both technology and consumer demand are simply too dynamic. Congress should provide some basic guidelines for the Federal Communications Commission ("FCC") to use in establishing the service definition and in reviewing the definition over time to ensure that it reflects current technology, economic conditions, and consumer requirements.

I think that "international competitiveness" should be a principal guideline for use by the FCC in defining the precise service that should be universally available. In other words, universal service should encompass the basic communications services all Americans need access to in order to maintain full employment and a rising level of real income. Today, this basic service package might include digital, interactive voice and data, as well as one-way broadcast video. As technology progresses, this definition could be enhanced to include interactive, integrated voice, data, and video.

In addition, care should be taken to define universal service as a "capability," not a "technology." The same service can be provided using a number of competing technologies. Carriers should be free to determine which technology most efficiently provides a given service.

Question. What level of competition can we expect to see in smaller states?

Answer. Under pro-competitive legislation like S.1086, it is reasonable to expect that local competition will evolve in all states, both small and large, urban and rural. Technological developments and innovations in the lab for wireless, wire, and optical transmission are changing the benefit/cost calculations, and thus, hold out the potential for competition reaching most areas of the United States. In fact, as has been the case with long distance, more than two additional providers are likely in most urban areas and many rural areas.

But there are no guarantees that competition will evolve everywhere. Therefore, the legislation must include provisions, as S. 1086 does, to allow for monopoly regulation in areas where competition does not evolve.

Question. What is the best way to ensure that we do not become a nation of information have's and have not's?

Answer. Basically, three elements must be included in legislation to ensure that the nation doesn't become bifurcated. First, we need to foster competition in local communication services. As we have already seen in the equipment and long distance markets, competition brings down the price of goods and services, making them more affordable for the disadvantaged and those in high cost areas.

Second, the FCC must be given clear authority to establish a rolling definition of universal service. As explained earlier, this authorization should include guidelines for the FCC to ensure that all Americans have access to the services they need to maintain full employment and a rising level of real income.

Finally, a financing mechanism needs to be established that requires all competitors to contribute their fair share to the goal of providing universal basic service as envisioned in S. 1086. This financing mechanism needs to be established on a national basis, not solely on a regional or statewide basis. A national plan is necessary to ensure that subscribers in poorer states have access to surplus resources generated in more wealthy states.

Question. How do current rules discourage investment?

Answer. The current local exchange monopolies face numerous regulatory complications as they attempt to invest in their networks. First of all, they are restrained in their investment behavior by state regulatory authorities that can simply choose to disallow investment. The cost of such disallowed investment must be carried solely by the shareholder. This creates a natural tension for the local exchange carrier in which it must decide how much regulatory risk to take. If it over-invests, its shareholders will pay. If it under-invests, subscribers will be under-served. It is a difficult dilemma within which to operate, one which results in under-investment.

Secondly, the current subsidy mechanism between business and residential service necessitates a low level of investment. Because such subsidies exist, local exchange carriers desperately try to keep the subsidies low by maintaining the lowest rates possible. This necessitates a lower rate of investment than would otherwise be the case.

Thirdly, local exchange carriers are prohibited from providing the one video service with demonstrated demand, video programming, which requires investment in fiber optics and other broadband infrastructure. If local exchange carriers can only provide narrowband service, why would they invest in broadband infrastructure?

Finally, state regulation inhibits and, in some cases, even prohibits new entry, and hence, competition in local telephone service. Without competition, incumbent providers have less of an incentive to invest. Moreover, the denial of entry to new providers certainly reduces investment.

Question. What effect would the changes you recommend have on universal service?

Answer. If legislation is enacted to create a competitive market for local service, in all probability, universal service will be improved. By enacting the statute, the Congress will be able to give clear direction to the FCC to improve the nature of universal service.

This will be particularly helpful for the rural areas. As you know, rural areas are oftentimes the last to benefit from investment in telecommunications infrastructure, particularly those areas that do not benefit from REA financing. Enacting legislation now would no doubt accelerate the pace of investment in rural areas because Congress would speak on the matter.

Prepared Statement of Consumer Federation of America

Because the Consumer Federation of America (CFA) is very concerned that local telecommunications competition will not develop evenly, or at all (particularly in rural regions), we urge the Commerce Committee to amend S. 1086 with a clearly worded "universal service" insurance policy and regulatory safeguards (like those proposed by Senator Inouye in S. 2112, introduced last congress). we believe that universal service involves preservation of the 60 year old declining-real-price trend for local telephone service and the regulatory flexibility to, from time to time, expand universality to include new services that become essential to day-to-day communications. In my written testimony, I propose modification of today's rural high cost fund, low-income lifeline fund, and carrier interconnection charges that contribute to local infrastructure costs, to protect consumers against local rate increases that exceed historical levels, as local telecommunications competition grows.

By locking in these local price protections, Congress can open the door to competition without jeopardizing the social benefits that our traditional universal service policy yielded from a monopoly environment. Also, by establishing price limits for local phone service, Congress can ensure that areas in which competition does not develop will be no worse off than they are in today's monopoly environment.

CFA believes that, before local telephone companies are allowed to expand into new markets—information, video or long distance[1]—consumers should have at least one other viable vendor to choose local phone service from, and strict regulatory safeguards must be in place to prevent anticompetitive practices. Unfortunately, S. 1086 does not provide adequate protections to ensure the maintenance or growth of competition in all telecommunications markets. We urge the committee therefore to add the strict separate subsidiary, cost allocation and other regulatory requirements

[1] While long distance prices are not "competitive" for the most popular consumer services (i.e., evening, night, and weekend discount rates have risen slightly in recent years), there is no indication that the Bell companies would be more likely than existing long distance carriers to reduce rates for the typical, small-volume consumer long distance customer.

proposed in S. 2112, to prevent anticompetitive practices[2] as the local exchange carriers are allowed to diversify.

QUESTIONS ASKED BY SENATOR PRESSLER AND ANSWERS THERETO BY MR. LASHER

LOCAL COMPETITION/UNIVERSAL SERVICE

Question. What services do you think should be universally available?

Answer. The question on what services should be universally available depends upon the entity asking the question. Does a household anywhere in the country need access to the same broadband data communications that the Sioux Falls school system, hospitals or the University of South Dakota may require? Probably not! But they should have access to advanced services such as two way interactive video which could be made available through upgrades to existing telephone or cable television wires. Wireless communications may prove more cost effective for delivery of service in rural and low population density areas for advanced services. If certain services are essential to day to day communications, such as availability to bank automated teller machines have become, then they should be available to all.

Question. What level of competition can we expect to see in smaller states?

Answer. The availability of services described in (1) will be dependent upon not just the consumer demand for such services but also the ease with which they can be deployed—often a regulatory decision. Long distance competition has flourished in South Dakota under the auspices of an enlightened regulatory agency. The Competitive Telecommunications Association Definitive List of Long Distance Carriers lists six (6) long distance companies headquartered and providing service in the state. This does not even include the services that are most likely available from more national carriers. Similar growth for competition among local providers of service could be experienced if the market was opened to full and fair competition as envisioned in S. 1086.

Question. What is the best way to ensure that we do not become a nation of information haves and have-nots?

Answer. The best way to ensure that we do not become a nation of information haves and have-nots is to continue to encourage the development of competition by moving regulatory barriers to entry and permitting access to essential public facilities on just and reasonable terms. In today's telecommunications linked world, geography is not a limiting factor, but availability to advanced systems could be. The decision by Citibank to locate its credit card servicing operation in Sioux Falls is witness to this fact that the rural/urban make up of a state need not be determinative in attracting business and sophisticated information users there.

Question. How can we ensure fair competition? Is there a danger that telephone companies could cross-subsidize competitive services with ratepayer revenues?

Answer. When any market that has been a monopoly is initially opened to competition, there is always a concern that the entrenched provider will take advantage of their superior knowledge of the customer base, including years of relationships, and ability to shift costs to still as yet noncompetitive market segments in order to thwart the development of competition. When AT&T was divested, the Court ordered and the FCC implemented a complex set of safe guards designed to ensure the development of a competitive long distance market place. These safeguards included the implementation of equal access and customer presubscription. It also included heavier regulatory oversight for AT&T than for the developing competitors, with requirements for AT&T such as tariffed offerings with cost of service justification. These mechanisms allowed the long distance market to develop into a competitive market. As portions of the market, such as business services, were deemed to have become sufficiently competitive, AT&T was relieved of its regulatory burdens.

In order to ensure fair competition in the local telecommunications market, we must set standards for interconnection that would be analogous to opening long distance to equal access, each entity providing service must be able to do so at the same cost of access and on equal technical conditions. Local telephone companies should, like AT&T before them, be subject to increased regulatory scrutiny until such competition develops. While ALTS has not espoused a position on the MFJ questions raised in S. 1086, believing that those issues are best argued by the parties most familiar with the ramifications of any changes, ALTS is very concerned about granting the local companies any additional regulatory or pricing flexibility

---

[2] Our written testimony cites previous documentation of Bell company competitive abuses and ratepayer rip-offs. This pattern of abuse has continued since the information services restriction was lifted, with ratepayers facing local phone bills that are inflated by billions of dollars above reasonable price levels.

pending the establishment of truly effective competition. To this end, we are attempting to develop a definition of effective competition to submit to the Subcommittee by Friday, October 1.

PROTECTION FOR RURAL AREAS

Question. What steps can state and federal regulators take to ensure that new services reach rural areas?

In addition to our comments contained in a previous question above, any government programs designed to incent new infrastructure investment in rural areas, such as the REA loans of past, should be available on an open and fully competitive basis.

PREPARED STATEMENT OF INTERMEDIA COMMUNICATIONS OF FLORIDA, INC.

Intermedia Communications of Florida, Inc. ("ICI") is pleased to endorse S. 1086 and its pro-competitive measures that will open the local telecommunications market to competition. As a leading provider of competitive telecommunications services, ICI believes S. 1086 will provide the momentum and catalyst for construction of an advanced telecommunications infrastructure throughout the United States.

ICI applauds you for assuming leadership on this critical Issue and for your foresight in recognizing that current restrictions on competition stand in the way of the full development of the advanced telecommunications infrastructure in this country. Your introduction of S. 1086 has advanced the debate on the role of competition in telecommunications. As S. 1086 appropriately recognizes, a national policy promoting competition is essential to break down the barriers that thwart competition in the local exchange market.

Free competition will do more than any government policy or regulation to foster the development of an advanced telecommunications infrastructure and spur new private investment in telecommunications and in our economy. The products of competition will be a new diversity of services and customer choice, as well as job creation and real economic growth.

While ICI is a firm supporter of S. 1086, ICI encourages you to consider amending the non-discrimination provisions of S. 1086 to require that all utilities and telecommunications carriers charge non-discriminatory pole attachment and conduit fees for all telecommunications carriers and cable television systems. Such an amendment would facilitate the development of competition and increase customer choice.

ICI—A LEADING COMPETITIVE ACCESS PROVIDER

By way of background, ICI is one of the leading providers of competitive access telecommunications services in the United States. ICI's fiber optic networks in Orlando, Tampa, Miami, Jacksonville, and St. Petersburg provide Florida customers a much-needed alternative to the local telephone companies for the transmission of voice, data and video signals. Through network route diversity, multi-vendor security, high-quality transmission, and sophisticated broad bandwidth products, ICI is meeting the needs of Florida businesses for competitive leading edge telecommunications services. ICI also assists telecommunications users in solving complex networking problems that enable them to reduce costs, improve efficiency and increase their competitiveness.

COMPETITIVE ACCESS PROVIDERS DEMONSTRATE THE PROMISE OF COMPETITION IN LOCAL TELECOMMUNICATIONS SERVICES

Competitive access providers ("CAPs") emerged in direct response to the consumer need for high capacity, reliable, dependable communications networks. CAPs provide an alternative to the local telephone company for end-user business customers and long distance carriers and provide a choice of vendors, innovative new services and superior service quality, using fiber optic cables which have nearly limitless capacity and unparalleled quality.

Today, however, competitive providers of local telecommunications services, such as ICI, are extremely limited by historical regulatory obstacles to local exchange competition that continue to exist but that have no place in today's advanced telecommunications age. Currently, ICI is permitted to compete for only a small portion of local telecommunications services. Passage of S. 1086 would alleviate this problem by permitting CAP entry into virtually all local service markets.

The primary services of CAPS today are: (1) dedicated transmission facilities between various point-of-presence of long distance carriers, providing backup facilities in case of service Interruption; (2) dedicated lines for large telecommunications users to connect with their long distance carrier, bypassing the local telephone company; and (3) enhanced products and services, such as interconnection of local area networks and enhanced transport services utilizing frame relay and asynchronous transfer mode technologies for corporate customers. The majority of local telecommunications services remain subject to the exclusive monopoly of the local exchange carriers. S. 1086, however, will remove antiquated state and local barriers to competition and permit ICI to offer its customers comprehensive local telecommunications services. The strong user response that CAPs have received in this limited area provides a glimmer of the benefits of full competition for all local exchange services.

NEED FOR NON-DISCRIMINATION IN POLE ATTACHMENT AND CONDUIT FEES

ICI encourages you to consider amending the nondiscrimination provisions of S. 1086 to prohibit utilities and telecommunications carriers from imposing discriminatory charges for access to conduit, pole attachments and rights-of-way. This amendment is consistent with S. 1086 and furthers its goal of eliminating barriers to competition and opening the local telecommunications market to full competition.

ICI has experienced first hand the excessive and discriminatory fees imposed by utilities for pole attachments and conduit occupancy. ICI currently leases facilities from a cable television company that utilizes pole attachments provided by Florida Power Corporation ("Florida Power"). Florida Power has imposed pole attachment fees for ICI's telecommunications services that are 576 percent higher than the price cable companies are charged for the same facilities, even when they offer common carrier services that compete with those CAP services of ICI. Specifically for the year 1992, Florida Power charged the following discriminatory pole attachment rates for cable companies and telecommunications carriers:

Annual pole attachment rate per distribution pole: Cable TV Company, $3.44; Telecommunications Carrier, $23.25—**Difference,** $19.81 (576 percent).

Annual pole attachment rate per transmission pole: Cable TV Company, $10.19; Telecommunications Carrier, 68.87—**Difference,** $58.68 (576 percent).

Recent developments appear likely to exacerbate the situation further. Several power utilities now are seeking to enter the telecommunications marketplace themselves, and when they do, they will have a strong incentive to charge unjust and discriminatory rates to their competitors. Tampa Electric Company ("TECO"), or example, recently applied for alternative access vendor ("AAV") status in Florida. If the Florida Commission grants its application, TECO would be able to utilize its own pole attachments at cost, while tending off competition with discriminatory rates.

A recent decision by the Court of Appeals for the D.C. Circuit has confirmed the ability of utilities to continue such discriminatory practices, specifically it has sanctioned the disparate pole attachment and conduit occupancy charges that a utility company assessed cable television companies and telecommunications providers. in *Texas Utilities Electric Co.* v. *FCC*,[1] the D.C. Circuit affirmed the FCC's decision that the federal Pole Attachment Act requires utilities to impose a single rate for all pole attachments by cable TV companies, even if the attachments are used by the cable television system to provide telecommunications services. The D.C. Circuit also held that its jurisdiction extends only to attachment rates charged to cable television systems and not to competitive providers seeking to provide telecommunications services that may compete with cable TV companies. The FCC's decision gives cable TV systems a competitive and discriminatory cost advantage over other competitive providers of telecommunications services such as ICI.

The FCC's decision was based primarily on its limited view of its statutory jurisdiction and calls out for a legislative solution to reflect current competitive conditions that can appropriately be provided by S. 1086:

> We are mindful of the competitive concerns raised by TU Electric. * * * However, we cannot refrain from fulfilling our statutory mandate to ensure just and reasonable pole attachment rates, terms and conditions for cable television systems operators, simply because our authority may not extend to every competitor in the data transmission market. We recognize that the result we reach today will afford cable television system operators an advantage over other competitors who are not entitled to regulated pole attachment rates under Section 224. * * * While the Commission favors opening markets to competition as fully as possible, Congress, by enacting Section 224, expressly limited its applicability to cable television systems.

---

[1] No. 92-1032 (June 25, 1993).

Federal legislation is the only way to ensure just, reasonable and nondiscriminatory conduit and pole attachment rates, terms and conditions for all entities providing local telecommunications services to the public. Equitable treatment of all local service providers is required for the promotion of robust competition.

ICI stands ready to assist the Senate as it considers S. 1086 and the benefits of competition in the local telecommunications market. We look forward to working with you on this important issue.

Questions Asked by Senator Pressler and Answers Thereto by Ms. Easterling

LOCAL COMPETITION/ UNIVERSAL SERVICE

Question. I am concerned not only with access to affordable phone service. I care about the quality and diversity of services available in small states like South Dakota. If access to the information superhighway becomes an essential means of fully participating in society, we may need to redefine universal service. What services do you think should be universally available?

Answer. Every actual or potential user must have at least dial tone access, provided for the foreseeable future by the local exchange telephone company, or "LEC." The debate over the last several years, in CWA's view, has shifted meanings of terms and the perceived basic needs of users.

The network should be set to offer all desired and required services; how these services are to be paid for requires close public agency monitoring, to ensure that the ordinary user who only wants and needs basic dial tone is not overcharged. (About a dozen years ago, much debate concerned the question of whether "touch-tone" service was part of the basic package or something beyond. That question became moot as local telephone exchanges were modernized, with the result that all access lines became "touch-tone"; to remove the "touch-tone" feature would have entailed extra cost.)

Beyond the discussion above on basic dial tone service, the response to the query is that the users should have as many service options as the providers can offer at reasonable and affordable (as determined on solid legal records) rates.

Question. What level of competition can we expect to see in smaller states?

Answer. In the smaller (population/rural) States, only modest degrees of competition can be foreseen. Low population density is a factor working against high degrees of competition. The Congress must keep in mind that the presence of competition in a given market often leads to easier entry into and exit from that market. Over the last two decades, competition has encouraged providers to abandon or "reprice" (i.e., raise prices) for some services and equipment. Apparent long distance toll reductions should not cloud this issue: end-user access charges usually are not factored into this part of the debate.

Question. What is the best way to ensure that we do not become a nation of information haves and have-nots?

Answer. The definitions of "information haves and have nots" never have been stated with clarity in the long policy debates of the last dozen years. Especially unclear is whether the ordinary (perhaps low-income) basic service user who only uses the telephone for voice messages is a "have not." Basic telephone service can arguably be taken as placing the user into the "have" category.

The question seems to be cast in the context of assuring continued universal service. Government can ensure that rate structures provide sufficient resources to help pay costs of basic local service affordable to all. All carriers and large customers using parts of the overall network must be required to contribute toward local service costs, even when "bypassing" certain facilities.

The "carrier of last resort" LECs must have enough revenue to continue their operations offering high-qualify service.

Question. S. 1086 would allow telephone companies pricing flexibility to respond to their new competitors. It is always a difficult regulatory problem to introduce competition into what was previously a monopoly markets We all know the FCC's resources are severely strained. State regulators also have limited resources. How can we ensure fair competition?

Answer. "Fair competition" can be ensured first by defining the term "competition." For the last 15 years, CWA has been publicly calling for the Congress and FCC to define the term in order to evolve rules. The September 8 hearing thus became a major vindication of CWA's long and lonely quest to divine the meaning of a key term.

A second needed step is to ensure that only one set of rules will apply to all providers; this means equal tariffing, disclosure and Section 214 requirements, in public files. Tariff "forbearance"is an improper public policy, since only some carriers are required to meet the full set of regulatory requirements. CWA urges the Congress to remind the FCC of the need for neutrality.

Fair competition then depends on the willingness of regulators and competitors to use enforcement tools available.

Question. Is there a danger that telephone companies could cross-subsidize competitive services with ratepayer revenues?

Answer. While some cross-subsidy danger may exist, regulatory actions are able to prevent improper flows of funds. Imposition of rules requiring the separate subsidiary device to cut down on the danger of cross-subsidy is itself an anti-competitive action, since only some—but not all—providers would be faced with such rules. In the example of the "non-dominant" carrier, ratepayer1subscriber revenues would be allowed to flow with little restriction; revenues of "dominant" carriers would be treated differently. Since the term "cross-subsidy" is so often used, it requires a definition.

PROTECTION OF RURAL AREAS

Question. S. 1086 requires regulators to ensure that rural customers have access to high-quality telecommunications services. The bill specifically directs the FCC to ensure that local telephone companies in the same geographic area engage in joint coordinated planning and design of the telephone network. However, the bill does not provide much further direction. In your opinion, what steps can state and federal regulators take to ensure that new services reach rural areas?

Answer. Regulators should require carriers to plan and design on a coordinated basis, using their current authority. Requiring the highest quality of basic services and a uniform set of tariffing and other rules will help ensure better service for rural areas. Competitors of conventional telephone companies must be required to contribute toward rural area costs.

MFJ ISSUES: LONG DISTANCE

Question. S. 1086 would amend the AT&T consent decree's (MFJ's) so-called "interLATA" (long distance) restriction to allow the Bell Companies to provide certain long distance services. Specifically, it would permit them to provide some cellular and cable television services across LATA boundaries. Essentially, the bill would codify limited waivers currently pending before the Department of Justice. I understand that the Bell Companies would like the waiver completely lifted. Their competitors may agree to these limited provisions, but want to make sure long distance relief goes no further. How much competition is there today in the long distance market?

Answer. Considerable competition exists today in both inter- and intra-LATA markets. Many States are in process of allowing expansion of intra-LATA offerings. The FCC Common Carrier Bureau's Industry Analysis Division periodically reports on the several hundred facilities-based and resale long distance providers.

Question. How would Bell Company entry affect that competition?

Answer. Bell company entry into inter-LATA services would considerably increase the supply and competition. The smaller resale carriers can be expected to exit the business in the presence of more large competitors.

MFJ: INFORMATION SERVICES

Question. The Courts have lifted the information services restriction. S. 1086 would not reimpose the restriction, but it would impose certain safeguards designed to prevent the Bell Companies from cross-subsidizing their "information services," and from discriminating against their competitors. It would also require the Bell Companies to use separate subsidiaries to offer electronic publishing. Are these safeguards necessary?

Answer. Current law and regulatory policy are sufficient to prevent the litany of imputed abuses often ascribed to the Bell companies. The "safeguards" of S. 1086 are superfluous, unnecessary. CWA opposes the bill in its present form.

Question. How effectively will the safeguards prevent cross-subsidization and discrimination?

Answer. As set out above, the legislation's "safeguards" are unneeded. Enforcement of current law and regulation will be sufficient. CWA views the "safeguards" as incomplete, with only selective enforcement included.

Question. The Bell Companies have been able to offer information services in the absence of these safeguards for almost two years. Do we have any evidence of the Bell Companies cross-subsidizing or discriminating in the provision of information services during this period? If not, do you think the safeguards are needed? Why or why not?

Answer.CWA is not aware of any evidence of abuses by Bell companies in offering and information services.

CWA does not favor imposing "safeguards" until such evidence establishes a pattern of abuse. If competition is to be the policy, those wanting to compete ought to be free to do so, with minimum and consistent restrictions.

In CWA's experience, many who call for "safeguards" are off the mark, by also telling the Congress that even with "safeguards" the user will not be protected. Some of these parties show themselves unwilling to work to evolve policies to help avoid the abuses they grimly predict. We believe a consistent set of rules for competition will be an effective solution to this question.

○

70-334 O - 94 (288)

TP
APEX
9780160460685